Series: Intelligent Technologies for Smart Precision Agriculture

Plant Disease Detection Using Machine Learning, Deep Learning, and Metaheuristics

Series: Intelligent Technologies for Smart Precision Agriculture

1. Plant Disease Detection Using Machine Learning, Deep Learning, and Metaheuristics
   Nebojsa Bacanin, Amit Chhabra, Satveer Kour and *Ghaith Manita (Editors), 2027*
2. Computational Intelligence for Plant Disease Detection: Models, Methodologies, and Emerging Trends
   Amit Chhabra, Nebojsa Bacanin, Satveer Kour and *Ghaith Manita (Editors), 2027*

Series: Intelligent Technologies for Smart Precision Agriculture

Plant Disease Detection Using Machine Learning, Deep Learning, and Metaheuristics

Editors:

Nebojsa Bacanin
Professor, Vice-Rector for Scientific Research and
Head of Applied Artificial Intelligence
Singidunum University, Belgrade, Serbia

Amit Chhabra
Department of Computer Engineering and Technology
Guru Nanak Dev University, Punjab, India

Satveer Kour
Department of Computer Engineering and Technology
Guru Nanak Dev University, Punjab, India

Ghaith Manita
Associate Professor, Faculty of Science of Tunis
University of Tunis, El Manar, Tunisia

CRC Press
Taylor & Francis Group
Boca Raton London New York

CRC Press is an imprint of the
Taylor & Francis Group, an **informa** business
A SCIENCE PUBLISHERS BOOK

First edition published 2027
by CRC Press
2385 NW Executive Center Drive, Suite 320, Boca Raton FL 33431

and by CRC Press
4 Park Square, Milton Park, Abingdon, Oxon, OX14 4RN

CRC Press is an imprint of Taylor & Francis Group, LLC

Library of Congress Cataloging-in-Publication Data (applied for)

ISBN: 978-1-041-28977-7 (hbk)
ISBN: 978-1-041-28999-9 (pbk)
ISBN: 978-1-003-76753-4 (ebk)

DOI: 10.1201/9781003767534

Typeset in Times New Roman
by Prime Publishing Services

Preface

Agriculture continues to be the main pillar of food security around the world, and early detection of plant diseases is vital to crop productivity and sustainability. In the last ten years, agriculture has undergone a major transformation with the advent of Artificial Intelligence (AI). Specifically, machine learning, deep learning and metaheuristic optimization techniques have provided new opportunities for early detection of plant diseases, aiding in better crop management and sustainable agriculture.

Although a lot of research work has been done in this field, it has been noticed that most of the work is directed towards specific techniques or specific applications. The importance of a single source that encompasses various computational methods and allows the reader to grasp the integration of these methods within a unified framework and their practical implications is increasing. This realization prompted us to put together the present volume.

This volume gathers together fourteen chapters by researchers from a variety of academic and research backgrounds. The chapters not only cover fundamental concepts but also recent developments, including machine learning techniques, deep learning architectures, metaheuristic optimization techniques and their hybridization for plant disease detection. An attempt has been made to bring in contributions which are balanced between theory and application so that the book will be useful for study and application.

The book is logically organized. It starts with an introduction to plant disease detection and basics of machine learning in agriculture. It then progresses to more specific areas such as deep learning for image analysis and data acquisition strategies and optimization-driven approaches with metaheuristics. A few chapters discuss cutting-edge deep learning models and examples, including comparisons of contemporary plant disease detection methods. Later chapters discuss a variety of significant issues, including limitations, challenges, economic and environmental impacts, and future research directions. The book ends with insights into future developments and the contribution of intelligent systems to sustainable agriculture.

The volume will be useful to researchers, post graduate students, academicians and professionals dealing with the interface of agriculture and artificial intelligence in general. We hope it will be used as a reference for current methods and as a starting point for new research ideas and collaborations.

We would like to thank all the contributing authors for their time, effort and valuable insights. This book is based on their contributions. We would also like to thank the reviewers for their insightful comments that have enhanced the quality of the chapters. Last but not least, we would like to express our gratitude to the editorial staff at CRC Press for their guidance and support in the development of this volume.

We hope that this book will make a small contribution – in some way, in some measure – to the continuing search for smarter and more sustainable agriculture and that it will inspire others to continue the search in this important area.

(*Editors*)
Nebojsa Bacanin
Amit Chhabra
Satveer Kour
Ghaith Manita

Contents

List of Contributors

A. Poonguzhali
Sri Sairam College of Engineering, Anekal, Bangalore, India.

A.S. Malleswari
Assistant Professor, Department of CSE, Aditya University, Surampalem, Andhra Pradesh, India.

Adarsh Kumar Pandey
Assistant Professor, Faculty of Computer Science and Engineering, Poornima University, Jaipur, India.

Allirani Saminathan
Department of Electrical and Electronics Engineering, Sri Ramakrishna Engineering College, Coimbatore, Tamil Nadu, India.

Amit Kumar Maurya
Teerthanker Mahaveer College of Agriculture Sciences, Teerthanker Mahaveer University, Moradabad Uttar, Pradesh-244001, India

Ande Disha
Ph.D. Scholar, Department of Plant and Soil Science, University of Kentucky, Lexington, KY 40506, United States.

Anuraj Singh
Department of Engineering Sciences, Atal Bihari Vajpayee Indian Institute of Information Technology and Management Gwalior, Gwalior, Madhya Pradesh, India.

Chidozie Ezekwem
Department of Mechanical Engineering, University of Port Harcourt, Port Harcourt, Nigeria

Cynthia Jayapal
RV University, Bengaluru - 560059, India.

Dadi Syamala
Assistant Professor, Department of Microbiology, Gayatri Vidya Parishad College for Degree and PG Courses (A), Visakhapatnam, Andhra Pradesh, India.

Damanpreet Kaur
University Institute of Computing, Chandigarh University, Mohali-140413, Punjab, India.

Divvela Srinivasa Rao
Associate Professor, Department of AI&DS, Lakireddy Bali Reddy College of Engineering, Mylavaram, Krishna District, Andhra Pradesh, India.

Gaddam Venu Gopal
Associate Professor, Department of CSE (AI & ML), B V Raju Institute of Technology, Narsapur, Medak, Telangana, India-502313.

Ganesh Datt Bhatt
Teerthanker Mahaveer College of Agriculture Sciences, Teerthanker Mahaveer University, Moradabad Uttar, Pradesh-244001, India.

Gaurav Kumar
Assistant Professor, Faculty of Computer Science and Engineering, Poornima University, Jaipur, India.

Hemlata Pant
Department of Zoology, CMP Degree College, University of Allahabad, Prayagraj, Uttar Pradesh-211002, India.

Kadraka Harshini
Student, Department of Microbiology, Gayatri Vidya Parishad College for Degree and PG Courses (A), Visakhapatnam, Andhra Pradesh, India.

Kannappan Manikandan
Ph.D. Scholar, Department of Agronomy, Faculty of Agriculture, Annamalai University, Chidambaram, Tamil Nadu, India-608 002.

Karthikeyan Palaniappan
Department of Electrical and Electronics Engineering, Kongu Engineering College (Autonomous), Perundurai, Erode, Tamil Nadu, India.

K. Ulaga Priya
Vels Institute of Science Technology and Advanced Studies, Chennai, India.

Lanka Sujatha
Head of the Department, Department of Microbiology, Gayatri Vidya Parishad College for Degree and PG Courses (A), Visakhapatnam, Andhra Pradesh, India.

Lekkala Venkata Ravishankar
Division of Plant Pathology, FoA, SKUAST, Jammu-180009, Jammu and Kashmir, India.

Mahesh Singh
Teerthanker Mahaveer College of Agriculture Sciences, Teerthanker Mahaveer University, Moradabad Uttar, Pradesh-244001, India.

Manjushree Nayak
Amity School of Engineering & Technology, AMITY UNIVERSITY, Raipur, Chhattisgarh, Manth (Kharora), State Highway 9, Baloda Bazar Road, 493225, India.

Mynavathi R.
Velalar College of Engineering and Technology, Erode, Tamil Nadu-638012, India.

Nebojsa Bacanin
Faculty of Informatics and Computing, Singidunum University, Belgrade, Serbia.

Prashant Atmakuri
Assistant Professor, Department of CSE, Prasad V Potluri Siddhartha Institute of Technology, Kanuru, Vijayawada, Andhra Pradesh, India.

Priyanka
University Institute of Computing, Chandigarh University, Mohali-140413, Punjab, India.

R. Jansi
Department of Electronics and Communication Engineering, Faculty of Engineering and Technology, SRM Institute of Science and Technology, SRM Nagar, Kattankulathur 603203, India.

Ravikirushna S.
Velalar College of Engineering and Technology, Erode, Tamil Nadu-638012, India.

Rituparna Seal
JECRC University (Sunstone), Jaipur, Rajasthan, India.

Rocky Kumar
Assistant Professor, Faculty of Computer Science and Engineering, Poornima University, Jaipur, India.

S. Divya Bairavi S.
Vels Institute of Science Technology and Advanced Studies, Chennai, India.

S. Haarish
Department of Electronics and Communication Engineering, Faculty of Engineering and Technology, SRM Institute of Science and Technology, SRM Nagar, Kattankulathur 603203, India.

Sabitha P.
Velalar College of Engineering and Technology, Erode, Tamil Nadu-638012, India.

Sachin Raj
University Institute of Computing, Chandigarh University, Mohali-140413, Punjab, India.

Sakshi
Apex Institute of Technology (CSE), Chandigarh University, Mohali-140413, Punjab, India.

Sakshi Sharma
Assistant Professor, Faculty of Computer Science and Engineering, Poornima University, Jaipur, India.

Sasthikha V.K.
Velalar College of Engineering and Technology, Erode, Tamil Nadu-638012, India.

Satender Kumar
Quantum School of Technology, Quantum University, Roorkee, India.

Satveer Kour
Department of Computer Engineering and Technology, Guru Nanak Dev University, Amritsar, Punjab, India.

Shree K.V.M.
Dhanalakshmi Srinivasan Engineering College (Autonomous), Perambalur.

Sridhar B.
Department of Electrical and Electronics Engineering, Sri Ramakrishna Engineering College, Coimbatore, Tamil Nadu, India.

Subhakanta Panda
Department of Computer Science and Engineering, Centurion University of Technology and Management (CUTM), Paralakhemundi, Odisha-761211, India.

Suresh Muthusamy
Department of Electrical and Electronics Engineering, Kongu Engineering College (Autonomous), Perundurai, Erode, Tamil Nadu, India.

Sushma Singh
Teerthanker Mahaveer College of Agriculture Sciences, Teerthanker Mahaveer University, Moradabad Uttar, Pradesh-244001, India.

Syed Mohammed Azmal
Assistant Professor, Department of Computer Science & Engineering, Koneru Lakshmaiah Education Foundation, Vaddeswaram, Guntur District-522502, Andhra Pradesh, India.

Vasanth Poomalai
Ph.D. Scholar, Department of Agronomy, Faculty of Agriculture, Annamalai University, Chidambaram, Tamil Nadu, India-608 002.

Vinny John
Department of Agriculture, Mangalayatan University, Jabalpur-483001, Madhya Pradesh, India.

Vivek John
Uttaranchal Institute of Technology, Uttaranchal University, Dehradun-248007, Uttarakhand, India.

Vivek Kumar
Quantum School of Technology, Quantum University, Roorkee, India.

1 Introduction to Plant Disease Detection in Agriculture

Lanka Sujatha[1*], Dadi Syamala[2] and Kadraka Harshini[3]

Plant pathogens are a significant obstacle to world food security resulting in massive economic damages in the agricultural industry. The quality of crop yields depends on the success of early and accurate diagnosis and detection of disease, which is inseparable in its success and development of precision agriculture. This paper provides an overview on traditional and advanced methods of plant disease detection. The traditional disease detection methods include manual inspection by the farmer or investigator followed by tedious and lengthy laboratory techniques that can be sometimes error prone when working in a large-scale farming. As the early symptoms are subtle, when unnoticed can lead to faster spreading of the disease due to delay in timely action.

The advanced methods include the technologies that are easy, accurate, automated and non-invasive in nature for plant disease detection. These applications of advanced methods have created a need to overcome the constraints of the traditional methodologies. Precision agriculture focuses on the adoption of new modern technologies like sensor-based methods, artificial intelligence, machine-learning, and deep-learning models for analyzing the data to detect disease symptoms of the plants with great speed and accuracy. The advantages of such modern technologies are addressed in the context of the improvement of antimicrobial effect, making it possible to monitor it in real time, contributing to the evidence-based decision-making to prevent crop losses, and sustainable practices. The presented insights are expected to enhance the comprehension of the emerging technologies of disease-detection, deal with the problematic issues, and orient future developments.

1. Introduction

Plant disease detection is a significant component of precision agriculture that is meant to guarantee minimal losses of crops and food security in the world.

1 Head of the Department, Department of Microbiology, Gayatri Vidya Parishad College for Degree and PG Courses (A), Visakhapatnam, Andhra Pradesh, India.

2 Assistant Professor, Department of Microbiology, Gayatri Vidya Parishad College for Degree and PG Courses (A), Visakhapatnam, Andhra Pradesh, India.

3 Student, Department of Microbiology, Gayatri Vidya Parishad College for Degree and PG Courses (A), Visakhapatnam, Andhra Pradesh, India.

Email: dsyamala@gvpcdpgc.edu.in[2]; 2023-2411006@gvpcdpgc.edu.in[3]

* Corresponding author: lsujatha@gvpcdpgc.edu.in

DOI: 10.1201/9781003767534-1

Pathogens such as bacteria, fungi, viruses, or environmental stress cause plant diseases which decrease the crop yield quantitatively and qualitatively, in many cases with a significant economic implication. The economic and environmental sustainability of the agriculture sector can be ensured by early detection of disease in the plants to reduce the losses. The detection of the disease in plants has various uses including enhancing yield, boosting the cost effectiveness and performance, and contributing to food security across the globe [1][2].

2. Types and Causes of Plant Disease

The Plant diseases are mainly divided into two types Infectious (Biotic) and Non-infectious (Abiotic) diseases, which depend on their cause of infection or spoilage, i.e., Living organisms cause Infectious disease, and non-infectious disease is caused by environmental or non-living factors [3].

2.1 *Infectious (Biotic) Diseases*

Pathogens that are living organisms cause these diseases and may be transmitted between a diseased and a healthy plant. The plant pathogens are fungi, bacteria, viruses, nematodes and parasitic plants.

2.1.1 *Fungi*

Fungi are the causative organisms of most of the plant diseases. They multiply by means of spores, and have an extremely varied range of symptoms, which include rusts (red/brown spots), smuts (black, sooty masses), mildews (powdery or downy growth), and other blights and rots. One such famous example is Potato Late Blight, which is an infection of the oomycete, *Phytophthora infestans*.

2.1.2 *Bacteria*

These are single-celled organisms that gain their way into a plant either by natural openings (such as stomata) or by wounds. They usually lead to such symptoms as spots of the leaves (they can be angular and water-wetted), wilts (they can block the vascular system), soft rots of fruits and storage organs. One of the leading bacterial diseases is Citrus Canker.

2.1.3 *Viruses*

These are mandatory intracellular parasites, made up of pure genetic material (DNA or RNA) that is covered with a protein coat. They have to utilize the cells of the host plant so as to replicate. The insect vectors such as aphids or leafhoppers are known to transmit viruses. Symptoms usually include the case of stunting; the leaves being curled and mosaic pattern (mottling) on the leaves.

2.1.4 Nematodes

These are non-segmented, round worms that are microscopic and inhabiting the soil. They feed on the roots of plants leaving symptoms on the surface which resemble nutrient deficiency like retarded growth and yellowish hue. Root galls develop or knots are formed as a result of their feeding.

2.1.5 Parasitic Plants

These are less frequent but important and are the plants which feed directly on one more plant. Some examples are dodder and mistletoe.

The plant pathogens bring about the disease shown in Figures 1, 2, 3, 4 [4].

2.2 Non-infectious (Abiotic) Diseases

These disorders are not transmissible from one plant to another. They are also known as physiological disorders and are caused by unfavorable growing conditions or environmental factors.

2.2.1 Environmental Extremes

Conditions outside the optimal range can severely stress a plant. This includes drought (leading to wilting), excessive water/flooding (causing root suffocation), temperature extremes (such as freezing or heat stress leading to sunscald), and insufficient light.

2.2.2 Nutrient Imbalances

The lack or excess of essential mineral nutrients (e.g., nitrogen, iron, calcium) can cause disease symptoms. For instance, Iron Deficiency leads to yellowing between the veins of new leaves (interveinal chlorosis).

2.2.3 Pollution and Chemical Injury

Plant damage can be caused by air pollutants (e.g., ozone), soil toxins, or the misuse of pesticides and herbicides.

3. Symptoms and Signs of Plant Disease

In plant pathology, a clear distinction is made between a symptom and a sign. Understanding this difference is the first step in correctly diagnosing a plant problem.

3.1 Symptoms

A symptom is the visible reaction or expression of the host plant to the presence of a pathogen or an abiotic (non-living) factor. It is the plant's way of showing it

Fig. 1 Bacterial Diseases.

Fig. 2 Fungal Diseases.

Fig. 3 Viral Diseases.

is ill. Symptoms are indirect evidence of disease. The major categories of symptoms are necrosis, hypoplasia, and change in color or appearance.

Root knot of vegetables

Ear cockle of wheat

Root gall of soyabean

Mature cysts of potato golden cyst nematode

Common dodder parasitic on shrubby host

True mistletoe plants parasitic on hardwood tree

Witchweed (*Striga* sp) parasitic on maize roots

Broomrapes parasitic on pepper

Red rust of tea

Fig. 4 Nematode Diseases.

3.1.1 Necrosis (Death of Tissue)

It is the mature death of the tissue.

- **Leaf Spots:** These are the dead areas, typically brown or black, of limited size on a leaf (e.g., Black Spot of rose).
- **Blight:** It includes the rapid and widespread killing of leaves, flowers, or entire shoots (e.g., Fire Blight of pear).

- **Canker:** A localized, sunken, necrotic area on a stem, branch, or trunk is known as canker.
- **Rot:** It is the decomposition and disintegration of plant tissue. Which can be often soft and watery or dry and firm, known as Soft Rot or Dry Rot respectively.
- **Dieback:** Progressive death of shoots, starting from the tip and moving toward the base.

3.1.2 Hypoplasia

It is an Underdevelopment or Stunted growth of any part of the plant.

- **Stunting/Dwarfing:** Failure of the plant or its parts to grow to normal size.
- **Chlorosis:** Yellowing of normally green tissue due to the destruction or failure of chlorophyll to develop. If the yellowing is between the veins, it is called interveinal chlorosis.
- **Rosetting:** Clumped rosette-like growth pattern is due to the shortening of internodes.

3.1.3 Hyperplasia/Hypertrophy

It is an overdevelopment or Overgrowth of any part of the plant.

- **Galls/Tumors:** These are the swellings or abnormal localized overgrowths on roots, stems, or leaves (e.g., Crown Gall).
- **Witches' Broom:** Excessive, bushy and localized growth of shoots from a single point.
- **Leaf Curl:** The uneven growth of leaves can cause abnormal curling, crinkling, or distortion.

3.1.4 Change in Color or Appearance

- **Mosaic:** Viral infection on leaves can cause characteristic patchwork of light green and dark green areas (ex-Tobacco Mosaic Virus).
- **Water-soaking:** The bacterial and water mold infection can lead to a translucent, dark, water-soaked appearance on tissues.

3.2 Signs

It is the direct evidence of disease shown by the causative organism or pathogen on or in the host plant. Common Examples of Signs are:

3.2.1 Mycelium

The visible, thread-like vegetative body of a fungus, that appears as white, gray, or black mold growth.

3.2.2 *Spores/Fruiting Bodies*

The reproductive structures of fungi (e.g., the bright orange or red pustules of Rust fungi or the white, powdery growth of Powdery Mildew are the pathogen's actual structures).

3.2.3 *Bacterial Ooze (Exudate)*

A slimy, viscous mass of bacteria and host cells, often exuded from cankers or cut vascular tissue.

3.2.4 *Sclerotia*

Hard, dark resting structures of certain fungi.

3.2.5 *Nematode Cysts*

The tough, protective bodies of certain female nematodes, visible on the roots [5] [6].

4. Evolution of Plant Detection Methods

The Traditional method of plant disease detection is based on the manual inspection by the experts. It is time-consuming, labor-intensive, often subjective, and can miss early-stage infections. These limitations have paved the way to modern agriculture which include the advanced technological methods that are fast, accurate, and automated detection systems.

The plant disease detection system in agriculture has three main historical phases:

4.1 *Ancient to Mid-19th Century*

This was the pre-scientific era that basically relied on observation and theory. For ages, farmers have closely observed their crops visually for a disease to detect the symptoms like rusts, mildews, and blights. In this era, it was believed that the diseases are caused spontaneously and it is an act of divine punishment. In the 17th century, the invention of the microscope allowed the visualization of microorganisms.

4.2 *Mid-19th Century to Late 20th Century*

This period is called the Laboratory era where plant disease was identified based on different laboratory techniques. In this era, Plant Pathology was grown as a science. In the 1850s, Heinrich Anton de Bary proved the Germ Theory of disease for plants where fungi were the cause of diseases like the Irish Potato Blight.

The pathogen was confirmed primarily by visual inspection by trained experts, followed by laboratory techniques like:

- **Culturing:** The pathogen was isolated from the diseased plant by growing on artificial media.
- **Microscopy:** This technique includes the use of a microscope to observe the morphology of a pathogen.
- **Chemical/Genetic:** It includes the molecular diagnostics methods like ELISA and PCR for accurate identification of the pathogen.

4.3 *Early 21st Century Onward*

This period was called the Precision era where digital and automated techniques were developed and used for the identification of a pathogen. It includes the integration of expensive cameras, drone/remote sensing, and powerful computing that led to automation in the agriculture sector like computer vision and machine learning.

- **Computer Vision:** The sensors-based techniques like RGB (red, green, blue), multispectral, and hyperspectral are used to capture images.
- **Machine Learning (AI):** It includes the use of Convolutional Neural Networks (CNNs), an algorithm that is trained on thousands of images to instantly detect, classify, and quantify disease symptoms in real-time.

5. The Early Detection of Plant Disease

It is important for the following reasons:

5.1 *Minimizing the Losses and Financial Impact*

Early and accurate diagnosis of the plant disease is the crucial factor to reduce the qualitative and quantitative losses in crop yield [7].

5.1.1 Containment of Epidemics

Plant diseases can spread rapidly. Identification, isolation, and treatment of the first signs of infection in a plant prevents the pathogen from spreading to the entire field or region, which is essential for maintaining crop productivity [8].

5.1.2 Financial Gains

Financial losses can be significantly mitigated by the timely actions of the farmers. Diseases can reduce income of the farmers and increase prices for consumers, potentially leading to food insecurity in vulnerable regions [9].

5.2 *Facilitating Effective and Targeted Management*

Proactive strategies can be followed when the symptoms are detected at the early stage, which leads to efficient interventions and disease management.

5.2.1 *Timely Intervention*

Many plant diseases are treated by antimicrobial agents like fungicides, bactericides, and biopesticides, which are most effective when they are applied at the very onset of the disease. It can be often before the symptoms are obviously visible to the naked eye.

5.2.2 *Precision Agriculture*

Usage of modern detection methods like advanced imaging or molecular diagnostics, provide the necessary information for precision agriculture. This allows the farmers to be target specific, i.e., the infected areas rather than treating the entire field. This saves resources, reduces waste, and makes treatments more effective.

5.3 *Promoting Environmental and Resource Sustainability*

Early detection of the disease can reduce the severity and spread of disease and is the key to sustainable farming practices.

5.3.1 *Reduced Chemical Use*

Early, accurate and targeted treatments can significantly reduce the use of pesticides and other chemical products. This decreases the risk of environmental contamination, minimizes impacts on non-target organisms, and slows the development of pesticide resistance in pathogens.

5.3.2. *Informed Decision-making*

Advanced systems provide farmers with the required knowledge to make the quick and smarter decisions that enhance the efficiency and long-term sustainability in agriculture [10].

6. Traditional Plant Disease Detection Methods

6.1 *Visual Inspection and Field Diagnosis*

The first step is most crucial in disease diagnosis, which includes the simple visual inspection of the field by the skilled diagnostician. It relies on the knowledge of symptoms, signs, and the environmental conditions.

- **Symptom Recognition:** It is the identification of the plant's reaction to the disease (e.g., wilting, leaf spots, galls, and chlorosis).
- **Sign Recognition:** It is the direct observation of the physical structures of the pathogen (e.g., fungal spores, mycelial mats, or bacterial ooze).
- **Disease Triangle:** It includes the correlation of the visual evidence with the presence of a susceptible host (a virulent pathogen) and favorable environmental conditions like temperature, moisture.
- **Pattern Analysis:** It is the observation of the disease distribution across the field. The analysis types of patterns include:
  a) A random pattern often indicates abiotic disorder such as nutrient deficiency or chemical damage.
  b) A clustered pattern often suggests an infectious pathogen spreading from plant to plant.

6.2 *Microscopic Examination*

Once a sample is collected, microscopy is used to confirm the presence of a pathogen's structure.

- **Bright-field Microscopy:** It is used to observe pathogen structures directly from infected tissue, such as fungal spores, hyphae, and nematodes.
- **Moist Chamber Incubation:** It is the process of placing infected tissue in a humid environment (moist chamber) for 12–48 hours, encouraging the sporulation or growth of many fungi, making them easier to observe and identify under the microscope.
- **Tissue Sectioning:** Preparing thin cross-sections of infected tissues (e.g., a stem with a wilt) to observe internal structures like plugged vascular bundles in wilt diseases or the location of the pathogen within the host cells.

6.3 *Isolation and Culturing*

The fungal and bacterial pathogens are identified by isolating them from the infected tissue and growing them on an artificial nutrient medium.

- **Aseptic Isolation:** Small pieces of infected tissue are surface-sterilized (usually with bleach) to kill superficial contaminants and plated onto an agar medium like Potato Dextrose Agar for fungi or Nutrient Agar for bacteria.
- **Identification by Morphology:** The pure culture grown on the culture media are identified based on the colony characteristics (color, texture, growth rate) and the microscopic morphology of its spores (for fungi) or cell shape (for bacteria).
- **Fatty Acid Analysis (FAME):** It is a semi-traditional method for bacteria identification, where the fatty acids in the cell membrane are analyzed for a characteristic profile, which aids in genus and species identification.

6.4 *Koch's Postulates (The Diagnostic Test)*

Developed by Robert Koch in the late 19th century, these are the fundamental criteria used to prove that a specific organism causes the disease.

1. The pathogen must be associated with all diseased plants and not with healthy plants.
2. The pathogen must be isolated from the diseased plant and grown in pure culture.
3. The isolated pathogen must be inoculated onto a healthy, susceptible host and must cause the same disease symptoms.
4. The pathogen must be re-isolated from the experimentally inoculated plant and shown to be identical to the original organism.

7. Advanced Detection Methods

It includes the use of modern approaches by integrating the laboratory and sensor-based techniques to detect plant diseases.

7.1 *Laboratory-Based Methods*

The accurate diagnosis can be achieved by the use of molecular, serological and microscopic techniques to detect the pathogen directly. Examples include:

7.1.1 Molecular Techniques

The DNA/RNA of the pathogen can be analyzed by Polymerase Chain Reaction (PCR).

7.1.2 Serological Techniques

The pathogen-specific antibodies or antigens are detected by various serological methods and mostly by Enzyme-Linked Immunosorbent Assay (ELISA).

7.1.3 Microscopic/Culture Methods

The specific pathogen is isolated by different cultural techniques and identified by the use of electron microscopes like TEM or SEM.

7.2 *Sensor-Based and Imaging Techniques*

These techniques are used to detect minute changes in physiology or appearance of the diseased plant even before the visual development of the symptoms. These are suitable for real-time, large-scale field monitoring and are often non-destructive. Examples include:

7.2.1 Optical Techniques

The changes in leaf color, texture, and light reflection are detected by the use of RGB imaging by standard cameras, hyperspectral/multispectral imaging and chlorophyll fluorescence.

7.2.2 Thermal Imaging (thermography)

The pathogen interference with stomatal function and transpiration can affect the leaf surface temperature and can be analyzed by thermal imaging.

7.2.3 Artificial Intelligence (AI) and Machine Learning (ML)

In large-scale fields, drones or IoT sensors are deployed to obtain vast image data of the experimental field. These vast image datasets are used to train the Deep Learning (DL) models like CNNs to automatically classify and diagnose diseases from acquired images.

The integration of these advanced technologies helps in targeted application of treatments that allows for more effective disease management strategies to support sustainable and productive farming practices [11] [12].

8. Sensor-Based Approaches for Plant Disease Detection

The advanced non-invasive and sensor-based technologies like remote and proximal sensing have revolutionized the monitoring and detection process of plant diseases. These methods are used to capture the physical and biochemical changes in plants under stress from a pathogen. The primary difference between the two approaches is the distance between the sensor and the plant.

8.1 Proximal Sensing or Close-Range Sensing

In this method, the sensor is placed very close to the target, often hand-held or mounted on ground-based vehicles like tractors or autonomous robots. This approach provides very high spatial and spectral resolution and makes it ideal for the early, asymptomatic detection of disease on individual leaves or plants [13] [14]. Technologies used in proximal sensing are:

8.1.1 Handheld and Ground-based Spectroscopy

It includes instruments like spectroradiometers that measure the detailed spectral reflectance of a leaf across hundreds of narrow bands. Since a plant's spectral signature changes even before visual symptoms appear due to altered pigment, water content etc. This is a powerful tool for early diagnosis.

8.1.2 *Chlorophyll Fluorescence Imaging*

This technique measures the light re-emitted by chlorophyll, which is directly related to the efficiency of photosynthesis (Photosystem II). In diseased plants, the pre-symptomatic stress can be detected by the drop of photosynthetic efficiency.

8.1.3 *Close-range Thermal Imaging (Thermography)*

High-resolution thermal cameras are used to measure the temperature of specific plant organs. The plant water regulation (transpiration), is disrupted during the infection stage, which causes the temperature increase in the stressed tissue. This change in temperature can be measured by thermal cameras.

8.1.4 *High-Resolution Imaging (RGB/multispectral)*

The detailed images of the plants are captured by cameras that are mounted close and analyzing the shape, size, and texture of lesions. This enables the automated counting and classification of disease spots.

8.2 *Remote Sensing or Large-Area Coverage*

The data is collected from a distance, usually from high altitudes using Unmanned Aerial Vehicles (UAVs or drones), airplanes, or satellites. This is the suitable and efficient method for monitoring large agricultural fields or entire regions efficiently. The resolution obtained in this method is generally coarser than proximal sensing. It provides the spatial context necessary for precision agriculture management [15]. Techniques used in remote sensing are:

8.2.1 *Multispectral Imaging*

The sensors capture the data across a wide band of the electromagnetic spectrum, particularly the visible, Red Edge, and Near-Infrared (NIR) regions. The overall plant health and greenness across the field is referred to as Vegetation Indices (VIs) (Normalized Difference Vegetation Index), which is calculated by the data captured by the sensors.

8.2.2 *Hyperspectral Imaging*

The data is captured by the advanced sensors across hundreds of narrow, contiguous spectral bands. It helps the researchers to build a highly detailed spectral signature which is used to distinguish between different types of diseases or abiotic stresses and also detect early infections.

8.2.3 Aerial Thermal Imaging

It helps to monitor canopy temperature over a large area with the help of thermal cameras mounted on drones or planes. By identifying these hot spot areas, farmers can take action against water scarcity or early signs of disease or outbreak.

8.2.4 RGB Imaging (Aerial)

In this method, the visible symptoms of diseased plants are detected by standard high-resolution cameras. The symptoms like yellowing (chlorosis), browning (necrosis), or canopy gaps are analyzed using image processing and machine learning techniques [16][17].

9. Machine Learning and AI for Plant Disease Detection

Machine Learning (ML) and Artificial Intelligence (AI) is changing the way plant disease diagnosis is performed, automating the identification and classification process successfully and rapidly, making the sphere of agriculture more manual than precise management. AI systems use the information that sensors (mostly images) capture to recognize unnoticeable symptoms, which may arise and subsequently provide immediate and localized treatment [18].

9.1 ML and AI in Plant Disease Detection

The essence of AI/ML application is to make a certain diagnosis using the models that are able to work with various types of data which are spectral, visual, thermal images depicted in Figure 5.

Traditional Machine Learning (ML)

This technique depends on a two-step process, i.e., feature extraction and classification.

9.1.1 Feature Extraction

To quantify disease characteristics, specific metrics are designed by human experts or algorithms. The metrics include the features like color (e.g., hue, saturation), texture (e.g., lesion smoothness or roughness), and shape (e.g., lesion area and perimeter).

9.1.2 Classification Algorithms

The extracted features are then passed into traditional machine-learning models, which classify the image into the appropriate disease category. The common algorithms used are Support Vector Machines (SVM), K-Nearest Neighbors, and Random Forest.

Fig. 5 Integration of AI and Machine learning in Agriculture [19][20]
(www.ucdavis.edu/food/news/big-data-comes-to-dinner & www.bitsathy.ac.in/blog/generative-ai-in-agriculture-revolutionizing-crop-management/)

- **Support Vector Machines (SVM):** It is used to create a clear separation between healthy and diseased classes.
- **K-Nearest Neighbors (KNN):** It helps to classify the data based on the majority class of its nearest data points.
- **Random Forest (RF):** This method uses a collection of decision trees to make predictions that are more stable and accurate.

9.2 Deep Learning (DL)

It is a subset of AI, which is mostly used for image-based plant disease detection. It performs end-to-end learning and eliminates the need for manual features [21][22].

9.2.1 Convolutional Neural Networks (CNNs)

CNNs learn key features straight from raw image inputs from RGB, multispectral, and hyperspectral data, without any manual feature design. They are used for three main tasks, namely, Image Classification, Object Detection, and Semantic Segmentation.

- **Image Classification:** The type of disease is identified from a single leaf or plant image (e.g., "Tomato Early Blight").
- **Object Detection:** It identifies and locates each disease spot or pest in a given field image and highlights the affected areas with bounding boxes.
- **Semantic Segmentation:** It precisely maps the diseased area at the pixel level and helps in accurately quantifying the severity of the infection.

9.2.2 Transfer Learning

In plant pathology, this step is very crucial as available labeled datasets are insufficient. A powerful AI model can be used which is expertized on a massive dataset and is trained on the smaller dataset of specific plant diseases. This will eventually reduce the training time and data requirements while enhancing the performance.

10. Predictive Modeling and Decision Support

In precision agriculture, AI is used for smarter analysis like identification, and higher-level decision-making [23].

10.1 Prognosis and Forecasting

It uses machine learning models to predict the timing and severity of plant disease outbreaks by utilizing historical disease data, weather patterns, and crop variety information.

10.2 Real-time Decision Support

The AI systems deployed on edge devices like drones and cameras can instantly diagnose and send alerts. It helps in site-specific treatment, optimization of fungicide uses, and reduces cost.

11. Data Pipeline: From Acquisition to Decision

In precision agriculture, plant disease detection includes the data pipeline which is a well-structured and multi-stage process. It transforms the raw data from sensors into a useful insight for the farmers to take decisions. It is the backbone of any data-driven management system.

The data pipeline structure includes four main phases, i.e., Acquisition, Processing, Analysis, and Decision or Action.

11.1 Data Acquisition (Ingestion)

This initial step involves the collection of high quality, relevant data from the field. The information obtained must accurately reflect the health status of the plant.

- **Sensors and Platforms:** Data of the diseased plants are collected by using the sensors that are placed close to the plants or from a distance through drones etc. Important types of data are Optical/Spectral Data, Thermal Data, Contextual Data and Data Ingestion.
- **Optical/Spectral Data:** RGB images, Multispectral data (e.g., NDVI), and Hyperspectral image measures change in pigmentation, structure, and water content of leaves with disease.
- **Thermal Data:** Infrared data are taken to monitor localized changes in temperature, which indicate impaired transpiration (early disease stress).
- **Contextual Data:** GPS position, time stamp and environmental information (weather, soil moisture) are essential for geo-referencing and contextual analysis.
- **Data Ingestion:** The data are moved (ingested) into a storage system which may be in the form of massive parallel file systems with raw remote sensing images or structured metadata in the form of databases.

11.2 Data Processing (Cleaning and Transformation)

Raw data sampled in the field is not always clean, distorted or unorganized and thus it takes a lot of cleaning and processing before it can be used to be analyzed.

- **Pre-processing (Cleaning):** This involves atmospheric and lighting effects correction of remote sensing data, pixel mapping to accurate field coordinates, and noise reduction. In the case of image data, it may include the resizing of images, their cropping, and elimination of undesired artifacts.

- **Feature Engineering:** The raw sensor data is converted to quantitative and meaningful features.
- **Vegetation Indices (VIs):** Computing multispectral indices to indicate the health and greenness of the vegetation.
- **Feature Extraction:** It involves the process of separating the crop and the background (soil, weeds) in the picture and deriving texture, shape and color features of visible lesions.
- **Data Storage:** Once the cleaning and processing is done, the data of the features is stored in a data warehouse where it is readily retrieved by the analysis models.

11.3 Data Analysis (Modeling)

This phase utilizes the advanced computing processes to identify the key patterns and classify the health condition of the plants.

- **Model Training and Validation:** ML or DL models, e.g., CNNs, are trained on the processed information. The model is trained to be able to categorize that a plant is healthy or diseased, or otherwise in a stress (e.g., nutrient deficiency).
- **Classification and Detection:** The model is trained and tested on the new and unknown field data. The resultant is a very precise prediction map that identifies the existence of a disease and usually gives a measure of the severity of the disease (e.g., area of infection).
- **Spatial Analysis:** Visualization and analysis of the spatial heterogeneity of the disease with Geographic Information System (GIS) tools are carried out to determine the high-risk areas, disease foci, and patterns of disease dissemination on the field.

11.4 Decision and Action (Output)

The last phase converts the analysis outcomes into a localized teaching that can be useful to the farmer.

- **Decision Support System (DSS):** A disease map is inputted into a DSS. The system suggests an action based on preset thresholds and economic models (e.g., cost of fungicide versus what size yield loss is expected).
- **Prescription Map Generation:** The output is a Variable rate Application (VRA) Prescription Map that displays the specific GPS coordinates and the required dose of treatments like use of fungicide, water, fertilizer etc. to be applied for each particular area on the field.
- **Action/Execution:** The prescription map is sent to a farm machine (e.g., a smart sprayer or drone) which then performs the application of the site-specific use of inputs, which means that the resources are applied where and when they are required. This completes the cycle of data driven accuracy management.

12. Practical Challenges in Plant Disease Detection

The main challenges include the data limitations, environmental variability, model generalization, and practical deployment [24][25].

12.1 Data and Annotation Limitations

The performance of AI models is heavily reliant on the quality and quantity of data, which is a major bottleneck.

- **Data Scarcity and Imbalance:** Comprehensive datasets are often limited, especially for less common crops, rare diseases, or specific local conditions. Furthermore, datasets are frequently imbalanced, with a large number of healthy samples compared to diseased samples, or more data for common diseases than rare ones.
- **Lack of Diversity:** Training data often lacks images taken under the full range of real-world field conditions, such as different lighting (shadows, low light, harsh sun), diverse geographical locations, varied crop varieties, or different growth stages.
- **High Cost of Labeling:** Acquiring and accurately labeling images requires expert knowledge (plant pathologists) to confirm the disease type and severity, making the process time-consuming and expensive.
- **Co-occurrence of Issues:** Datasets often focus on single diseases, but in the field, plants frequently suffer from multiple diseases, pests, or nutrient deficiencies simultaneously, which is complex to annotate and classify.

12.2 Environmental and Visual Variability

Field conditions introduce variability that can severely degrade the accuracy of laboratory-trained models.

- **Lighting and Weather Effects:** These environmental Changes in illumination like shadows, glare, time of day, cloud cover can dramatically alter the image features. This will lead to poor performance of computer vision models that were usually trained on controlled images.
- **Visual Ambiguity of Symptoms:** The automated systems face significant difficulty in accurate differentiation of different diseases, insect damage, and abiotic stresses like nutrient deficiencies or sunscald as they visually exhibit similar symptoms e.g., yellowing or leaf spots.
- **Early-Stage Detection:** Diseases in their early, subtle stages often present minimal or hidden symptoms, which can be easily obscured by natural field conditions like dust, water on leaves, or the plant canopy itself.
- **Viewpoint and Clutter:** Images captured in the field contain background clutter (soil, weeds, and adjacent plants) and complex canopy structures that can hide the affected area or confuse the model.

12.3 Model Generalization and Robustness

Even the models themselves are not always able to perform under the conditions that are not within the training environment.

- **Poor Generalization:** It is found that models that are trained on data related to a single type of crop, or a region, or controlled conditions in a laboratory may later perform much worse when applied in new, unseen conditions (low field-deployment accuracy).
- **Resolution and Sensitivity:** Systems must be able to identify symptoms at different scales, i.e., tiny spots on a single leaf to an outbreak over a whole field (i.e., drone-based vs. smartphone-based images).
- **Computational Constraints:** Deep learning models can be very computationally intensive, and therefore demand a lot of processing power to be trained. This is a realistic constraint to be deployed on resource constrained devices such as low-cost smart phones or cheap edge computing devices in the field.

12.4 Practical Deployment and Adoption

The implementation of the technology into a workflow of a farmer establishes non-technical barriers like

- **Accessibility/Usability:** The last tool, e.g., a mobile application or drone system has to be convenient, very reliable, and give non-expert farmers clear and actionable recommendations (diagnosis and treatment recommendation).
- **Infrastructure Requirements:** Cloud-based processing often requires a good internet connection or cellular connection which may be a problem in isolated rural agriculture areas.
- **Interoperability and Integration:** New detection systems have to be interconnected to the current farm management systems, precision agriculture equipment, and processes of farmer decisions.
- **Cost-Effectiveness:** The overall price of the automated system including software, hardware, cameras, drones, sensors, maintenance, and training should provide a tangible and significant payoff to the average farmer.

Modern research in this area usually entails the cooperation of computer scientists, plant pathologists and agricultural engineers. Two prominent personalities related to pioneering activity in the area of utilizing deep learning in recognizing plant infection by using large publicly accessible datasets comprise:

- **David P. Hughes and Marcel Salathe:** These are the authors who published a groundbreaking 2016 article on the application of deep learning for classification of 26 diseases in 14 crop species on a publicly available set of more than 50,000 images and showed the potential high accuracy of the technique [26]. Their work acted as a critical point of reference and resource to other researchers.

A different scientist that worked on autonomous detection systems is:

- **Tyr Wiesner-Hanks:** A former Cornell graduate student, he developed a study to show a proof-of-concept showing an autonomous phenotyping system which combined a deep learning method with Unmanned Aerial Vehicles (UAV) or drone-based imagery to effectively detect the symptoms of maize disease in the field. This underscores the shift towards real world and in-field applications of these technologies. In one of his works, Tyr Wiesner-Hanks illustrates one of the main applications of AI and UAV technology in detecting plant disease in one of the primary crops [27].

13. Conclusion

The detection of plant diseases in agriculture has been redefined with the integration of advanced technology. This has paved a way for representing industries to be beyond the limitation of manual scouting, subjective, as well as delayed issues. The research clearly indicated that the future of crop protection is predictive, precise and decentralized.

The current immediate technologies like DL and computer vision have proved to create a superior impact by providing a high level of accuracy to both the farmers and the agronomists. Such precision may be done under the support of quick diagnostic instruments that in reality can easily identify diseases based on the provided simple visual information. The challenges such as the necessity of fast inference on the device and confidentiality of the data preservation can certainly be tackled through the development of the lightweight model architectures and federated learning. The most crucial agricultural resilience plan is the digitalization of plant disease detection and limited traditional approaches.

Thanks to its ability to introduce timely and selective treatment, it decreases the losses in crop yields, decreases the application of broad-spectrum pesticides, and enhances sustainability and economic feasibility of agriculture.

It is the need to promote further partnership of inter-disciplinary research among plant pathologists, computer scientists and agricultural engineers that will assist in setting up potent instruments that can easily identify illnesses long before they occur to ensure the realization of solutions to healthy food consumption globally.

References

[1] Barbedo, J.G.A. (2019). A review on the use of deep learning for the diagnosis of plant diseases. *European Journal of Plant Pathology*, *154*(1), 173–189.

[2] Agrios, G.N. (2005 or current edition). *Plant Pathology*. Academic Press.

[3] Shaad, N.W., Jones, J.B. and Chun, W. (2001). *Laboratory Guide for Identification of Plant Pathogenic Bacteria. American Phytopathological Society. Paul, MN, USA.*

[4] Sharma, P.D. (2018), "*Plant Pathology*", Second edition, Rastogi Publications.

[5] Lucas, J.A. and Jones, P.L.P. (1998 or current edition). *Plant Pathology and Plant Pathogens.* Blackwell Science.

[6] Westcott, Cynthia. (2001 or current edition). *Westcott's Plant Disease Handbook*. Springer.
[7] Mahlein, A.K. (2016). Plant disease detection by imaging sensors: Parallels and specific demands for precision agriculture and plant phenotyping. *Plant Disease*, *100*(2), 241–251.
[8] Kamilaris, A. and Prenafeta-Boldú, F.X. (2018). Deep learning in agriculture: A survey. *Computers and Electronics in Agriculture*, *147*, 70–90.
[9] Gautam, V., Mishra, A.M., Kaur, P., Singh, M.P., Singh, P., Diwakar, M. and Gupta, I. (2025). Composite deep learning model for characterization of tomato leaf disease. *Multimedia Tools and Applications*, 1–30.
[10] Raza, A., Razzaq, A., Mehmood, S.S., Zou, X., Zhang, X., Lv, Y. and Xu, J. (2019). Impact of climate change on crops adaptation and strategies to tackle its outcome: A review. *Plants*, *8*(2), 34.
[11] Bock, C.H., Poole, G.H., Parker, P.E. and Gottwald, T.R. (2010). Plant disease management: A perspective on the role of new technologies. *European Journal of Plant Pathology*, *128*(1), 1–18.
[12] Amrita, Nayak, S.K., Kumar, R. and Tomar, S. (2025). Plant Disease Detection Using Deep Learning. *Journal of Mechanics of Continua and Mathematical Sciences*, *20*, 134–149.
[13] Mahlein, A.K. (2016). Plant disease detection by imaging sensors—Review of close-range sensing techniques. *Italian Journal of Agronomy*, *11*(2), 87–99.
[14] Pethybridge, S.J., Nelson, A. and Mahlein, A.K. (2020). Proximal and remote sensing for plant disease detection. *Annual Review of Phytopathology*, *58*, 249–275.
[15] Gogoi, N.K., Deka, B. and Bora, L.C. (2018). Remote sensing and its use in detection and monitoring plant diseases: A review. *Agricultural Reviews*, *39*(4).
[16] Bendig, J., et al. (2017). Combining UAV-based imaging and machine learning for early detection of plant stress and disease. *Remote Sensing*, *9*(3), 263.
[17] Thandapani, S.P., et al. (2019). "Decision Support System for Plant Disease Identification. In: *International Conference on Advanced Computing and Intelligent Communication* (pp. 539–548). Springer, Singapore.
[18] Priyadharshini, S. and Arivazhagan, S. (2020). A survey on plant disease detection using image processing and machine learning techniques. *International Journal of Modern Agriculture*, *9*(3), 143–155.
[19] https://www.ucdavis.edu/food/news/big-data-comes-to-dinner.
[20] https://www.bitsathy.ac.in/blog/generative-ai-in-agriculture-revolutionizing-crop-management/.
[21] Ferentinos, K.P. (2018). Deep learning models for plant disease detection and diagnosis. *Computers and Electronics in Agriculture*, *145*, 311–318.
[22] Chen, J., Zhang, D., Sun, Y. and Nanehkaran, Y.A. (2020). Using deep transfer learning for image-based plant disease identification. *Computers and Electronics in Agriculture*, *173*, 105393.
[23] Holst, N. and Baret, F. (1998). "Decision support systems for pest and disease management in agriculture: A review. *Agricultural Systems*, *57*(2), 237–252.
[24] Perini, C.R., et al. (2023). Current and emerging trends in techniques for plant pathogen detection. *Frontiers in Plant Science*, *14*, 1120968.
[25] Fry, W.E. (2008). *Principles of Plant Disease Management*. Academic Press.
[26] Mohanty, S.P., Hughes, D.P. and Salathé, M. (2016). Using Deep Learning for Image-Based Plant Disease Detection. *Frontiers in Plant Science*, *7*, 1419.
[27] Wiesner-Hanks, T., Stewart, E.L., Kaczmar, N., DeChant, C., Wu, H., Nelson, R.J. and Gore, M.A. (2018). Image set for deep learning: Field images of maize annotated with disease symptoms. *BMC Research Notes*, *11*(1), 440.

2 Machine Learning for Agricultural Application

Manjushree Nayak[1*] and Subhakanta Panda[2]

Despite the fact that farms are critical to sustaining food security in the world, it is experiencing growing challenges due to a lack of resources, population expansion as well as climate change. Machine learning (ML), one of the branches of artificial intelligence (AI), has brought a revolutionary technology to address these concerns by helping to make key agricultural decisions based on data.

Precision agriculture: This is a method based on sensor and satellite-data that optimizes the irrigation, fertilization, and pesticide application, reducing costs, and minimizing environmental impact; disease and pest identification, where machine learning models of images are used to detect the earliest indications of infections; and prediction of crop yield, which helps farmers to allocate resources efficiently. In addition, ML enables the optimization of a supply chain, identification of weeds, and animal monitoring, which enhance the overall sustainability and efficiency of the agriculture industry. The ability of ML to monitor and automate in real time is also upgraded by the advancements in deep learning, computer vision and integration with the Internet of Things (IoT). The adoption is however not done in large scale due to challenges such as the inadequate access to quality datasets, high cost of implementation, and the failure of farmers to have technical knowledge, despite its benefits. Researchers, legislators, and tech companies should collaborate in order to establish machine learning systems that are scalable, easy to use, and inexpensive and tailored to different agricultural applications. Machine learning can revolutionize agriculture to a significant extent by making agriculture resilient, sustainable, and productive. Through continuous development, it can have a major impact on the future of smart farming and ensure food security to the growing population of the world.

1 Amity School of Engineering & Technology, AMITY UNIVERSITY, Raipur, Chhattisgarh, Manth (Kharora), State Highway 9, Baloda Bazar Road, 493225, India. ORCID (https://orcid.org/0000-0001-6383-780X)

2 Department of Computer Science and Engineering, Centurion University of Technology and Management (CUTM), Paralakhemundi, Odisha-761211. ORCID (https://orcid.org/0009-0006-1862-5684)
Email: subhakanta.panda@cutm.ac.in[2]

* Corresponding author: manjushreesai44@gmail.com

DOI: 10.1201/9781003767534-2

1. Introduction

Agriculture has been the foundation of the entire human civilization as it has made food, raw materials, and offered employment to billions of people throughout the world. In the developing world, such as India, agriculture does not only sustain the economy but also a larger population in the rural areas. The traditional form of farming was carried out on the basis of the empirical knowledge, local experience, and seasonal cycles. But with the world population increasing continuously (estimated to hit over 9 billion by the year 2050), the agricultural industry is under tremendous pressure to be more productive and at the same time to be sustainable [1]. Contemporary agriculture is forced to face a host of thorny issues: the decreasing agricultural land area, global warming, incessant precipitation and soil erosion, and the increasing water shortage. There has also been over-exploitation of fertilizers and pesticides which have resulted in poor quality of soil and low-level biodiversity. There are labour gaps, unstable market requirements, and increased costs of inputs, which, in turn, have complicated farmers in making a living. These issues highlight the imperative of seeking new and sustainable solutions which will ensure that agricultural yields are optimized without a danger to the environmental well-being [1][2].

Conventional agricultural practices despite being time-tested are becoming inadequate to deal with the dynamics of the modern agricultural environment. Implementation of new technology in agriculture, commonly known as smart farming or precision agriculture, has become an important agenda leading to sustainable development. The technological intervention assists in keeping track of the soil conditions, to maximize irrigation, predicting the weather pattern and more effectively dealing with pest and disease outbreaks. The technologies produce large volumes of data on the soil moisture, temperature, nutrients content, crop health, and weather dynamics. The difficult part however is in interpreting this complicated data in order to make actionable decisions. In this case, the role of the Machine Learning (ML) is transformational in that it analyses extensive data, finds latent patterns, and creates predictive information to aid in evidence-based decision-making among farmers and policymakers.

Soil makeup and harvesting output are all measurable and assessable, as well as market prices and weather patterns. Everything about agriculture can now be quantified and analysed. Through successful data utilization, we can have precision agriculture whereby the application of inputs like water, fertilizers, pesticides, etc. are used in the optimal amounts at the right time and place. Fields are fitted with sensors that monitor temperature, nutrient content, and moisture of the soil in real time. Satellites and drones offer high-resolution images that can be used to monitor the stress of crops, outbreaks, and weeds. Meteorological information gathered in weather stations assists in the short- and long-term forecasts. Therefore, data-driven agriculture will decrease waste, boost productivity, and support sustainability. Artificial Intelligence (AI) and ML have transformed various sectors, and agriculture is not an exception. AI is defined as the systems that can do the tasks that ordinarily need human intelligence whereas ML is defined as algorithms that learn and get better with time without being explicitly programmed [1].

ML can be used in agriculture to do automation and intelligent decision support. Image recognition algorithms can detect crop diseases using images of the leaves, regression models can estimate yields depending on soil and weather conditions, clustering algorithms can cluster fields based on their productivity, and reinforcement learning (RL) can optimize irrigation and fertilization decisions. As more and more agricultural data is becoming accessible and computing power becomes more affordable, there has been a surge of new applications of ML making the divide between conventional farming and digital advancement smaller. In this chapter, the author is trying to discover the ways ML can tackle the burning issues of contemporary agriculture through transforming data into actionable insights. It starts with the introduction of the key principles of ML, then moves on to the data sources, tools, and important agricultural applications, including prediction of yields, disease detection, and optimal use of resources. Reality case-studies, the existing issues, and the future of the field are also brought out in the chapter. Finally, it is aimed at showing how ML can be used as an agent of sustainable and efficient farming that will enable farmers to make informed choices, reduce losses, and adjust to environmental changes. With the combination of ML with other newer technologies such as IoT (internet of things), robotics, and remote sensing, agriculture will be able to move to a smarter, more durable, and data-driven future.

2. Fundamentals of Machine Learning

ML is a branch of AI, and it is associated with creating algorithms that can learn based on data and base predictions or choices without being explicitly written. The notion of ML is that even complex patterns in massive data sets can be recognized by the computers and can be utilized by them to make predictions based on the data. In agriculture, computers can train on a vast amount of data, including soil properties, weather, satellite images, sensor measurements, and others, to forecast such things as crop yields, disease outbreaks, and irrigation demands with the help of ML. This learning and adaptation ability is what makes ML a major facilitator of precision agriculture whereby all decisions are made out of precise, data-driven information [1][2][3].

2.1 Types of Machine Learning

There are four major categories of ML algorithms, which are supervised learning, unsupervised learning, semi-supervised learning, and RL.

They have various functions based on the character of the problem and the accessible labelled data.

2.2.1 Supervised Learning

In supervised learning, algorithms are trained with labelled-data sets, i.e., data in which the inputs are known and the desired result is known. The model gets to learn

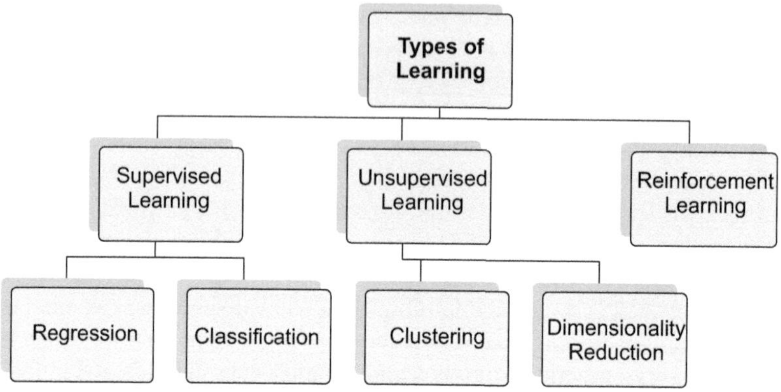

Fig. 1 Classification of Machine Learning.

the association of inputs and output, and is capable of making predictions of new data. Agricultural examples include modelling crop yield with soil and weather data, identifying crop diseases through images of labelled leaves, and estimating nutrient content from soil samples. Common algorithms are Linear Regression, Support Vector Machines (SVM), Decision Trees, Random Forests, and Neural Networks.

2.2.2 Unsupervised Learning

Unsupervised learning concerns unlabelled information, in which the algorithm determines trends or frameworks within the data, but without definite results.

Agricultural uses: grouping of soils according to their fertility, defining similar areas of crops or grouping bands of plants by satellite. Standard algorithms K-Means Clustering, Hierarchical Clustering, and Principal Component Analysis (PCA).

2.2.3 Semi-supervised learning

A semi-supervised learning is a middle ground method that involves a small quantity of labelled data and a huge quantity of unlabelled data. Agricultural data (such as a disease image that has to be labelled manually) can be a particularly valuable area since this process can also be time-intensive and costly, which is why semi-supervised learning can come in handy [9].

2.2.4 Reinforcement Learning

The idea behind RL is the interaction between the agents and the environment to accomplish a goal. The algorithm is taught by trial- and- error by being fed by rewards or punishments.

Agricultural examples: RL can be used to improve irrigation schedules, pesticide spraying, and temperature control in a greenhouse. The system continuously learns the best way to maximize harvests while using fewer resources.

3. Data Sources and Technologies in Agriculture

3.1 IoT and Sensor-Based Collection of Data

IoT has become a major facilitator of smart and data-driven agriculture. IoT is a system of connected devices that are integrated with sensors to gauge, transfer, and analyse real-time information about the physical world. In farming, IoT applications measure the state of soil, crops, and the environment in real-time, allowing making decisions in a timely manner and with high accuracy [13].

Fields fitted with sensors can be used to measure soil moisture, pH, temperature, salinity, as well as nutrient content. Such readings are sent to cloud services or edge computing and processed and analysed. As an illustration, irrigation systems can be programmed to automatically switch irrigation systems on when the moisture content in the soil is below the optimum level, so irrigation can be avoided by either over- or under-irrigating the soil. On the same note, the nutrient sensors have the potential to guide the application of fertilizers so that the crops get the appropriate quantity at the appropriate time.

In addition, the weather stations installed on IoT networks also gather information about rainfall, humidity, and temperature, which predictive models can be used to predict disease outbreak or irrigation requirements. The real-time data stream of the IoT devices forms the basis on which ML algorithms can be applied to determine trends, correlations, and anomalies that may be overlooked by human eyes.

3.2 Remote Sensing and Satellite Imagery

Remote sensing technology is the technology which deals with taking data on the surface of the earth using an aircraft or satellite without physical contact with the ground. It has transformed original agricultural observations by offering extensive, intermittent, and multi spectral data that may be used to evaluate the health of crops, soil, and land-use patterns.

Current satellites like Sentinel, Landsat, and MODIS record pictures in different spectral bands, which includes visible band, infrared, and thermal bands [1]. It is the spectral images, which are used to identify the subtle changes in vegetation indices like the normalized difference vegetation index (NDVI), which is associated with plant vigour, chlorophyll content, and biomass. Models based on ML can be used to detect stressed areas in a farm, early stages of a disease, or even the type of crops. Moreover, the combination of satellite data, weather, and soil data helps to improve yield prediction models. As an illustration, multi-temporal remote sensing imagery combined with regression algorithms can be used to estimate the crop growth stages and ultimate yields way before the harvesting period. Remote sensing is therefore important in creating scalable ML applications in managing agriculture at regional and national scales.

3.3 UAVs (Drones) and Ground Robots

Precision agriculture has become unimaginable without Unmanned Aerial Vehicles (UAVs), also referred to as drones. Drones can also take highly detailed images of particular farm plots any time, unlike satellites which are restricted by [12] the orbital schedule, and cloud cover. Imaging with drones is in support of a number of ML-driven applications, such as:

- **Agricultural disease and pest detection:** Convolutional Neural Networks (CNNs) have the potential to evaluate aerial imagery and identify discoloured or damaged leaves.
- **Weed detection:** Image segmentation using the ML is able to differentiate crops and weeds, and instructs selective spraying of herbicides.

Ground robots, conversely, are used to support drones, which complete tasks on a field-level, including weeding, harvesting, and soil sampling. These robots are able to respond to varying field conditions, navigate autonomously, and make precision manoeuvres with minimal human oversight with the implementation of ML integration.

3.4 Weather and Climate Data

Among the most significant aspects in agricultural productivity are the weather conditions. Crop growth periods and vulnerability to pests or diseases are dependent on the variables of rain, temperature, humidity, and wind velocity. Precise weather prediction is hence important when making plans to carry out agricultural activities. ML models utilize meteorological data to a considerable extent in terms of short and seasonal forecasting. Local weather stations, national weather services, or international databases such as the European Centre of Medium-range Weather Forecasts (ECMWF) and POWER data at NASA can provide the data. Agricultural data analysis has taken new dimensions with climate change. ML models made on the historical data about the climate can model future conditioning to enable the policymakers to be able to develop adaptive strategies concerning sustainable farming. As an example, predictive analytics may detect changes in the time of the planting season or recommend other crops that would not be lost in changing the conditions.

3.5 Soil and Crop Databases

Soil is among the most important resources of agriculture production. Knowledge of its physical, chemical, and biological characteristics allows it to be managed well in terms of resources. The parameters usually comprised of soil data are soil texture, organic matter, nutrient availability (N, P, K), soil moisture, and pH level [4].

Across the globe, government departments and research institutions have large databases containing information about the soils, including the National

Bureau of Soil Survey and Land Use Planning (NBSSLUP) in India or the USDA Soil Survey Geographic (SSURGO) database in the United States. These datasets are used in the creation of ML algorithms that predict soil fertility, types of soil, and recommend crops. Crop databases, conversely, put together data on species, varieties, requirements to grow, and susceptibility to diseases. By combining soil and crop data, ML systems can align the empirical type of crop to the particular condition of soil and climatic environment and enhance productivity and minimize resource wastage [3].

3.6 Data Integration Problems and Open Datasets

Availability of large high quality datasets is important to the success of ML models in agriculture. Most open-source agricultural datasets have been made available to researchers and developers in the past few years. Examples include:

FAO and World Bank data: They are data sets that present agricultural statistics, climate, and soil maps in the whole world. Although these developments are made, merging different data sets is a significant challenge. There are many different heterogeneous information providers of agricultural data such as satellite sensors, field devices, laboratory experimentation, and surveys, and they can vary in terms of resolution, scale, and format. Harmonization of data and preprocessing is time-consuming to ensure that they can be trained with an ML model.

Also, there are data privacy, ownership, and standardization problems that are an impediment to massive data sharing. Farmers might be reluctant to provide proprietary field data and the national policies of data complicate the collaboration further. To solve these issues, there should be open data standards, the development of secure sharing models, and the inclusion of farmers in data governance models. Data and emerging technologies are integrated in order to enhance the quality of education programmes delivered by the university.

3.7 Data and New Technologies Integration

Agriculture has moved to a new stage in the integration of various sources of data to form integrated platforms. The integration of IoT, remote sensing, and cloud computing makes possible the development of so-called digital twins of farms, virtual models that can mimic real-time farming. Also, edge computing enables data processing either locally on IoT devices or on local servers, which shortens latency and will not require Internet connectivity at all times. This will be quite useful in rural or remote areas with agricultural regions and limited network access. The blockchain technology has also demonstrated the prospects of protecting the integrity of data as well as transparency in agricultural supply chains [8].

The quality, diversity, and accessibility of data are the core elements of making ML in agriculture a successful tool. Data is the cornerstone of smart decisions, whether it is the IoT sensors and drone images or soil maps and climate records. Nevertheless, the challenges of successful integration of all these different sources are warranted by the need to have solid data management frameworks and sophisticated computational tools. The current convergent development of IoT, remote sensing, and ML is slowly turning agriculture into a data-driven environment that can be real-time monitored, predicted, and optimized.

4. Major Applications of Machine Learning in Agriculture

Big Data: Big uses of machine learning in agriculture

ML has become a revolutionary technology in every step of the agricultural production process, including analysis of soil before sowing, and quality control of the harvest. ML helps farmers and agronomists to make data-driven decisions that will increase productivity, lower costs, and improve sustainability by revealing the hidden relationships between complex datasets. This chapter discusses the key areas that have used ML to attain fruitful results in agriculture.

4.1 Crop Yield Prediction

Crop yield prediction is one of the oldest and the most popular uses of ML in the agricultural industry. The old system of yield estimation uses manual sampling and statistical models which are constrained in large scale and accuracy [2][3].

Random forest, SVM and Gradient Boosting are some of the ML algorithms that can train on large volumes of data consisting of soil parameters, weather patterns, satellite images, and crop management data to predict yields. These models are capable of learning nonlinear correlations among variables with greater accuracy as compared to the traditional regression techniques. As an illustration, yield prediction systems in India and the United States utilize meteorological records and remote sensing information in predicting the crop output of regions weeks before harvest. Such insights can then be used by the farmers to optimize the fertilizer levels, irrigation programmes, and price policies [10][11].

4.2 Disease and Pest Detection of Crops

Among the top causes of the loss of yields around the globe are the crop diseases and pest infestations. Timely treatment and prevention of outbreaks is important to ensure that damages are minimized through prevention and early identification. ML and especially Deep Learning (DL) with CNNs have demonstrated impressive results in the automation of the process of disease and pest detection by analysing images.

CNNs are able to pick out small differences in colour, texture, and shape which cannot be easily noticeable by a human eye. As an illustration, time-series ML models forecast the occurrence of pests using weather and crop development. These predictive tools contribute to the integrated systems of pest management where excessive use of pesticides is not necessary and environmentally friendly methods of farming are encouraged [8].

4.3 Weed Detection and Management

Weeds will compete with crops in terms of nutrients, water, and sunlight, and in most cases result in huge productivity losses. Traditional methods of controlling weeds are dependent on the use of chemical herbicides which may have adverse effects of damaging the environment and increasing the cost of production. ML-based weed detectors can manage weeds on the site by detecting and localizing weeds in agricultural fields [7].

This can be sprayed in particular places, removed mechanically, or automated weeding by robotic arm. ML is therefore promoting sustainable agriculture by lowering the amount of herbicides used as well as saving soil health.

4.4 Soil Health and Nutrient Analysis

Good soil is the key to successful cultivation, and in large regions, it is difficult to track the properties of soils. ML methods help to forecast the soil properties including moisture, organic content, and nutrient concentrations based on few samples taken in the field, as well as satellite or sensor observations [6].

The algorithms of regression and classification, such as the Random Forests, K-Nearest Neighbors (KNN), and Artificial Neural Networks (ANNs), are commonly applied to predict the soil fertility and identify the deficiencies in the nutrient level. As an example, ML models can predict the level of nitrogen or phosphorus using the spectral reflectance values obtained with remote sensors. Moreover, it is possible to combine soil data with crop performance data to develop soil recommendation systems. Such systems will guide farmers on the best fertilizer application and crop rotation procedure, thereby cutting down on wastage and ensuring the conservation of soil in the long run [4].

4.5 Water and Irrigation Optimization

Water management is one of the most significant elements of sustainable agriculture especially in regions that have been affected by drought and depletion of groundwater.

Regression models, decision trees, and the RL algorithms can be used to predict the crop water requirements and automate irrigation scheduling. As an example, the IoT devices with ML can check the soil moisture levels and activate irrigation pumps when necessary. Such a process prevents the losses in yields caused by water stress as well as in saving water. Based on temperature, humidity, and solar radiance data, sophisticated ML models also predict evapotranspiration (ET) rates, which is a vital measure of water loss at the plant and soil. These projections help farmers to develop efficient irrigation programmes that are specific to some crops and weather conditions.

4.6 Crop Type Mapping and Land Use Classification

National food security planning, crop insurance programmes, and sustainable land management require proper mapping of the type of crop. To classify land cover and identify the types of crops that are cultivated at some locations, multi-temporal satellite images are analysed using ML algorithms. To this end, supervised classification algorithms such as Neural Networks, Decision Trees, and SVMs are commonly used. The phenological patterns, vegetation indices, and spectral signatures of sensors, such as Sentinel and Landsat are analysed.

Indicatively, in India, ML-based classification models that are developed with the assistance of Google Earth Engine can be used to distinguish between sugarcane fields, wheat fields, and rice fields with more than 90% accuracy.

4.7 Surveillance of Health of Livestock

ML is increasingly gaining significance in non-crop management in livestock. Wearable sensors and technology are put on animals that are then able to continuously monitor their movement, temperature, feeding habits, and milk production.

An example is that using anomaly detection models to identify early signs of mastitis in dairy cows, the models can observe an abrupt drop in the production of milk or a change in the body temperature. Similarly, computer vision systems are used to check on the posture and gait of animals so as to ascertain their welfare and detect lameness. The advantages of ML-powered livestock monitoring to animal welfare, reduced veterinary costs, and enhanced productivity are all positive factors that would benefit sustainable animal farming.

4.8 Post Harvest Quality Assessment

Some of the post-harvest activities that are critical to market value and customer satisfaction include sorting, grading, and quality assessment. These are normally manual, subjective, and labour-intensive jobs. ML and computer vision are automated to ensure that the procedure is efficient and consistent. ML algorithms are also able to evaluate the characteristics like colour, size, shape, and surface flaws and that is done by training models on labelled photos of fruits or vegetables. CNNs are commonly deployed to analyse the quality of products, such as mangoes, tomatoes, and apples, based on their pictures.

These technologies enable farmers and processors to meet high standards of quality through reducing losses after harvesting.

4.9 Precision Agriculture and Intelligent Farming

Precision agriculture systems utilize real-time field data analysis to make decisions regarding pest management and irrigation timing, and planting density. ML models are techniques that use multiple data, including drone pictures, weather prediction, and soil sensors, to generate prescriptive analytics, which suggest not only what is happening but also what is to be done. In addition, autonomous farm machinery controlled by ML algorithms is able to carry out field operations like plowing, seeding, and spraying with high precision, thereby minimizing their dependence on labour and operational costs.The numerous applications of ML in the field of agriculture demonstrate that it can be groundbreaking to the entire agricultural value chain. ML can improve the farming process by making it smarter, more efficient, and sustainable because it allows prediction of production and detection of diseases, optimization of irrigation, and post-harvest quality assessments. Besides enhancing productivity, these solutions address pressing international problems such as a lack of resources, food security, and weather change. ML is expected to keep changing agriculture because computing technology and availability of data are increasing to allow intelligent, automated, and ecological-friendly agriculture systems.

5. Case Studies and Real-World Implementations

ML has already moved past the stage of theoretical studies and is already applied in agriculture, showing quantifiable productivity, efficiency, and sustainability improvements. This chapter offers major case studies and real-life applications involving the way ML is changing agriculture in different contexts.

5.1 Satellite Data to Predict Crop Yield: India

India is another giant agricultural economy in the world with recurrent problems of unpredictable monsoons and soil erosion. In response, the Indian Space Research Organisation (ISRO) and the Indian Council of Agricultural Research (ICAR) worked together to come up with ML-based models of yield prediction using remote sensing and weather data.

Using the satellite values of the NDVI coupled with historical yield data, models such as the Random Forest Regression and SVMs can predict the yields of key crop products, such as wheat, rice, and maize. The predictions assist the government and farmers in planning to procure, to control storage, and to stabilize the prices in the market.

As an example, in such states as Punjab and Haryana, a combination of ML-based estimates of yields with early-season satellite images enables officials to predict production patterns several months before harvest time, greatly enhancing the efficiency of the supply-chain and avoiding food waste.

5.2 Pest Detection with Deep Learning: PlantVillage Project (Africa)

The application of ML to detect pests and disease in Africa can be illustrated by the PlantVillage Project which is a worldwide project of Pennsylvania State University and the UN Food and Agriculture Organization (FAO). Smallholder farmers are usually faced with disastrous losses of crops by late detection of diseases like cassava mosaic virus or maize leaf blight.

To solve this, scientists created a mobile app that is based on DL and is trained on thousands of labelled crop images. The system operates on the basis of CNNs in order to detect diseases with an accuracy of more than 90%. All that farmers do is take a picture with their smartphone of an infected leaf, and the app automatically diagnoses the issue and gives remedies.

Through the project, millions of farmers in Kenya, Uganda, and Nigeria have been empowered with the low-cost, accessible tools employed to monitor the health of their crops; thereby reducing their reliance on the agricultural extension officers and minimizing the crop loss through early interventions.

5.3 Accurate Irrigation and Water Optimization: California, USA

Water shortage is one of the factors that are highly threatening agricultural sustainability, particularly in dry areas like California.

ML algorithms are making predictions of the actual water needs of various crop zones based on the data on soil moisture, temperature, humidity, and evapotranspiration. SmartIrrigate systems are being used to dynamically modify irrigation programmes in order to accomplish optimal water delivery and to avoid over-irrigation using RL.

In the commercial vineyards and almond farms, these models have decreased water usage by up to 25%, hence the strength of the ML in ensuring a sustainable use of resources. Moreover, predictive analytics enable the water managers to allocate resources to a greater extent during the drought periods.

5.4 Soil Fertility Mapping and Recommendation System: Netherlands

An ML-based soil health assessment system designed by the Wageningen University and Research (WUR), in the Netherlands, was the first attempt to increase nutrient management and decrease the misuse of fertilizers. Based on the KNN and the Random Forest algorithms, the soil samples were examined in terms of their pH, organic matter, and nutrient concentrations at numerous regions.

The model produced soil fertility maps whereby specific nutrient application based on crop and soil type are fully recommended. The method resulted in high savings and yields when it was incorporated in a Decision Support System (DSS) to assist farmers.

This case indicates the strength of ML to convert complicated soil chemistry data to practical recommendations, which foster precision farming and environmental preservation through the minimization of fertilizer runoff and pollution.

5.5 Automated Fruit Grading and Sorting: Japan

The agricultural sector of Japan is experiencing the problem of labour shortage and therefore automation is introduced in the post-harvest operations. Kyoto University and FANUC Corporation created a fruit grading system based on DL and computer vision which operates with AI. The system records a high-resolution image of the fruit like apples, oranges, and peaches on the conveyor belts. The ML model is capable of making training data more effective in all types of lighting conditions and fruit varieties unlike the traditional systems that are rule-based and therefore do not improve.

This automated grading has also 40% faster processing speed and has more than 95% accuracy. It has already become an example of intelligent post-harvest management, minimization of human error, and uniform quality of the product.

5.6 *Predictive Farming to Smarter Help the Climate: Australia*

Droughts and extreme weather are common phenomena that put the productivity of agriculture in danger in Australia. Through the use of time-series forecasting and ensemble learning models, such systems forecast variations in rainfall, alterations in soil moisture, and variations in temperature many months ahead. Mobile apps provide localized advice to farmers, who may then decide on drought-resistant types of crop varieties or change the date of sowing.

Moreover, risk mapping devices, based on ML, can detect the areas that are the most susceptible to heat stress or lack of water, and assist policymakers in adopting specific climate adaptation tools. The innovations are some examples of how ML can assist in individual decision-making and also planning of large-scale agriculture.

5.7 *Animal Health in Wearable Sensors: Denmark*

Companies like Smartbo and Delaval equip cows with temperature sensors, RFID (radio frequency IDentification) tags, and accelerometers to collect information about their physiology and behaviour. ML algorithms can detect abnormalities in movement, food, and activity over time that may indicate stress, illness, or estrus. For example, if anomaly detection models identify mastitis or lameness several days before symptoms appear the veterinarian can take action. These techniques have improved milk production and animal comfort while reducing disease-related losses by 30%. Recent successes in these applications may raise awareness of the use of ML in precision livestock farming, aiming to balance productivity and welfare.

5.8 *Smart Farm Robotics: Israel and India*

The integration of ML and robotics has revolutionized the farming operation in technologically developed nations such as Israel, and are currently spreading to India. There are startups like TartanSense (India) and Blue River Technology (USA, acquired by John Deere) that have created AI-based agricultural robots that are able to weed, spray, and plant plants without needing human supervision. The robot TartanSense, which has a name Bringi, identifies weeds and sprays herbicide in specific spots using computer vision and DL, thus saving 80% of chemicals. Equally, the See and Spray system of Blue River uses ML models, which identify crops and weeds in real-time, and provide specific treatment.

These inventions are some of the examples of how ML is transforming the future of autonomous farming and making it less reliant on labour, more efficient with inputs, and more productive in the field.

5.9 Supply Chain Optimization and Predictive Market Analytics

In addition to optimization of agricultural supply chains (predicting demand, minimizing waste, and stabilizing prices), ML is also implemented on the farm. Some websites like the IBM Watson Decision Platform in Agriculture apply ML to combine market prices with weather predictions and logistics to give traders and policymakers real-time advice.

In developing countries, such ML systems assist the cooperatives to deal with storage and distribution, minimizing wastage of post-harvest due to inaccurate demand forecasting. The advantage of this is that the farmers get to know when and where to sell their produce at the highest price, which helps the farmers in maintaining an income stream.

The case studies above demonstrate that ML is not a visionary technology but a dynamic force of change in the world of agriculture. It can be disease detection in Africa, smart irrigation in California, or soil mapping in Europe, but in any case, the applications of ML are proving to have real economic and environmental advantages.

ML provides the bridge of the gap between wisdom in traditional farming and the effectiveness of technologies by transforming data into knowledge. The next step will be how to scale these innovations to the advantage of smallholder farmers to ensure that data-driven agriculture is a source of global food security and climate resiliency

6. Benefits and Impact of ML in Agriculture

ML is transforming the conventional farming methods as it allows making smarter and data-driven decisions in agriculture. The positive and negative effects of ML in agriculture are multi dimensional, i.e., economic, environmental, social, and technological. These dimensions are discussed herein.

6.1 Economic Benefits

6.1.1 Greater Crop Productivity and Yield Optimization

The ML algorithms can process soil quality, weather trends, crop development, and pest infection to prescribe the best farming measures. Predictive models enable farmers to know the most appropriate time to plant, the rate of applying the fertilizers, and when to irrigate their lands, and this maximizes their crop production.

Indicatively, ML-based yield prediction models in combination with remote sensing data offer practical information several weeks before harvest. This enables farmers to have a better plan on how resources should be allocated, labour and sales in the market. ML positively affects the income of farmers and national food security by decreasing losses of crops, and maximizing their productivity.

6.1.2 Cost Reduction

Efficiency in the utilization of inputs like fertilizers, pesticides, and water will lead to reduced operation costs. An example of such an irrigation system is ML-based irrigation systems, which apply sensor data and predictor models to apply the exact quantity of water needed by crops. Likewise, site-specific application of pesticides minimizes the amount of chemicals spent as well as environmental degradation. This is a precision farming method, which will guarantee that resources are utilized effectively and reduce costs as well as increase profitability.

6.1.3 Optimization of the Market and Supply Chain

ML models have the ability to predict crop supply, demand, and prices in the market giving farmers and stakeholders the ability to make appropriate decisions regarding sales, storage, and distribution. Predictive analytics minimizes after harvest losses due to market changes or overproduction. This economic optimization empowers the lives of farmers as well as stability of the entire agricultural market.

6.2 Environmental Benefits

6.2.1 Sustainability Resource Management

ML helps in precision farming, lessens excessive use of water, fertilizers, and pesticides. Among others, these include soil sensors, predictive irrigation models which use the required levels of water depending on the water scarcity in a region and crop nutrient prediction models which avoid unnecessary use of fertilizers. These are some of the practices which reduce soil erosion, groundwater contamination, and groundwater runoffs.

ML helps in promoting sustainable farming by directing the utilization of resources to enhance environmentally friendly farming that is sensitive to both agricultural output and environmental sustainability.

6.2.2 Climate Change Adaptation

The agricultural industry is very prone to climate change (change in the rainfall, temperature variation, and extreme weather conditions). Real-time observations can be used in ML models to predict crop stress, drought risks, or pest outbreaks using historical information with real-time data. This then allows farmers to implement adaptive management like planting of hardy crop species or changing the planting time, hence minimizing losses due to climatic conditions.

This prediction ability makes the agricultural systems more resilient and sustainable, which facilitate long-term environmental protection.

6.2.3 Lessening Carbon Footprint

Automation based on ML decreases the necessity of human work and interventions with a high energy consumption. Robotic weeders, autonomous tractors, and optimal irrigation save energy and greenhouse emissions. Also, reducing the amount of chemicals used, ML helps to reduce the emission levels caused by fertilizer manufacturing and chemical pollution of the soil, and thus, makes agriculture more climate-aware.

6.3 Social Benefits

6.3.1 Empowerment of Farmers

ML tools enable farmers to make informed decisions instead of just using intuition or traditional practices to make decisions by offering actionable insights. Smart applications and online tools based on ML transform the ability to access agricultural knowledge into the hands of smallholder farmers in isolated areas, allowing them to enjoy the latest technological innovations. This empowerment makes them more productive, builds confidence, and eliminates the need to consult with intermediaries or extension officers.

6.3.2 Improved Food Security

ML helps in stable food production by increasing crop yields, decreasing losses associated with diseases or pests, and streamlining supply chains. Predictive modelling also makes sure crops are made at the appropriate time and quantity to avoid the risk of scarcity. This contributes directly to food security in the world, especially in areas where agriculture produces are important to the sustenance of livelihoods.

6.3.3 Job Creation in Agri-tech

Although ML automatizes some activities in the agricultural industry, it, at the same time, creates new jobs in the agricultural technological field such as data analysis, drone piloting, sensor repair, and software development. The implementation, control, and maintenance of ML-based systems can be done together by the farmers, technicians, and researchers, which contributes to new economic prospects.

7. Challenges and Limitations

ML has transformed the process of farming by bringing on automation, predictive analytics, and precision farming. Nevertheless, its use in farming encounters serious problems which impede the large-scale adoption, especially in the developing areas where farming is heterogeneous, the resources are scarce, and the technological literacy is poor (Figure 2).

This chapter is a critical analysis of the large-scale technical, infrastructural, socio economic, ethical, and environmental issues that influence the application and scalability of ML to agriculture.

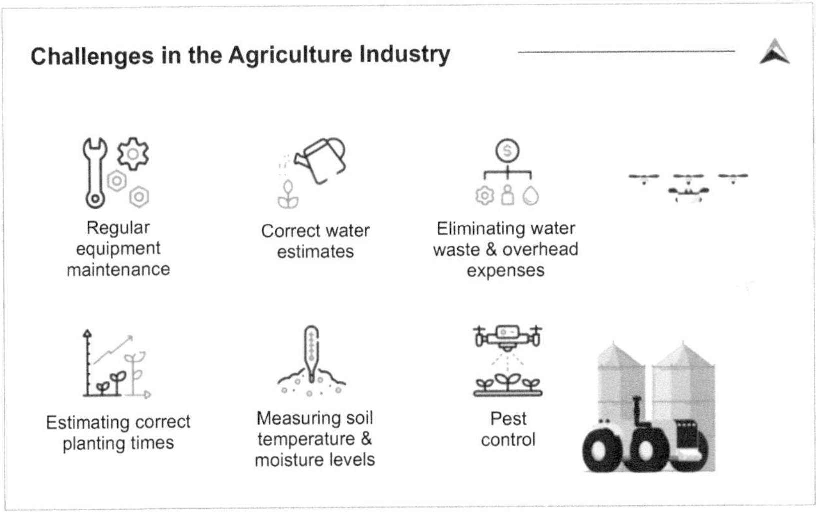

Fig. 2 Challenges in The Agriculture Industry.

7.1 *Data-Related Challenges*

7.1.1 *Data Insufficiency in Quality and Quantity*

ML models are largely dependent on masses of big, varied, and quality data to perform accurate predictions. Information in the agricultural sector is usually incomplete, patchy, or missing. Farmers do not have any digital systems of records and data collected with sensors or satellites can be costly or inaccurate. Smallholder farmers who constitute a large portion of the world agriculture and farming sectors do not use digital tools but usually depend on manual observation. Poor quality information, particularly the data on soil health, diseases in crops, and weather conditions prevents the formulation of precise models. Low data quality results in inaccuracies in the models which may result in misguided forecasts in the yield estimation or disease identification.

7.1.2 Standards of Data and Interoperability

There are various sources of agricultural information: IoT sensors, satellite images, drones, weather stations, and market databases. These data sources, however, tend to have varying formats and standards, and thus it is not easy to integrate them.

The standardization of data does not exist and makes training models and their extrapolation to different regions difficult. One example is that soil measurements in one nation will not be consistent with the measurements in a different country, so the reproducibility of ML models in the global context is poor.

7.1.3 Bias and Representativeness of Data

Bias happens when ML models are trained on non-representative datasets. The models that are trained with large commercial farms will not be effective with smallholder or organic farms where there is a variation in the types of crops and the management procedures. Such discrimination may cause unfair advantages in the distribution of technology as those with resources end up benefiting more than their counterparts in developing or resource deficient areas.

7.2 Technological and Infrastructural Problems

7.2.1 Small Internet and Connectivity in Rural Areas

The majority of the sophisticated ML systems rely on online data transmission and cloud computing. Nevertheless, poor internet connectivity and digital infrastructure plague a lot of rural agricultural areas particularly in Africa, South Asia, and Latin America. Farmers lack access to real-time data services (satellite imaging, weather forecasting, or automated irrigation systems) due to subpar connections. This digital gap is still one of the greatest obstacles to the use of ML in agriculture.

7.2.2 High Cost of Implementation

The implementation of agricultural systems based on ML involves the investment in sensors, drones, IoT, and computer equipment. The initial setup cost together with the cost of data storage as well as maintenance of the models may prove to be prohibitive to the small- and medium-scale farmers. Access is restricted even in the case of the availability of cheap instruments due to the absence of financial support mechanisms, including government subsidies or microloans.

7.2.3 Constrained Computational Resources

Most developing areas do not have access to highly computing systems to train more complicated ML systems, including deep neural networks. Lack of edge computing infrastructure causes a problem of processing the data at the point of origin which means having to rely on cloud servers which are expensive and subject to fluctuating network conditions.

7.3 Human and Skill-Based Limitations

7.3.1 Farmers Lack Technical Knowledge

Poor digital illiteracy among farmers is another major limitation to the implementation of ML in agriculture. As urban technologists create advanced algorithms, a majority of farmers are not equipped to comprehend data analytics, operate through digital interfaces, and diagnose technical malfunctions. Even the well-designed ML tools can be left unexploited or improperly implemented without the relevant education and extension programs [14].

7.3.2 Lack of Educated Professionals

The building, sustenance, and expansion of agricultural systems that operate by means of ML necessitates cooperation among data scientists, agronomists, and software engineers. But it lacks interdisciplinary professionals that are familiar with both agricultural systems and data science. This skill shortage slows the innovation process and limits the regionalization of ML models. The potential of change in technology resistance is examined in the second section of the questionnaire. Older farmers might be reluctant to accept digital interventions because of their traditional farming practices. There is trust that is important in the adoption of technology. Farmers may also be unwilling to adopt the use of ML models, particularly in cases where the decision made will have a direct impact on such important matters as irrigation or pest management, as long as they do not fully grasp the functioning of the models and their credibility.

7.4 Algorithms and Model Limitations

7.4.1 Model Interpretability and Transparency

The black-box character of many algorithms, particularly DL models, is one of the most controversial questions of ML. Farmers and policy makers usually need to be given clear explanations why a model suggests this or that [19].

Farmers require rational arguments when an ML model predicts poor yield or indicates to alter the dosage of fertilizer. The absence of transparency lowers the level of trust and implementation, particularly in risk-sensitive sectors such as agriculture.

7.4.2 Over-fitting and Lack of Generalization

The agricultural conditions are highly diversified based on the types of soils, climate, and systems of farming. A model that is trained using data in one region might not perform well in the other because of over-fitting when the model picks up too specific patterns based on the form of the training data.

The creation of generalizable ML models that can work both under varying field conditions is one of the key technical problems [2][10].

7.4.3 Processing Constraints in Real-time

Real-time decision-making is needed in applications of ML, including pest detection or autonomous harvesting. Nevertheless, when using remote cloud servers and large model sizes, systems with high latency or size of models do not allow processing and acting on information in time.

7.5 Socio Economic and Ethical Issues

7.5.1 Data Privacy and Ownership

With ML systems gathering an enormous quantity of agricultural data (soil parameters, payment history of farmers, etc.), the question of the owner of the data and its administration emerges. Most of the time, the information gathered by the private firms or government sites is not brought to the attention of the farmers who produce it. This casts ethical issues of privacy, abuse, and commercialization of data of farmers. It is essential to come up with solid and transparent data governance mechanisms.

7.5.2 Disparity in Technological Access

Large agribusinesses and other players that can afford technology have been benefiting more than the smallholders in agriculture as a result of ML-driven agriculture. This electronic gap threatens to increase socio economic inequalities in the farming industry. Otherwise, it might result in dependency on the corporate-controlled platforms and the loss of farmer autonomy.

7.5.3 Algorithmic Bias and Fairness

Trained on biased data, ML algorithms can make discriminatory predictions, e.g., inaccurate predictions of crops in the minority farming areas, or incorrectly underreporting some types of crops [4]. It is essential to have fairness in the design of the algorithm to make sure that the benefits of ML are equally shared by gender, geographic region, and the breadth of operation.

7.6 Environmental and Contextual Limitations

7.6.1 Agricultural Ecosystem Variability

Agricultural systems are highly dynamic unlike the controlled industrial environments and they are affected by various unpredictable variables like pests, rainfall, and temperature changes.

When the model is trained in a particular environment and then subjected to drastic changes in the environment (droughts, floods, etc.), the model can fail. Such dependency on the context thus makes it hard to sustain a uniform model performance.

7.6.2 Reliance on External Technology Ecosystems

ML systems usually rely on the other complementary technologies such as cloud storage, GPS, and IoT devices. The failure of any of these systems, i.e., sensor malfunction or connection problem can interrupt the whole ML pipeline. Reliability and redundancy planning are, therefore, necessary but not followed up [7][13].

7.7 Institutional and Policy Issues

7.7.1 Inadequate Government Policies

There are no clear policies in many regions on how to encourage digital agriculture or control data collection and sharing. Farmers are not given directions or protection in adopting ML tools without national plans or programmes to put the funds in place.

7.7.2 Lack of Cooperation and Information Sharing

Lots of agricultural data is commonly dispersed between research facilities, commercial institutions, and government bodies. The absence of shared data ecosystems constrains innovation and supports researchers to come up with holistic models [17].

7.8 Plans to Overcome Difficulties

Some of the strategies that can be applied in order to reduce these challenges include:

Information normalization and open-source systems to enhance the quality and interoperability of data; training farmers and other agricultural stakeholders on capacity-building to promote digital literacy; incentives and subsidies offered by government agencies to alleviate the economic pressure of the adoption of ML technologies; cultivation of inexpensive ML-based agricultural services that would be tailored according to local requirements through public-private partnerships [19]. Although the potential offered by ML to revolutionize the agriculture sector is substantial, its success is pegged on overcoming a variety of related issues, including, but not limited to, data quality and technological infrastructure, ethics, and social equity. The solution to these gaps is a multi-stakeholder approach that requires the government, researchers, technology firms, and farmers to close these gaps. The future of ML in agriculture is not only in the innovativeness of algorithms, but also in the construction of inclusive ecosystems that incorporate digital intelligence and human wisdom. By defeating these problems, sustainable, equal, and smart farming systems that will be profitable to society as a whole will emerge [5].

8. Future Trends and Research Directions

One of the most significant changes in contemporary food production is the introduction of ML within the agricultural industry. However, with the development of technologies and the changing conditions on the global scale, ML keeps changing and developing new horizons. Innovative agricultural systems that will be autonomous, sustainable, transparent, and inclusive are the next stage of innovation in agriculture. The chapter examines the major future trends and research priorities as well as new technologies that will shape the future of ML-powered agriculture [14].

8.1 Intelligent and Autonomous Agriculture Evolution

8.1.1 A Precision to Predictive and Prescriptive Agriculture

The new generation of ML applications will go beyond precision agriculture, which aims to maximize the inputs and outputs, to predictive, prescriptive agriculture, where algorithms are not only predictive but also prescriptive, i.e., recommend certain actions. In the case of pest outbreaks, as an example, forecasting the outbreak of the pests will take place weeks before, and the biological pest control will automatically be scheduled. Equally, ML systems will recommend precise ferment mixes, depending on actual soil chemistry, health of crops, and weather forecasts.

Such a shift will result in the formation of entirely smart farming ecosystems that will dynamically respond to changes in the environmental and economic conditions.

8.1.2 Autonomous Farming Systems

ML, robotics, and the IoT will result in the emergence of autonomous farms. These farms will have a low human element in operation; drones will be used to check the health of the crops, there will be self-driving tractors to plough and plant crops, and harvesting can be done by robotic hands [9].

Self-operating systems using the ML algorithms will improve efficiency, diminish the need to depend on labour and allow the 24-hour work of the farms. The studies will be based on RL methods that enable machines to learn optimal behaviours with experience to enhance the abilities to adapt to a variety of agricultural environments.

8.2 Embarkation of Emerging Technologies

8.2.1 Artificial Intelligence of Things (AIoT)

The intersection of AI and IoT, the so-called AIoT, will transform agricultural monitoring and control. IoT devices will learn the ML algorithmically [14] to analyse data collected by sensors by edge computing instead of depending solely on cloud servers.

This will enhance the response time considerably and make the ML applications more affordable to smallholder farmers.

It could be assumed that the future research will be dedicated to creating energy-efficient AI chips, efficient sensor fusion algorithms, and low-cost IoT networks that can be applied in agricultural settings.

8.2.2 ML and Blockchain Integration

Blockchain technology provides a transparent and secure data exchange and traceability to agricultural value chains. It can be paired with ML to guarantee predictive systems have integrity, trust, and accountability of their data [13].

As an example, verified data by blockchain can be used to train ML models to predict yields, making sure that it is a genuine one and minimizes any manipulation. Besides, the ML can analyse blockchain log entries to identify the supply chain inefficiencies or fraud.

Such synergy will be critical in the establishment of transparent, data-driven agri-economies safeguarding the rights of the farmers and promoting fair trade.

8.2.3 Cloud and Edge Computing

Cloud computing will remain to serve the large-scale training of the ML models, whereas on-site processing will be addressed by edge computing to take immediate action. Hybrid computing models (i.e., combined centralized cloud-based analytics and decentralized edge-based decision-making) will be used in future agricultural systems to balance between speed, accuracy, and energy efficiency [17].

8.3 Data-Driven Sustainability and Climate-Resilient Agriculture

8.3.1 Climate Adaptation and Modelling

With increased climate change, ML will be important in the modelling and reduction of the effects of climate change on agriculture. The models of the future will combine satellite, weather forecast with crop physiology to forecast droughts, floods, or heat stress [19].

Climate adaptation systems that resemble those of the MLS will advise farmers on seeding robust crop varieties, seeding dates, and adaptive irrigation methods.

The other area of research will be long-term environmental modelling, in which ML is used to discover sustainable agricultural methods that will minimize greenhouse gas emissions and preserve biodiversity [17].

This action plan aims to implement sustainable resource management.

8.3.2 Sustainable Resource Management

ML will also be used in an ever-smaller amount of resources in agriculture, which will reduce waste and maximize productivity. Nutrient management systems and predictive irrigation systems will optimize resource allocation according to environmental feedback and the needs of the plants [16].

The developing fields of study are carbon sequestration modelling, biodiversity monitoring, and estimating the water footprint based on DL and remote sensing data.

ML will be a facilitator of the UN Sustainable Development Goals (SDGs) associated with food security, water management, and climate action by enhancing sustainability.

8.4 Elaborate ML Techniques to Agricultural Innovation

8.4.1 Explainable Artificial Intelligence (XAI)

The complexity of the algorithms used as black boxes has been declared as one of the biggest criticisms of ML in agriculture. The next generation of studies will focus on Explainable AI (XAI), i.e., methods of making ML models transparent and readable [15].

Farmers and policy makers can know the reasoning behind the models making certain predictions, which will enhance trust and adoption.

As an example, AI predictions can be more comprehensible and credible with the help of visual explanation tools that indicate which characteristics (e.g., soil moisture or the colour of leaves) impacted the disease forecasting.

8.4.2 Federated and Transfer Learning

In the traditional ML, centralized data is necessary, in federated learning models are trained collaboratively with several devices or farms without accessing raw data. This ensures privacy and helps ML to operate efficiently in decentralized agricultural systems [12].

Transfer learning will also be critical, and the models trained in one location or crop can be used in new settings with limited data. This saves much time and expenses of implementing ML systems all over the world.

8.4.3 Generative Artificial Intelligence and Generative Data

Generative AI is a research direction that is becoming more popular and can be used to address data scarcity by generating synthetic data. By applying Generative Adversarial Networks (GANs), researchers are able to recreate images and soil profiles of real crops, weather conditions to train powerful ML models.

This method will improve the variety of data, particularly in rare occasions like pest outbreak or severe droughts [10].

8.5 Social and Ethical Research Frontiers

8.5.1 Inclusive Technology Among the Smallholder Farmers

Inclusiveness and affordability should be the priority of future ML studies. Small-scale farmers do not have the means to reach the digital tools or the bandwidth that is rather expensive. The digital divide will be overcome with the creation of lightweight ML models based on mobile devices, which operate in local languages and offline mode. Community-based data platforms that are collaborative will also be used to ensure that smallholders may actively engage in data collection and model insights to benefit [14].

8.5.2 Agricultural Ethical and Responsible AI

The ethical issues of data privacy, algorithmic bias, and corporate control will come into the foreground of the research in the future. Good AI systems will be responsible to guarantee transparency, equitability, and data sovereignty within agricultural systems.

Governments and global bodies will be required to put in place AI ethics policies to control the use of data, defend the rights of farmers, and advance fairness in adopting ML technologies [11].

8.5.3 Human-AI Collaboration

Instead of ousting human expertise, the ML systems of the future will enhance human decision-making. The cooperation between farmers, agronomists, and AI systems will be aimed at the interpretation of data and optimization of results [8].

The human-AI interaction models that increase the trust, usability, and co-learning between the intelligent systems and farmers will be investigated.

8.6 Policy Research Directions and Institutional

8.6.1 Open Data Ecosystems

The global and regional open agricultural data repositories are needed in order to exploit the full potential of ML. Standardized data sharing platforms will help to speed up innovation at the same time guaranteeing privacy [7].

Governments, research institutions, and private companies must be encouraged to share data to open-access data ecosystems in the future, which will make the data transparent and interoperable.

8.6.2 Policies in Digital Agriculture

Governments will be instrumental in setting up AI governance in agriculture. The policies should strike a balance between innovation and ethical control, i.e., how to tackle the problem of data ownership, liability in autonomous systems, and fair distribution of technology. The partnerships between the government and the business sectors will also play a crucial role in financing research, creating infrastructural development, and advancing capacity-building projects to farmers and rural populations [15][14].

8.7 International Research Opportunities Interdisciplinary

ML studies in agriculture will progressively become interdisciplinary, which is to say involving computer science, agronomy, ecology, economics, and social sciences.

The future research opportunities are promising, and are:

- **Agriculture-based bioinformatics:** Combining ML and genomics to create stress-resistant crops.
- **Agroecological modelling:** The combination of ecological and environmental information with the management of land.
- **Behavioural analytics:** Examining how farmers make decisions with the help of ML to create an efficient advisory system [16].
- **Economic forecasting:** Using ML to predict the world with commodity prices and optimizing the trade. This integration of fields will result in a holistic agricultural intelligence system that would be both productive and sustainable.

Intelligent, connected, and sustainable agriculture is the future, and it uses ML and data-based innovation to achieve that vision. In the next decades, the farms will become self-adjusted to the environmental factors and continuously learn, making predictive decisions to maximize the yields and save resources [18].

The main directions to be taken in the future are:

- Creation of Explainable and Ethical AI in order to black box the trust and transparency.
- Edge AI and IoT start-ups.
- Inclusion of block chain to have secure transparent supply chains.
- Expand federated and transfer learning to facilitate global learning.
- Development of inclusive AI solutions to smallholder farmers.

Finally, the future of ML in agriculture is not only in new sophisticated algorithms but in human-focused innovation, i.e., to make sure that technology does not replace traditional wisdom and is accessible to all society tiers. ML can contribute to the creation of a resilient, productive, and sustainable global food system able to feed future generations by engaging in responsible, inclusive, and interdisciplinary research.

9. Conclusion

ML is turning agriculture as a traditional and labour-intensive industry into a contemporary, data-driven environment. Precision, predictive, and prescriptive farming has been made possible through the integration of ML with new technologies (IoT, drones, remote sensing, and robotics). These systems increase productivity, maximize the resources used, decrease the environmental impact, and reinforce the food security in the world. ML models enable farmers to make good decisions about crop selection, irrigation, fertilization, pests, and post-harvest activities through the review of historical and real-time information. Predictive analytics allow the initial identification of crop diseases and pests and prescriptive models suggest specific actions to minimize losses and optimization of yields. The implementation of these smart systems has seen some quantifiable economic, environmental, and social returns, such as lower input prices, better resource utilization, greater profitability, and management of the food chain. Although these developments have taken place, the widespread use of ML in agriculture is associated with a number of challenges. Information constraints such as lack of data, heterogeneity, and quality challenge is also a major setback especially to the smallholder farmers. The deployment is also impeded by infrastructure limitations including lack of internet connectivity, inaccessibility to computational resources, and an expensive cost to implement. Also, the black-box character of most of the ML models generates trust and interpretability problems, and ethical issues of data property, privacy, and fair access also remain. Inclusive technological design, capacity-building, and favourable policy frameworks are needed to address such challenges.

The future of ML in agriculture is bright and it tends to have autonomous, sustainable, and intelligent systems in agriculture. The new trends are XAI to make decisions transparently, edge computing to support real-time analytics, block chain to provide secure and traceable data management, and the federation and transfer learning solutions across regions. The field of research is now becoming more interdisciplinary as agronomy, computer sciences, ecology, and socio economic studies are combined to create holistic, human-oriented solutions. This will make sure that besides improving productivity, ML will also lead to environmental sustainability, social equity, and economic resilience. ML is a paradigm change in the world of agriculture. It presents the possibility to transform the agricultural practices, enhance the livelihoods, and help create a strong and sustainable food system that can support the needs of an increasing population. The potential can, however, be achieved by governments, researchers, and technology providers working hand-in-hand with farmers in forming inclusive, transparent, and adaptive agricultural ecosystems. With its ability to overcome technical, socio economic, and ethical issues, ML can become a powerful resource of changing agriculture into a smart, efficient, and sustainable business, which will guarantee food security and environmental care to the generations yet to come.

References

[1] Araújo, S.O., Peres, R.S., Ramalho, J.C., Lidon, F. and Barata, J. (2023). Machine Learning Applications in Agriculture: Current Trends, Challenges, and Future Perspectives. *Agronomy, 13*(12), 2976. https://doi.org/10.3390/agronomy13122976.

[2] Botero-Valencia, J., García-Pineda, V., Valencia-Arias, A., Valencia, J., Reyes-Vera, E., Mejia-Herrera, M. and Hernández-García, R. (2025). Machine Learning in Sustainable Agriculture: Systematic Review and Research Perspectives. *Agriculture*, *15*(4), 377. https://doi.org/10.3390/agriculture15040377.

[3] Gobierno de Perú. MIDAGRI Fortalece Transferencia De Tecnologías De Agricultura De Precisión Para Mejorar La Calidad De La Producción Agrícola. National Aeronautics and Space Administration.

[4] Goldenits, G., Mallinger, K., Raubitzek, S. and Neubauer, T. (2024). Current applications and potential future directions of reinforcement learning-based Digital Twins in agriculture. *Smart Agric. Technol.*, *8*, 100512.

[5] Vinayaka, and Prasad, P.R.C. (2024). AI-Enhanced Remote Sensing Applications in Indian Sugarcane Research: A Comprehensive Review. *Sugar Tech.*, *26*, 609–628.

[6] Cravero, A., Pardo, S., Sepúlveda, S. and Muñoz, L. (2022). Challenges to use machine learning in agricultural big data: A systematic literature review. *Agronomy*, *12*, 748.

[7] Araújo, S.O., Peres, R.S., Barata, J., Lidon, F. and Ramalho, J.C. (2021). Characterising the Agriculture 4.0 Landscape: Emerging Trends, Challenges, and Opportunities. *Agronomy*, *11*, 667.

[8] De Clercq, M., Vats, A. and Biel, A. (2018). Agriculture 4.0: The future of farming technology. In: *Proceedings of the World Government Summit, Dubai, United Arab Emirates*, 11–13 February 2018 (pp. 11–13).

[9] Zambon, I., Cecchini, M., Egidi, G., Saporito, M.G. and Colantoni, A. (2019). Revolution 4.0: Industry vs. agriculture in a future development for SMEs. *Processes*, *7*, 36.

[10] Liu, Y., Ma, X., Shu, L., Hancke, G.P. and Abu-Mahfouz, A.M. (2020). From Industry 4.0 to Agriculture 4.0: Current Status, Enabling Technologies, and Research Challenges. *IEEE Trans. Ind. Inform.*, *17*, 4322–4334.

[11] Zhai, Z., Martínez, J.F., Beltran, V. and Martínez, N.L. (2020). Decision support systems for Agriculture 4.0: Survey and challenges. *Comput. Electron. Agric.*, *170*, 105256.

[12] Trendov, N.M., Varas, S. and Zeng, M. (2019). *Digital Technologies in Agriculture and Rural Areas*. Briefing paper; FAO: Rome, Italy.

[13] Rose, D.C., and Chilvers, J. (2018). Agriculture 4.0: Broadening responsible innovation in an era of smart farming. *Front. Sustain. Food Syst.*, *2*, 87.

[14] Ahmed, M., and Pathan, A.S.K. (2018). *Data Analytics: Concepts, Techniques, and Applications*. CRC Press: Boca Raton, FL, USA.

[15] Liakos, K.G., Busato, P., Moshou, D., Pearson, S. and Bochtis, D. (2018). Machine learning in agriculture: A review. *Sensors*, *18*, 2674.

[16] Mahesh, B. (2020). Machine learning algorithms: A review. *Int. J. Sci. Res.*, (*IJSR*), *9*, 381–386.

[17] Moher, D., Liberati, A., Tetzlaff, J. Altman, D.G. and Group, P. (2009). Preferred reporting items for systematic reviews and meta-analyses: The PRISMA statement. *PLoS Med.*, *6*, e1000097.

[18] PRISMA. (2020). *Prisma Transparent Reporting of Systematic Reviews and Meta-Analyses*.

[19] Li, Z., Liu, F., Yang, W., Peng, S. and Zhou, J. (2021). A survey of convolutional neural networks: Analysis, applications, and prospects. *IEEE Trans. Neural Netw. Learn. Syst.*, *33*, 6999–7019.

3 Disease Detection of Mango Leaf using Deep Learning: YOLOv11 Versus YOLOv12

Vivek Kumar[1*] and Satender Kumar[2*]

Mango (*Mangifera indica* L.) is one of the most important fruit crops in India, with regions such as Saharanpur in western Uttar Pradesh contributing significantly to large-scale production. However, mango cultivation is often affected by various foliar diseases that negatively impact plant health, fruit quality, and overall yield. Early and accurate detection of these diseases is essential for effective management and timely intervention. Traditionally, disease identification relies on manual inspection by farmers or experts, which is time-consuming, subjective, and prone to human error. These limitations highlight the need for automated, reliable, and scalable solutions. This chapter presents a deep learning-based object detection system for mango leaf disease identification using two advanced models: YOLOv11 and YOLOv12. A custom dataset containing 3,000 annotated images from mango orchards was collected and enhanced through data augmentation techniques such as flipping, scaling, and color transformation to improve model robustness. The dataset was divided into training, validation, and testing sets in an 8:1:1 ratio. Both models were trained using transfer learning with pre trained weights to improve convergence and detection accuracy. Experimental results showed that YOLOv11 outperformed YOLOv12, achieving a precision of 0.964, recall of 0.976, mAP@0.5 of 0.983, and an F1-score of 0.97. The trained YOLOv11 model can also be deployed on mobile devices, enabling farmers to detect diseases in real-time and supporting sustainable agriculture practices.

1. Introduction

Mango (*Mangifera indica* L.) is a very important crop in the Indian agricultural and economic scene. Being one of the most cultivated fruit crops, it gives a significant contribution to the livelihood of millions of smallholder farmers. In some parts of the world like Saharanpur, which is in the western section of Uttar Pradesh, mango production is not only a source of income but a crucial socio economic activity that determines the development of rural areas. Nonetheless, mango orchards continue to face numerous threats in terms of the sustainability and productivity due to the

[1,2] Quantum School of Technology, Quantum University, Roorkee, India.
* Corresponding author: vksingh087@gmail.com; satenderkumar.cse@quantumeducation.in

DOI: 10.1201/9781003767534-3

extensive range of the foliar diseases, which concern the quality of the leaf health, canopy density, fruit production, and the general quality of the crop.

Mango leaf diseases are an important part of orchard management that must be identified. Manual visual inspection has been the conventional method through which farmers and agricultural experts detected such symptoms as lesions, discoloration, fungal development, and wilting. Although diseases can effectively be diagnosed by a trained pathologist, manual diagnosis is slow, subjective, inconsistent, and unscalable, particularly when dealing with large orchards, or where agricultural specialists are inaccessible [1]. Late interventions are usually caused by a misdiagnosis and late detection which generates drastic losses of crops, poor quality of fruits, pesticide overuse and economic returns. As a result of the emergence of precision agriculture, the development of modern technologies like computer vision and artificial intelligence (AI) has become a force that can be used to address the old agricultural problems [2]. The object detection systems and convolutional neural networks (CNNs) have demonstrated good results in detecting diseases in crops including rice, wheat, tomato, grapes, and apple leaves with deep learning. The use of these methods on mango leaves creates new prospects of proper, fast, and automated detection that can provide farmers with real-time solutions [3].

This chapter of the book meets this requirement, providing a sophisticated deep learning-based object detection framework in the comparison of two latest versions of the YOLO family including YOLov11 and YOLov12 to detect mango leaf diseases. The two models use state-of-the-art architectures with the ability to identify lesions of disease in complicated backgrounds. The chapter provides an extensive knowledge of 3,000 annotated mango leaf images gathered on the orchards of Saharanpur to test a model developed using custom data, training, augmentation, evaluation measures, and the viability of the model in the field.

2. Related Work and Literature Landscape

The study on identification of plant diseases has experienced a major transformation in the last decade with the continued reliance on computational tools to aid agriculture. Early studies were based on hand developed image-processing algorithms, with color histograms, texture indicators and edge patterns extracted manually and used to distinguish between healthy leaves and diseased ones. Blazing new grounds, these classical approaches were intrinsically finite as the ability to hand-craft characteristics could not convey the intricacy of natural variations on a range of leaf surfaces, lighting situations, and sounds. Consequently, the transition to machine learning and, more precisely, deep learning has been a landmark event in the automation of agriculture. With the introduction of CNNs, accuracy levels were highly obtainable because these models learned successful hierarchical features directly in images without manual feature engineering.

Research on the detection of diseases in crops with CNNs has demonstrated impressive advancement in a wide range of crops. Scholars have investigated

Table 1 Leaf Disease with Visual Pattern.

Leaf Disease Name (Scientific Disease Name)	*Visible Pattern*	*Infected Image*
Bacterial Black Spot (*Xanthomonas campestris pathovar mangiferaeindicae*)	The leaves may curl and break, and there are dark spots and lesions on them.	
Cottony Cushion Scale (*Icerya purchasi*)	The existence of reddish brown insects that have egg sacs that are cottony in appearance.	
Sooty Mould (*Capnodium spp.*)	On leaf surfaces a fungal covering that is either velvety, powdery, or black in color, could have a dirty appearance.	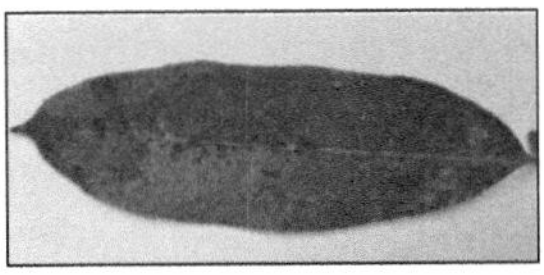
White Scale (*Aulacaspis tubercularis*)	On the leaves, there are little insects that are round or oval in shape and have a white or light gray scale.	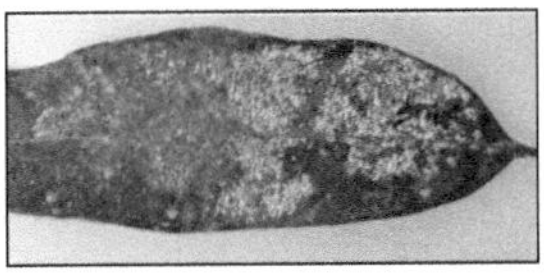
Dieback (*Botryodiplodia theobromae*)	Shoot tips have died and dried out, becoming brown and brittle.	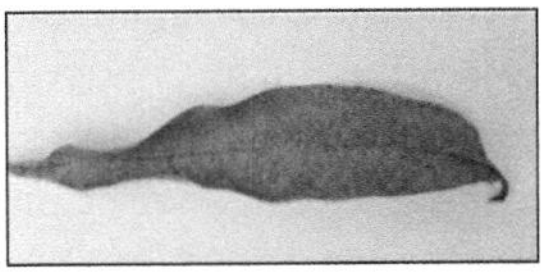
Healthy	Uniform dark green, smooth and glossy surface	

tomato pathogens, apple scab lesions, grape mildew, sugarcane diseases and rice blight and found that CNN-based classifiers have over 95% accuracy in controlled datasets [4]. Nevertheless, the weakness of the CNN models in the early days is that they did not always do classification but only the type of disease present, not its location. Leaf diseases tend to manifest as small spots or lesions, and spatial localization does not allow the classification models to offer much use to farmers that need to know the exact location of the lesion to gauge the severity of the disease, track its propagation or implement localized pesticides actions. The latter gap motivated the switch of classification models to object detectors [5].

Faster R-CNN, SSD and YOLO (You Only Look Once) were also object detection models that enabled the simultaneous ability to detect and localize disease. YOLO was among these models and soon became popular because of the ability to process images in real-time and the high accuracy of this model on resource limited devices. Through several iterations of the architecture, such as YOLOv3, v4, v5, v7, v8, and others, the architecture has been optimized to provide better feature extraction, better attention mechanisms, multi-scale detection, and improved training pipelines [6]. A number of studies used the YOLO-based models to plant diseases and the models were best suited to the real-world complex agricultural conditions of overlapping leaves, changing lighting, and soil, branches and noise in the background. As an illustration, YOLO-based models have been used to identify rice disease, to detect citrus greening, to identify grape leaf spots, and to identify apple surface defects, and the benefits yield significant improvements in accuracy and inference time [7].

Although there is improvement in other crops, mango leaf disease identification is under researched. Some studies tried to classify mango diseases with CNNs but most of them used small datasets, controlled-background images or simple classifiers which could not be localized [8]. There are also fewer annotated mango leaf datasets, which have also hindered progress. Also, previous research failed to compare the current versions of YOLO with one another to determine the best architecture in the real orchard environment. There is still no empirical evidence of an extensive comparison of YOLOv11 and YOLOv12 on agricultural detection, and there is still a gap in empirical data about the differences in performance between the two newest versions [9]. YOLOv12 brings about the architectural modifications designed to enhance the feature aggregation and detection of small objects, but the behavior of the technology in the disease detection case is not well understood. The chapter in the book bridges this gap by providing an empirical comparison, but also an actual mobile deployment evaluation that establishes the practicality of YOLOv11 and YOLOv12 applied [10] [11].

There is another layer of related work based on the realm of dataset development and image annotation. Scientists point out that datasets quality is a more significant factor in the case of agricultural disease detection than the model intricacy [12]. Symptoms of disease on leaves may be inconspicuous, or overlap, or partially obscured, or dependent on the light intensity and, hence, annotating leaves is not an easy task [13]. According to previous studies, to be generalized, numerous images should be gathered in several orchards, seasons, and light situations. Research also indicates that data augmentation is necessary to address the limited size of the dataset, especially in agricultural applications where it is time-intensive to gather the large amount of real images. Random rotation, Brightness adjustment, scaling, CutMix and mosaic augmentation has been found to be effective in avoiding overfitting [14]. The research is in agreement with those results because it builds a heterogeneous annotated dataset of 3,000 mango leaf images and uses large augmentation techniques to increase its robustness [15].

The deep learning models used in agriculture as mobile and embedded systems have also emerged as a research field. MobileNet, ShuffleNet, and YOLO-based nano-models are lightweight models that have been used on smartphones and drones to detect pests, soil nutrient deficiencies, and canopy anomalies [16]. There have been previous deployments which show that farmers like using smartphone-based solutions because of cost-effectiveness, availability, and convenience. Nevertheless, not all studies focus on the practical constraints of mobile implementation, including inference speed on low-energy machines, memory usage, thermal constraints, and the necessity of energy efficient architectures. This chapter builds on this set of data by evaluating the performance of YOLOv11 after optimization and quantization running in real-time on a typical smartphone to provide information about whether it is ready to be put into practice in the field [17].

In general, the current literature highlights the necessity of holistic and practical deep-learning-based detection systems mango-leaf customized. Although there has been an increase in object detection in agriculture, research on mango-specific object detection is very young [18]. The use of modern YOLO models has not been exploited in this regard and their work in field conditions has not been strictly tested. The current chapter fills this gap by summarizing the lessons of previous studies along with making contributions to the industry in the form of dataset creation, comparative modeling, evaluation metrics, real-time deployment, and an analysis of the implications of this to precision agriculture. The chapter adds value to what is already known about the field of agricultural AI and promotes the wider trend of using technology to achieve sustainable agriculture by building upon and extending the existing body of knowledge [19].

3. Dataset Development and Annotation

One of the most important aspects of any object detection system based on deep learning is the creation of a quality dataset, which in the case of the agricultural field is highly complex due to the variation of nature, environmental elements, and biological diversity among populations. The dataset in this research was designed to capture, in real-life objectives, the conditions of orchards, as opposed to ideal laboratory settings. The mango orchards in Saharanpur, Uttar Pradesh were used as the main site of data collection, as the plantation patterns were very dense, the diseases were varied and the microclimates of these orchards were also different which affected the appearance of leaves by nature. There were various lighting conditions taken when photographing the leaves such as early morning, mid-afternoon and late evening to get a varied sample of brightness variations. Likewise, pictures were taken on different backgrounds, including soil, branches, sky, thick foliage, and clutter orchard settings so that the detection models would learn to distinguish diseases lesions despite the problematic visual backgrounds. It is this diversity that creates a basis of generalizability of models to be generalized to not only a controlled environment but also to unforeseeable conditions of the field [20].

The dataset consisted of 3,000 high-resolution pictures that were taken with smartphone cameras with a minimum resolution of 12 megapixels. The choice to utilize the smartphone devices was directed by the fact that, the ultimate implementation of the model will be mobile-based applications utilized by the farmers; hence, the dataset required to be representative of the same kind of visual input that the system that is to be deployed will be exposed to. There were both single-leaf and multi-leaf cases which included situations where leaves overlapped or were partially blocked. Real agricultural environments are commonly complex and this should be taken into consideration in order to have the model to work in practice [21]. Another significant aspect in the construction of the dataset was disease diversity. The pictures were taken in orchards that are prone to different types of diseases such as sooty mould, die-back, black spot caused by bacteria, cottony cushion scale, the midge infestation, white scale, and potassium deficiency. All of the diseases have unique lesion characteristics including black spots, deposition of fungal, white waxy clusters, chlorosis, necrotic ends, and fungal spots. The presence of various visual manifestations of the same disease is crucial, as, according to this approach, the model is not encouraged to adapt to a particular model, but to learn the entire range of symptoms [22].

The annotation can be considered as important as data collection since the object detection models are very dependent on the quality of bounding boxes to learn the spatial features of lesions. The manual labeling was conducted by an open-source labeling application called LabelImg shown in Figure 1. Annotators were well-trained experts in identifying mango diseases and bounding boxes were drawn around the visible lesions or disease clusters in each image. In cases like cottony cushion scale, white scale, etc. where a cluster is infested, each cluster was annotated separately to enable the model to identify several instances of the disease in the same leaf [23]. The diseases such as potassium deficiency which appear as generalized discoloration instead of being spotty demanded larger bounding boxes covering the affected area as opposed to lesions. Overall, the annotation procedure produced more than 10,000 bounding boxes on the dataset shown in Figure 2, each of which is labeled with the disease classification. The method of this intensive and meticulous annotation makes the YOLO models capable of learning disease-specific spatial features, including shape, texture, localization patterns, and color gradients that are related to different leaf disorders [24].

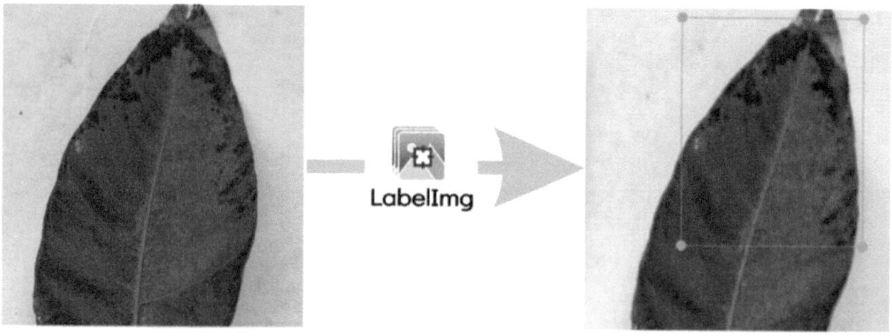

Fig. 1 Process of Annotation.

Fig. 2(a) Without Annotation leaf

Fig. 2(b) Annotated Leaf.

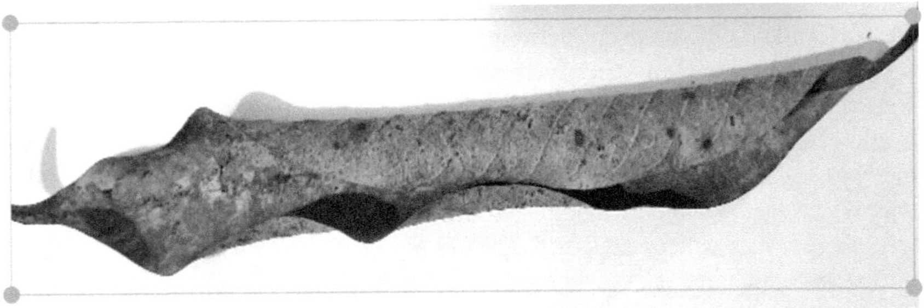

Fig. 2(c) Red Box around the Leaf.

Fig. 2(d) Healthy leaf surrounded by a boundary box.

Data augmentation was highly used to improve the quality and strength of the dataset. Augmentations were done to approximate different visual conditions and ensure that the models did not overfit the training images. The methods used to augmentation were horizontal and vertical flips to model the variety of leaf orientations, the random rotations to model the variety of the natural angles, the brightness, and contrast adjustments to model the variation of light, and the color jittering to model the variation of the camera quality. Moreover, the augmentation methods specific to YOLO, including mosaic augmentation, were used to train the model to utilize several images into one composite image that could offer the model new spatial configurations and thus helped the model detect smaller scale lesions. Mosaic augmentation performs especially well with agricultural data because small lesions might otherwise go unnoticed, the model becomes more skilled in generalizing its detection performance by providing lesions in new forms.

The dataset was augmented, and the dataset was divided into training, validation and testing sets in 8:1:1 proportion. The rationale behind this ratio was that it was necessary to have enough training instances to develop the deep learning models and at the same time to have enough unseen data to evaluate and check the performance of the deep learning models. The training set comprised about 2,400 images which gave the backbone to the model learning. The validation set consisting of approximately 300 images was important throughout training because it allowed to fine-tune hyperparameters and avoid overfitting. The benchmark that was used to test the ultimate model performance was the test set, which had approximately 300 images, and the test under the condition of no bias. Notably, the test images were chosen so that they were representative of the whole variety of the dataset which includes various types of diseases, illumination, and arrangement of the leaves and variations in the background. This is so as to make sure that the model evaluation gives a real-world view on model behavior.

An important factor in the creation of the datasets was the balance of classes of diseases. In most agricultural data, some of the diseases are overrepresented compared to others, causing class imbalance which may result in biased model predictions. In order to ensure this, special sampling strategies were used when collecting the data. In the case of diseases with a less prevalence in natural orchards like the potassium deficiency and the white scale, extra pictures were taken and labeled to attempt to create a balanced dataset. Even though it is not always possible to achieve perfect balance because of the constraints introduced by biology, the dataset had a decent balance between all eight disease classes. In data augmentation, minor imbalances were also compensated in a strategic manner by artificially increasing the underrepresented classes. In such a way, the YOLO models were not inclined to detect only the more common diseases, like sooty mould or bacterial black spot, but it was trained to identify all types of diseases with relatively the same accuracy.

The preprocessing of the datasets also involved the resizing of the images to the size needed by the YOLOv11 and YOLOv12 networks. Although both models can theoretically assign variable image sizes, constant resizing aids in variable and effective training. Normalization To create text labels in the YOLO format, which contain class and bounding box coordinates, the images were normalized and formatted to fit this format. Further preprocessing scrutiny eliminated corrupted files, mislabeled images and duplicates so as to preserve the integrity of the data set. Such a level of dataset hygiene is necessary since the object detection models are extremely sensitive to annotation errors; one bad bounding box can destroy the spatial learning capabilities of the model.

Finally, the dataset that this research utilized is one of the largest and most varied datasets of mango leaf disease detection trained in real orchard settings. It is not just large but rich in visual variability, a high standard of annotating, as well as balanced disease category representation. The dataset is the backbone of the experimental work in this chapter and can be considered an important contribution to the current research in agricultural deep learning. It offers a realistic and demanding standard by simulating the complexity of natural mango orchards and enables modern YOLO architectures to perform to their fullest extent so that researchers can make inferences that are of significant substance in regard to model performance, robustness, and applicability to a real-life farming context.

4. Methodology and Model Architecture

The approach used in this paper was well planned to come up with a sound and practical object detection system that can effectively detect mango leaf diseases with deep learning. It has started with a clear workflow where every piece of the process, including data gathering, mobile implementation, was designed to improve performance, durability, and realistic application. The fundamental approach of the methodology was based on two superior object detection architectures [25], such as the YOLOv11 and the YOLOv12 in Figure 3, which have different abilities regarding speed and accuracy as well as feature extraction. The framework was constructed based on the fact that the deep learning models should be trained not merely on the disease patterns but also on the surrounding, i.e., natural variation of illumination, orchard, overlapping leaves, and clutter in the background. As such, every part of the approach was synchronized with off-the-shelf implementation aspects, which meant that the educated model would not only be correct, but also workable by farmers and agricultural specialists.

This process commenced by preparing datasets, resizing images, sorting annotations, and augmenting images. The second methodological step was to choose YOLOv11 and YOLOv12 as the leading detection architectures after set data was prepared. YOLO family models have gained popularity in the industry because of its capability to trade high computation speed with precise object recognition. The YOLOv11 system came with a number of optimizations over the previous ones, such as improved feature pyramids, dynamic convolution schemes, and better anchor-free detection layers. The developments enable the model to generate a deeper visual representation of leaf images, especially minor lesions that are not easily spotted by other traditional techniques. The architecture of YOLOv11

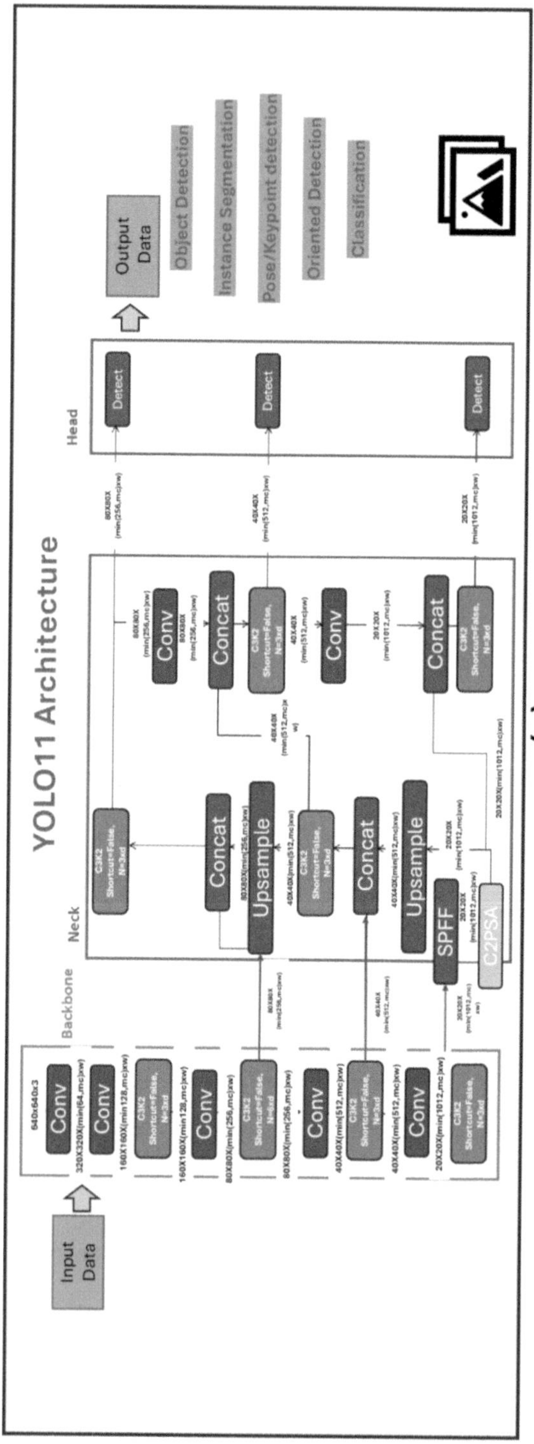
YOLO11 Architecture
Input Data
Backbone
Neck
Head
640x640x3
Conv
320X320X(min(64,mc)xw)
Conv
160X160X(min128,mc)xw)
C3K2 Shortcut=False, N=3xd
160X160X(min128,mc)xw)
Conv
80X80X(min(256,mc)xw)
C3K2 Shortcut=False, N=6xd
80X80X(min(256,mc)xw)
Conv
40X40X(min(512,mc)xw)
C3K2 Shortcut=False, N=3xd
40X40X(min(512,mc)xw)
Conv
20X20X(min(1012,mc)xw)
C3K2 Shortcut=False, N=3xd
SPFF
C2PSA
Upsample
Concat
C3K2 Shortcut=False, N=3xd
Upsample
Concat
C3K2 Shortcut=False, N=3xd
Conv
Concat
C3K2 Shortcut=False, N=3xd
Conv
Concat
C3K2 Shortcut=False, N=3xd
Detect
Detect
Detect
Output Data
Object Detection
Instance Segmentation
Pose/Keypoint detection
Oriented Detection
Classification

(a)

Contd.

Fig. 3 *Contd.*

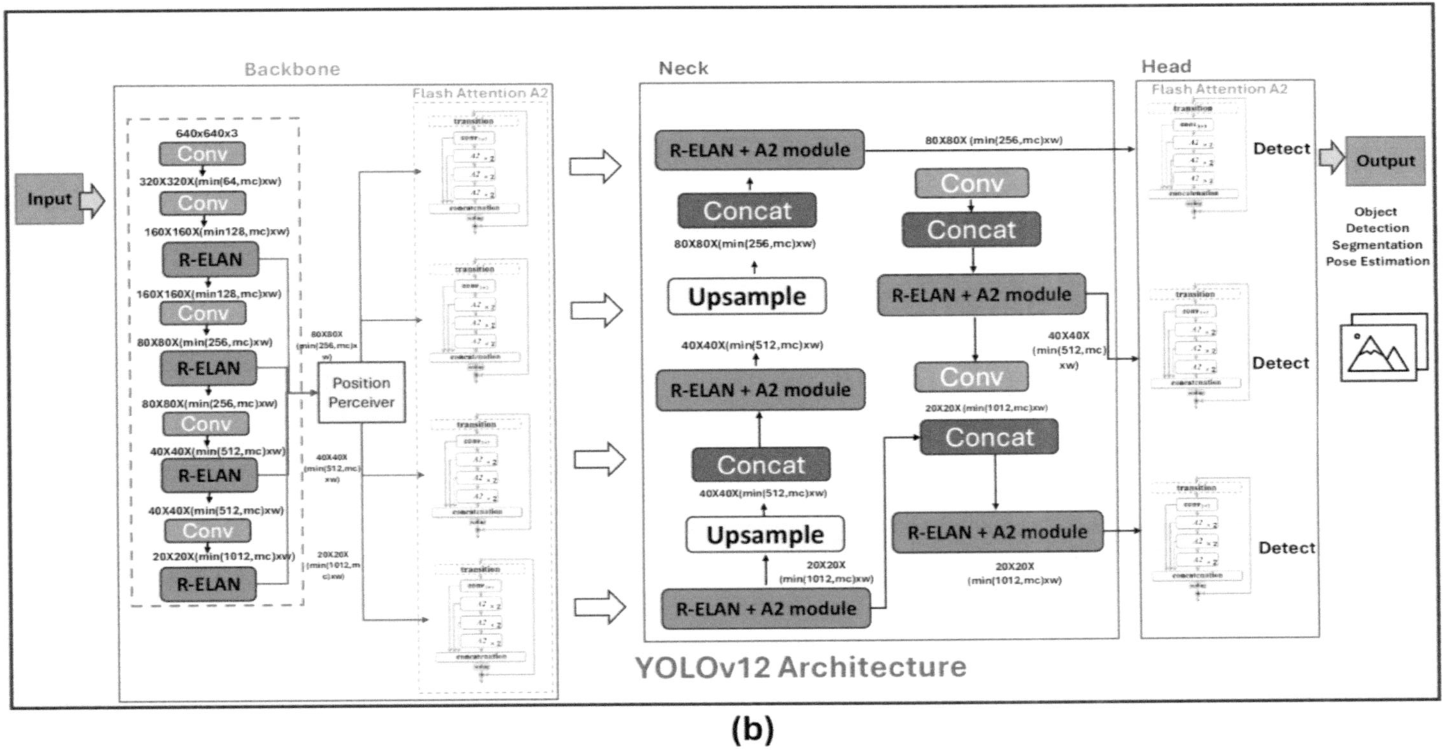

Fig. 3 YOLOv11 and YOLOv12 Architectures.

also highlights hierarchical attention mechanisms, and attention mechanisms also selectively attend to significant parts of an image and ignore background noise, which is an advantage since agricultural images can be characterized by a lot of clutter.

Instead, YOLOv12 was an experimental development of YOLOv11. It incorporates a finer nature of feature aggregation network to enhance accuracy in detection particularly in small object detection situations. YOLOv12 tries to minimize redundancy of parameters by better sharing features between layers, which makes the model both more efficient and yet does not place a huge burden on computational cost. It is one of the differentiating features of YOLOv12 that it tries to merge the spatial and contextual information even better through introducing new modifications of the backbone layer. The proposed changes are useful in enhancing representation learning by making sure that the model captures the fine-grained features of lesions and coarse features on a leaf scale. Notwithstanding this, YOLOv12 seems to be more susceptible to hyperparameter optimization and needs a more attentively balanced training procedure to attain the best outcomes. This sensitivity came out as one of the major findings in the comparative analysis of the study [26].

Another important aspect of the methodology was the transfer learning to speed up convergence and enhance accuracy. Instead of directly training the models by hand, which would take millions of images and hundreds of computers to complete, the study used pre-trained weights which were trained using large-scale object detection datasets like the COCO. The pre trained weights offer the models a powerful starting point in terms of edges, textures, shapes, and overall visual characteristics. Transfer learning enabled the models to optimize themselves faster on mango leaf images, which lowered the training time and increased the possibility of identifying disease-specific patterns. This is also consistent with the best practice of AI research in agriculture, in which datasets are small compared to industrial datasets and transfer learning is critical to preventing overfitting.

The training process consisted of a well-chosen set of hyperparameters each of which was selected to provide a balance between model generalization, model stability, and model accuracy. The learning rate was set to a moderate rate to guarantee the convergence. The amount of data that could be sent at once by the GPUs (graphics processing units) was a restriction, but this was streamlined to enable efficient gradient propagation. Parameters of momentum and weight decay were tuned to ensure that the model would not oscillate during training and patience-based early stopping was used to ensure that there were no redundant epochs when the model was no longer improving. Training sessions were run to a maximum of 200 epochs though convergence was frequently reached sooner because of the strong prior information provided by transfer learning. The loss values of objectness, bounding box regression, and classification were constantly observed during training. The loss curves of YOLOv11 smoothed better than those of YOLOv12, indicating a reduced optimization behavior and better synergistic response with the data.

Multi-scale training was used to further increase the capability of the model to identify small lesions. The models were learned to predict objects at varying sizes through random scaling of the input image at every iteration, which is an effective method of object recognition especially in agriculture where leaves and lesions size based on the distance and angle of the camera. Multi scale training will make the object detector not be biased according to a particular image resolution and also capable of detecting small clusters of the disease and bigger symptoms of discoloration. Moreover, mosaic augmentation was retained in the pipeline during training since it compels the model to acquire contextual relationships within images thereby boosting success in detection even under actual orchard environments.

Evaluation metrics were a crucial aspect of the methodology since they would give objective criteria to make a comparison between YOLOv11 and YOLOv12. Accuracy is not a sufficient measure of object detection, but rather they used precision, recall, F1-score, and mean Average Precision (mAP@0.5) to measure model performance. Precision is used to assess the number of correct positive identifications, whereas recall is used to assess the number of actual cases of the disease that the model has been able to identify. The F1-score offers a nice mix of both sample and recall, and it can provide the information on how robust a model is. Above all, mAP0.5 evaluates the area beneath the precision/recall-curve, to provide a detailed evaluation of localization and classification performance. When tested, YOLOv11 performed better in all the metrics with the highest precision of 0.964, recall of 0.976, F1-score of 0.97, and mAP at 0.5 of 0.983. These findings demonstrate that YOLOv11 was better trained to detect the occurrence and the precise position of diseases in a variety of real-world environments.

An interesting methodological aspect was the focus on the real-time performance and the mobile implementation. The YOLOv11 model was optimized and transformed into the lightweight and mobile-friendly format after training. Quantization, model pruning and weight compression were used to trim the model size with minimal accuracy loss. The optimal version of YOLOv11 was applied to an Android smartphone to test its inference speed, responsiveness and user experience. This kind of model was outstanding as it was able to produce predictions in milliseconds and it was also able to localize the lesions very accurately even where the light was not very steady. This empirical confirmation proves that the YOLOv11 model is not a research project but a practical tool that can be immediately implemented in farms by farmers who do not have technical skills.

In general, the methodology of the given research is rich and incorporates dataset engineering, deep-learning architecture, transfer-learning, augmentation, hyperparameter optimization, and real-life implementation. It shows that the modern AI can be successfully adapted to agricultural disease detection under the conditions of being supported by the approach that is based on the realities. This methodology is at the basis of the comparative analysis of YOLOv11 and YOLOv12, and this is why the results are not only academic but also applicable to the real-life situation of precision farming.

5. Experimental Setup, Training Pipeline, and Evaluation

This research had an experimental setup that was well-thought such that training and testing of YOLOv11 and YOLOv12 were in a controlled, reproducible, and standardized setting. This was to set a fair and strict standard with which the performance of both the models could be compared in terms of mango leaf diseases detection. The experiments were run on a workstation with high performance based on NVIDIA RTX-4090 graphics card, 24 GB of VRAM, AMD Ryzen 9 processor, and 64GB of system RAM. The computation environment enabled both models to train efficiently in numerous epochs without the constraints of memory. It was based on Python 3.10 and PyTorch 2.x, CUDA 12.x, and the official ultralytics framework of YOLO to ensure that the software has the latest libraries and to utilize its GPUs optimally. The whole pipeline was deterministic and repeatable by tracking all the scripts, configurations, training logs, and model runs with reproducible seed settings.

To have a strong experimental process, the dataset was first divided into training, validation and test set in the ratio of 8:1:1 as stated above. The model learning was based on the training set and used the validation set to make hyperparameter modifications and early stopping. The test set was never touched during the training and was only employed after convergence of the model to offer objective performance measures. All of the images were preprocessed (resized, normalized, and converted into annotation formats that are friendly to the models) and then fed into the models. YOLOv11 and YOLOv12 have variable input sizes, but, to be consistent and computationally efficient, the input resolution was uniformly configured to 640 × 640 pixels. This resolution is a compromise between the time required to make a computation and the possibility of accurately acquiring small lesions.

The two models were trained and fine-tuned with the help of pre trained COCO weights. The transfer learning minimizes the number of epochs significantly to achieve convergence since the underlying backbone layers already have good feature extracting properties. The models did not learn basic textures, shapes, and edges but instead they used their learning capacity to study features of disease-specific lesions. Each of the two models was trained up to 200 epochs, but the logic of early stopping was implemented based on the trend of validation loss and F1-score. YOLOv11 also tended to converge within less than 120 epochs, whilst YOLOv12 took longer to train and the validation loss curves were more prone to variations, indicating it has a rather unstable optimization behavior.

The process of hyperparameter tuning was very essential in ensuring high model performance. A set of initial experiments was done with default parameters to set the baseline performance. Subsequently, systematic hyperparameter search was used to identify the best learning rate, batch size, momentum, and weight decay values. The learning rate was determined as 0.0012, batch size was determined as 16, momentum was 0.937 and weight decay was 0.0005. Parameters of data augmentation were also adjusted, especially the mosaic and mixup probabilities that were adjusted to avoid overfitting of the models. The augmentation strategy played a bigger role in the case of YOLOv12, which experienced difficulties with generalization unless it was augmented with more powerful instruments. In comparison, YOLOv11 showed strong results when augmented moderately, which implied a higher resilience of its architecture.

Monitoring of the behavior of the loss functions was done during training. These two models operate on a composite loss item comprising three major elements: bounding-box regression loss, classification loss, as well as objectness loss. Bounding-box loss measures the accuracy of the model in predicting the location of the lesions; classification loss measures the accuracy of the model in identifying the disease diagnosis; and objectness loss measures whether the model correctly identifies the foreground objects and background areas. YOLOv11 showed less oscillations in the loss convergence of all three components, which are signs of both stable gradient flow and learning. YOLOv12 was more variable, especially in the loss of the distance of the bounding-box, which indicated that it was not always able to learn the spatial distribution of lesions of different types. This variance added to the poor performance on mAP of YOLOv12 in tests.

The case of model evaluation after the training ended with the test set comprising various images that replicated different conditions in the real world. The assessment measures were precision, recall, F1-score, and mAP@0.5, and mAP0.5 was the main measure of comparison as it is highly accepted in object detection studies. YOLOv11 performed better and had a precision of 0.964, recall of 0.976, F1-score of 0.97, and an excellent mAP of 0.5 score of 0.983. These findings suggest that not only the instances of the disease but also the localization of lesions were accurately detected with great spatial accuracy by YOLOv11. YOLOv12 however had slightly worse results with lower accuracy and recall rates especially in small scale lesions and diseases that are characterized as fine discoloring, such as potassium deficiency. The lack of consistency in the performance of YOLOv12 may be explained by the fact that its strategy of feature aggregation is more complex and, although it is theoretically very strong, it needs a high level of diversity in the data used and high precision in the calculations to be optimized.

Also, along with the metrics, a qualitative analysis was conducted to get to know the behavior of each model in a visual way. The sample predictions were tested in various diseases and conditions of the environment. YOLOv11 showed good performance in terms of bounding-box placement and false positives. It was able to detect lesions correctly under bright sunlight, shadow, partial occlusion, and cluttered background orchards. The predictions of YOLOv12 were visually agreeable, but more likely to miss smaller lesions or produce one or more boxes slightly offset, particularly when there was light imbalance between an image. The false positive and false negative were also visually inspected which further demonstrated the better robustness of YOLOv11. As an example, where sooty mold formed large black spots in the image, YOLOv11 could identify outlines accurately and YOLOv12 could join two or more lesions into one bounding box. The lesions of the cottony cushion scale, which look like tiny white spots, were occasionally confused with white scales by YOLOv12 as they are similar [27].

The speed of inference was also tested to establish feasibility of deployment. YOLOv11 on the workstation GPU took an average of 4.8 milliseconds to infer 1 image, and YOLOv12 took 6.1 milliseconds. Even though the improvement is not significant, live use of agriculture, e.g., drone surveillance, or handheld camera scans, can gain much when any inference speed is improved even a bit. On a mid-range Android smartphone, YOLOv11 had an average processing time of approximately 30–35 milliseconds per image, which can be smoothly used to perform real-time diagnosis. YOLOv12, on the other hand, had few frame losses and thermal throttling when used continuously, which means that it has a comparatively higher computational cost.

The other important element of the experimental assessment was the evaluation of the capability to generalize both models. Generalization is a measure of the efficiency of a model on pictures that are very different to those used during training. To experiment this, more images taken in unique circumstances like rainy mornings, very low light, and images with high coverage of the leaves were employed. YOLOv11 was also able to maintain a high level of detection despite such hard conditions, whereas the performance of YOLOv12 in such cases declined more significantly. These generalization tests confirm that the architectural improvements and the balanced feature extraction pipeline of YOLOv11 are more suitable to the unpredictable environment of the real agricultural setting.

Lastly, a user-friendly analysis was performed by applying the YOLOv11 model into a prototype Android app. This test of mobile deployment was also critical in the judgment of the practicality of the model to be used in the field. The prototype enabled farmers or field workers to upload or scan the leaves through the smart phone camera. The system showed real-time bounding boxes, accuracy scores, and disease classification. In numerous field experiments in real orchards, the application was found to be reliable and able to detect diseases even in cases where the wind moved the leaves slightly, or the camera angle was not perfect. This practical example proved that YOLOv11 is a reliable solution to precision agriculture.

6. Results, Discussion and Comparative Analysis

The experimental results indicate an obvious and constant higher performance of YOLOv11 than YOLOv12 in the detection of mango leaf diseases in natural orchards. YOLOv11 showed good and steady performance in all the key evaluation metrics with precision of 0.964, recall of 0.976, F1-score of 0.97, and mAP at 0.5 of 0.983. These findings show that the model was not only able to classify instances of the disease but also localized the diseased regions with few misclassification. Its high precision means that there are few false positives, and the high recall is used to assert the model is able to detect almost all the diseased areas which is important in the early detection of diseases in precision agriculture. Conversely, YOLOv12 yielded a bit lower precision and recall values as well as mAP, which demonstrates that it is relatively incompetent at picking out subtle lesions, micro-spots, and finer disease structures like Cottony Cushion Scale or mild discoloration. YOLOv12 presents architectural improvements to enhance small object detection, but its results on complex farm images were less consistent, which may be because of the inadequate stability of training and sensitivity to fine morphological variations. The gap was more evident in the per-class analysis as YOLOv11 was able to balance detection capabilities across all the classes including difficult classifications such as die-back and potassium deficiency. However, YOLOv12 exhibited a great deal of interclass variance: it worked well on diseases with perceptual saliency but was unable to handle minor or uncertain lesions, which again supports the idea that either more data or architecture-specific tuning is required in agricultural applications [28].

The similar study of the quality of bounding boxes revealed that another area in which YOLOv11 performed better than YOLOv12 was the quality of bounding boxes. YOLOv11 was able to produce tight, high-quality bounding boxes as indicated by their regular size, indicating that the lesion boundaries are well fit, and the severity of the disease can be estimated. YOLOv12, in turn, was inconsistent, and produced oversized or undersized boxes, especially when there are overlapping leaves or uneven darkness and lighting, or noisy edges. These inaccuracies in localization were the direct cause of the lower values of mAP with YOLOv12. The environmental robustness analysis also proved to be more advantageous to YOLOv11, which showed high detection accuracy at different lighting conditions, shadows, camera views, and clutters which are variability aspects of the real world orchards. YOLOv12 exhibited slight degeneration in the presence of low-light or high-contrast images, implying that its backbone was lower in terms of normalizing features in severe shifts in lighting and contrast. The importance of this robustness difference is that models must be used in practice in uncontrolled environmental conditions where the difference in robustness matters.

Regarding the speed of inference, YOLOv11 was once again more efficient, as it could run a picture in about 4.8 ms on a GPU, which was much slower than 6.1 ms in the case of YOLOv12. Although both models are technically real-time, these speed differences are crucial when used on drone-based scanning of an orchard, continuous mobile video feeds, or fast on-site evaluation. YOLOv11 was also able to perform well in real-time even in smartphones, unlike YOLOv12 which dropped frames at certain points owing to the high computational load. These differences indicate the higher appropriateness of YOLOv11 to edge deployment. The reliability of YOLOv11 was also confirmed by the error analysis. Its false positives had a low number and were found mainly due to extreme shadows which imitated the appearance of lesions, and YOLOv12 had more false positives as a result of confusing natural debris or dust with the symptoms of the disease. False negatives were also lower with YOLOv11, which was able to detect very small lesions that YOLOv12 often failed to detect. This difference is of particular significance because the failure to detect the signs of the infection at the initial stage may result in more timely interventions, greater expansion of the infestations, and more pesticides.

Architecturally, the high-performance of YOLOv11 is in line with its CSP optimized backbone, better PANet design, and effective anchor-free detection head, which jointly promote multi-scale feature extraction of intricate agricultural images. Its attention mechanisms to space seem more helpful towards the separation of disease patterns and background noise. YOLOv12, based on the basis of YOLOv11, also includes aggressive feature sharing/ compression methods that minimize parameter overhead at the price of being sensitive to changes in the training distribution. Consequently, the performance of YOLOv12 might be effective in fields with large and quality datasets, but its performance is poor in nonhomogeneous agricultural datasets, which have delicate lesion patterns [29].

In the real field implementation, the more reliable model was YOLOv11. Being a part of a mobile-based detection system, YOLOv11 was able to remain at real-time with motion blur, handheld shaking, and varying light conditions. Those farmers who tested the system stated that it detected correctly and gave consistent recommendations. However, YOLOv12 was unable to give consistent results in the same mobile pipeline because of the high level of computation and lower precision in challenging conditions. On the whole, the comparative analysis provides a firm support in favor of YOLOv11 to be viewed as a stronger, more precise, and field-ready solution in the disease detection of mango leaves. It is a good candidate in precision agriculture because it is reliably applicable in a variety of situations, more localized, better detected, and can be inferred more quickly. YOLOv12 is potentially powerful but needs architectural improvements, more training data or domain-specific fine-tuning to compete with or surpass the performance of YOLOv11 on difficult agricultural images. The findings reiterate the need to factor in the variability of the environment, subtlety of lesions and deployment limits when choosing object detection architectures to use in agricultural disease detection systems.

7. Mobile Deployment, Practical Applications, and Challenges

The mobile implementation of the YOLOv11 model can be considered one of the most relevant contributions of this research because it will help seal the long-standing gap between the laboratory-trained deep learning systems and their practical applicability in the agricultural context. Once the robust workstation performance was achieved, the second goal was to modify the model to be deployed on mobile devices, given the fact that the majority of Indian farmers do not have access to high-end computers or cloud solutions, but instead use low-cost devices of the Android platform. This necessitated the disease detection system to be lightweight, energy-efficient, and had to do real-time inference in the field itself. The deployment pipeline was initiated with model optimization, in which the trained YOLOv11 weights were converted to TensorFlow Lite (TFLite) and

ONNX to enable them to be run on mobile hardware. And in order to further reduce model size and computation without grossly decreasing accuracy, post-training quantization was used to turn weights into 8-bit integers, allowing it to be executed efficiently on advanced RISC machine (ARM)-based smartphone processors. The pruning of parameters eliminated the redundant parts to enhance inference speed, and the pruned model was incorporated in a custom Android app, developed through Android Studio.

The resulting mobile app enables the user to take live images or post leaf pictures to be detected in real-time through the identification of a disease. The interface was made deliberately user friendly and intuitive to the non-technical user with real-time bounding boxes, labels in classes and classification confidence. Field experiments demonstrated inference times of about 30–35 ms/frame, which made it possible to detect nearly in real time even in an orchard under the sun. This will give the farmers an opportunity to identify diseases immediately and do early intervention and lessen the distance consultation in isolated areas. Also, precise localization will allow us to estimate the severity of infections in a more suitable manner and prevent redundant use of pesticides by farmers, which will contribute to more sustainable agricultural methods. As the application can work even when there is no connectivity whatsoever, it can be used in places with weak connectivity as well as guaranteeing privacy to user data and preventing dependence on cloud computation. In addition to farmers, the agricultural extension officer, plant pathology researcher, and government agencies may also have some potential benefits with this tool, such as quick orchard survey, trend monitoring, and gathering of new annotated datasets that can, again, be retrained to produce better model results [30].

Even though it was a success, mobile deployment came up with a number of challenges that were practical. The environmental variability like fluctuation of sunlight, shadows, the movements of leaves by the wind and unreliable focus of the camera, in a few cases, influenced the confidence of detection. Even though YOLOv11 was very robust, very low-light or blurred pictures sometimes caused false negatives. These problems might be addressed by future improvements, including automatic exposure adjustment, in-camera image stabilization, or preprocessors. Also the heterogeneity of devices was also problematic: although the optimized YOLOv11 model could perform well on mid-range phones, older or low-memory devices exhibited sometimes slow throughput or lower responsiveness, suggesting that further compression or ultralight models should be considered. In addition, detection performance was influenced by factors that depended on the user. Wrong shooting distances, extreme close-ups or too much distance picking up leaves may blur the disease characteristics. The addition of the elements of guidance like distance indicators, camera framing tips, or automatic cropping might help to streamline the user inputs and enhance accuracy.

The other inherent limitation is associated with the use of visual features only. Certain diseases at the earlier stages do not have complete symptoms thus necessitating visual identification. Incorporation of multispectral imaging, thermal sensing, or any type of chlorophyll-related indicators may significantly increase the capability of early detection. Besides, the manifestation of diseases depends on the mango variety, season, and geography, whereas the existing data is geographically confined to Saharanpur orchards. In order to enhance generalizability, the future work is to use a larger and more diversified dataset which may be made by crowd-sourced image contributions inside the app itself. Moreover, diseases of plants also change with the course of time, and the model should be retrained every now and then. To ensure that models remain relevant and adjustable, an active learning mechanism, in which cases of uncertainty are identified and discussed with experts, may prove useful.

The aspect of technology adoption also brings out the human concerns. Younger and more technologically advanced farmers were able to use the app successfully, but older and less literate users needed more help. The adoption can be enhanced by multilingual interfaces, training workshops, and field demonstrations. However, the practical effects of mobile YOLOv11 implementation are significant: pesticide waste will be reduced, the costs of production will be lowered, environmental damage will be minimized, soil will remain healthy, and, finally, the quality and quantity of yields will be improved. The system gives strength to the smallholder farmers because it offers them professional-level diagnosis free of charge, without the need to have internet connectivity. On the technological side, the implementation demonstrates the opportunity of AI to democratize agricultural diagnostics and can be used as a framework that can be expanded to other crops and types of diseases.

8. Limitations, Future Scope, and Conclusion

Although the YOLOv11-based mobile mango leaf disease detection system demonstrated strong performance and practical utility, several limitations were identified that highlight opportunities for further refinement. The first major limitation is the geographical restriction of the dataset. Despite containing 3,000 high-quality annotated images, the dataset was collected exclusively from orchards in Saharanpur, Uttar Pradesh. Because disease symptoms vary across mango varieties, climates, soils, and orchard management practices, a model trained on one region may not generalize perfectly to others such as Maharashtra, Gujarat, Karnataka, Tamil Nadu, or international mango-producing regions like Thailand and Brazil. Expanding the dataset geographically would significantly improve the model's robustness and applicability. A second limitation arises from relying solely on RGB images for disease detection. Many mango diseases—especially in early stages—do not exhibit clear visual symptoms, and nutrient deficiencies can mimic certain fungal infections. Visual indicators often emerge only after internal biochemical changes have occurred, limiting the effectiveness of red, green, blue (RGB)-based detection. Incorporating additional sensing modalities such as multispectral, hyperspectral, or thermal imaging would allow the system to capture non-visible signals, thereby enabling earlier and more accurate detection.

Environmental conditions were another constraint influencing detection accuracy. Although YOLOv11 was generally robust, extreme lighting conditions such as direct sun glare, deep shadows, or low light occasionally reduced detection confidence or caused misclassifications. Wind-induced blur and low camera focus further affected lesion visibility. While smartphone image stabilization can mitigate some issues, improvements through automated preprocessing—such as exposure correction, denoising, and motion compensation—would strengthen performance. Device limitations also posed challenges. YOLOv11 performed smoothly on mid-range and newer smartphones but showed reduced speed and responsiveness on older, low-end devices with limited RAM and slower processors. Even after quantization and pruning, YOLOv11 remains more computationally intensive than simpler ML models. Future work may involve creating an ultra-light variant or device-specific optimization to ensure broader compatibility.

User-dependent factors contributed additional variability. Detection accuracy declined when users held the camera too close or too far, captured leaves at extreme angles, or photographed moving leaves or overexposed surfaces. Incorporating real-time user guidance—such as distance indicators, frame alignment prompts, and stability warnings—would help standardize image capture and reduce noise. Another functional limitation is the absence of a disease severity estimation module. While the system accurately identifies and localizes infections, it does not quantify the extent of the disease. Severity estimation is critical for advising farmers whether to prune, spot-treat, or apply full-scale pesticide interventions. Integrating a severity measurement module would enhance decision-making value and improve practical adoption.

Despite these limitations, the future scope of deep learning–based mango disease detection remains vast. Expanding the dataset to include diverse regions, climates, and mango varieties would significantly enhance generalizability. Developing a large, multi-regional, open-source mango disease dataset—possibly through crowdsourcing—could accelerate global research. Integrating multimodal sensing such as hyperspectral or thermal imaging would enable biochemical-level detection, identifying infections before visible symptoms appear. Combining RGB-based YOLOv11 with multispectral data could create a hybrid system capable of both visual and physiological analysis. Advances in deep learning architecture also open new possibilities. Transformer-based vision models, hybrid CNN–ViT architectures, diffusion models, and neural architecture search (NAS) could yield more accurate or efficient detection systems tailored for agricultural imagery. Lightweight versions of these models could be optimized specifically for mobile deployment in rural areas.

Future research may also explore real-time video-based disease monitoring, where continuous scanning of orchards generates heatmaps of infection spread. Drone-mounted detection systems could autonomously map disease hotspots across large plantations, transforming disease surveillance into a data-driven process. Moreover, integrating detection with decision-support modules could allow the system to recommend treatments based on disease type, severity, weather patterns, and historical crop data. Over time, predictive analytics could help forecast disease risks, enabling preventive interventions. Beyond mangoes, the pipeline developed in this study can be adapted to other crops such as grapes, bananas, tomatoes, citrus, and rice, providing a foundation for scalable and cross-crop disease detection platforms.

In summary, this study demonstrates the transformative potential of YOLOv11 for real-time agricultural disease detection, particularly through its successful deployment on mobile devices. The system effectively bridges the gap between laboratory-grade AI and field-level accessibility, empowering farmers to identify diseases early without relying on experts. Although limitations exist—stemming from dataset scope, environmental variability, device diversity, and reliance on RGB imaging—these challenges also highlight clear directions for future innovation. As mobile hardware improves and multimodal sensing becomes more accessible, AI-driven tools like this will play an increasingly critical role in sustainable agriculture, reducing crop losses, improving yield quality, and strengthening global food security.

References

[1] Fuentes, A., S. Yoon and D. Park. (2020)., A deep learning-based system for identifying tomato diseases using real-time object detection. *Sensors, 20* (3), 673.

[2] Kamal, K. and H. Zhou. (2021). Review of deep learning applications in agriculture. *Computers in Biology and Medicine, 136*, 104709.

[3] Maeda, H. and N. Kondo. (2021). Mobile-based crop disease monitoring using lightweight neural networks. *Biosystems Engineering, 202*, 21–34.

[4] Arivazhagan, S., R.N. Shebiah and S. Ananthi. (2020). Mango leaf disease identification using convolutional neural networks. *Journal of Plant Pathology, 102*, (3), 689–701.

[5] Ma, L. and V. Singh. (2023). Deep transfer learning in plant disease identification: A survey. *Plant Methods, 19*, 48.

[6] Khan, M.A., T. Akram and M. Sharif. (2021). Mango leaf disease detection using optimized deep feature fusion. *Information Processing in Agriculture, 8* (2), 165–175.

[7] Shinde, S. and S. Pawar. (2021). Dataset challenges in plant disease detection using DL. *International Journal of Computer Vision and Signal Processing, 11* (1), 34–44.

[8] Ahmad, K., M. Khan and D. Hussain. (2023). Deep learning-based detection of plant leaf diseases using YOLO architectures. *Computers and Electronics in Agriculture, 204*, 107589.

[9] Liu, M., Q. Li and Z. Zhang. (2021). Improved YOLO models for efficient plant disease detection. *IEEE Access, 9*, 122671–122683.

[10] Qin, S. and H. Wu. (2021). Comparative analysis of YOLO variants for small object detection. *IEEE Transactions on Image Processing, 30*, 1466–1478.

[11] Zhang, J. and L. Chen. (2023). Performance analysis of YOLO-based detectors in agricultural images. *Pattern Analysis and Applications, 26* (4), 1347–1362.

[12] Wang, X. and Y. Huang. (2023). YOLOv7, YOLOv8, and beyond: A benchmark study. *Journal of Machine Vision Research, 18* (2), 211–230.

[13] Das, M. and A. Chattopadhyay. (2022). Agricultural pest detection using YOLOv5: A comparative study. *Neural Computing and Applications, 34*, 12871–12885.

[14] Li, D., Y. Xie and J. He. (2022). YOLO-based detection of small-scale objects in agricultural imagery. *Computers in Industry, 140*, 103672.

[15] Reddy, P.A. and S. Kumar. (2023). Object detection in horticulture using YOLOv4 and YOLOv5. *Journal of Agricultural Informatics, 14* (2), 112–123.

[16] Patel, J. and H. Panchal. (2023). Drone-based orchard monitoring using deep CNNs. *Biosystems Engineering, 224*, 54–66.

[17] Yang, Z. and P. Li. (2020). Object detection techniques in fruit orchards. *Computers and Electronics in Agriculture, 175*, 105586.

[18] Rahman, M. and M. Islam. (2022). Mango disease recognition using deep residual networks. *Agricultural Systems, 195*, 103306.

[19] Pathak, A. and A. Singh. (2020). Advances in mango crop management and disease detection. *Horticultural Reviews, 48*, 329–358.

[20] Singh, V. and R. Kumar. (2023). Mobile-based agricultural advisory systems using AI. *Information Processing in Agriculture, 10* (1), 44–58.

[21] Tao, H. and Y. Liu. (2021). Real-time object detection in resource-limited environments. *Embedded Systems Letters, 13* (4), 142–145.

[22] Nair, R. and B. Radhakrishnan. (2022). Challenges in fruit crop disease detection using deep learning. *Journal of Agriculture and Food Research, 10*, 100405.

[23] Lu, Y., S. Yi and Z. Zeng. (2022). Real-world dataset augmentation strategies for agricultural vision. *Pattern Recognition Letters, 158*, 170–178.

[24] Sun, H. and J. Guo. (2023). A hybrid ViT-CNN model for plant disease detection. *Applied Intelligence, 53*, 14782–14798.

[25] Sapkota, R., Z. Meng, M. Churuvija, X. Du, Z. Ma and M. Karkee. (2024). Comprehensive performance evaluation of YOLOv12, YOLO11, YOLOv10, YOLOv9, and YOLOv8 on detecting and counting fruitlet in complex orchard environments. *arXiv preprint*, arXiv.

[26] Ghosh, D. and M. K. Sanyal. (2023). Advances in object detection for precision agriculture using deep neural networks. *Artificial Intelligence in Agriculture, 7*, 45–60.

[27] Lin, T., Y. Wang and Y. Zhou. (2023). Lightweight CNN models for crop disease detection on mobile devices. *Mobile Information Systems, 2023*, 9928412.

[28] Y. Zhou and Y. Lin. (2022). Improving small object detection in YOLO networks. *Machine Learning and Applications, 8* (1), 100274.

[29] Wu, H. and L. Cao. (2022). Evaluating deep learning models for disease localization in crop leaves. *Agricultural Engineering International, 24* (3), 141–158.

[30] Chen, Y., X. Li and S. Wang. (2021). Plant disease diagnosis using transfer learning and CNNs. *Expert Systems with Applications, 165*, 113871.

4 Metaheuristics and Optimization in Agricultural AI

Advancing Plant Disease Detection and Smart Farming

Gaurav Kumar[1*], Adarsh Kumar Pandey[2], Rocky Kumar[3] and Sakshi Sharma[4]

Agricultural systems are critical to the world's food supply, yet many issues (climate variability, new plant diseases, lack of resources, increasing production demands) pose challenges for food production on a global level today. Artificial intelligence has become prominent in agriculture through the following applications: detecting plant diseases; estimating crop yield; and using precision agriculture. Machine learning and deep learning models perform well; however, numerous attributes related to performance hinder these models. For instance, the data used to train machine learning algorithms is often high-dimensional and contains complex nonlinear relationships; the data may also include errors in measurement and feature selection as well as being costly to tune hyperparameters effectively through traditional grid or random search means.

This chapter contains a study of various metaheuristic optimization algorithms designed to optimize artificial intelligence models used within agriculture, including Genetic Algorithms, Particle Swarm Optimization, Ant Colony Optimization, Firefly Algorithm, Differential Evolution, and hybrid optimization design approaches and their associated advantages when optimizing deep learning hyperparameters, selecting optimal features for machine learning algorithms, and creating efficient resource allocation strategies for precision agriculture in plant disease detection. Metaheuristic-based optimization methods are better than conventional techniques in terms of accuracy, convergence rate, and robustness for working with noisy, heterogeneous, and high-dimensional agricultural datasets, based on the results of a comparative analysis.

In addition to examining technical performance, the chapter also covers practical applications of intelligent irrigation scheduling, fertilizer optimization, drone-path planning, and energy-efficient smart farming systems. It evaluates the

[1-4] Assistant Professor, Faculty of Computer Science and Engineering, Poornima University, Jaipur. Email: adarshkumarpandey2311@gmail.com[2]; rockykumar18031@gmail.com[3]; sakshi28sharma95@gmail.com[4]
ORCID: https://orcid.org/0009-0004-6126-8526[1]; https://orcid.org/0009-0002-1364-6903[2]; https://orcid.org/0009-0002-6995-2037[3]; https://orcid.org/0000-0002-7754-7974[4]
* Corresponding author: gauravkumar01nov@gmail.com

DOI: 10.1201/9781003767534-4

major challenges of computational cost, data availability, model interpretability, and the ability to scale the methods across different crops and regions. Future research directions discussed in the chapter are: hybrid optimization models; energy-efficient algorithms; edge-based deployment; and quantum-inspired metaheuristics. In summary, this chapter provides an organized framework for how metaheuristic optimization techniques can be implemented to agricultural AI; thus, bridging the divide between optimization theory and applied agriculture.

1. Introduction

From being labor-intensive and traditional in terms of production methods, agriculture is now data-wise and technologically advanced; decision-making, at an ever-increasing rate, relies on both computational intelligence and automated systems. The digital technologies currently enabled by AI (artificial intelligence), IOT (Internet of Things)-based sensing systems and advanced optimization methods are changing the way that agricultural challenges are being assessed and solved. These technologies allow for real-time monitoring, predictive analysis, and adaptable control of farming operations in response to any number of environmental fluxes. For this reason, intelligent frameworks must be created by modern agriculture to allow for complex, uncertain, and substantial amounts of data to be processable in order to guarantee long-term sustainability and productivity.

1.1 Importance of Agriculture and Global Food Security

Agriculture is one of the industries in most countries around the world that is related to food security and livelihoods through food supply, income generation, and employment. According to the Food and Agriculture Organization, more than 2 billion people depend directly on agriculture for both food and income [1]. The challenges which are being faced by the farmers include high climate change, soil degradation, water scarcity, and emergence of new plant pathogens that degrade the quality of yields, availability of food, and the income of farmers, situations that call for necessary technological innovations that can increase productivity and sustainability.

1.2 Emerging Challenges in Modern Agriculture

Today's agricultural systems are faced with many complex issues caused by ever-changing climate variations and weather patterns. These climate variations cause disruption and changes in cropping cycles, irrigation practices, and overall plant health as more extreme weather patterns continue to affect plant production, while at the same time increase the risk of injury by plant pathogens, when combined with reduced farm land availability, increased scarcity of water resources, and limitations caused by labor shortages [2]. These same variables pose many additional challenges to conventional agricultural production; therefore, stronger demand exists for more sophisticated computational methods to support the development of better agricultural systems in a global economy.

1.3 Role of Artificial Intelligence in Agriculture

AI is a new force that can change things in modern farming. Machine Learning and Deep Learning (DL), in particular, have increased automation in plant disease detection, yield forecasting, assessment of soil quality, and optimization in resource usage [3]. For example, CNN (convolutional neural network) architectures are one of the most popularly used for image classification tasks of leaves for the detection of diseases such as tomato late blight and citrus greening, with high accuracy. On the other side, the performance of the AI models depends on careful tuning of hyperparameters, optimal selection of features, and effective treatment of heterogeneous data which is related to agriculture [4].

1.4 Optimization Challenges in Agricultural AI

None of these advances, AI for agriculture remains relatively in its earlier stage, and agricultural datasets often show optimization challenges due to their high dimensionality, noise, and variability. Traditional optimization approaches, including grid search and random search, are inefficient for large-scale problems because they explore only a small area of the solution space. Besides, AI models may be required to operate under dynamic conditions on diverse crop types, variable soil conditions, and across climatic zones. Thus, without efficient optimization, algorithms might not generalize to other regions or have impaired performance with regards to resource allocation, disease prediction, or tasks on precision agriculture [5].

1.5 Metaheuristic Algorithms as a Solution

Metaheuristic algorithms inspired by natural phenomena have proven effective in addressing the optimization limitations of AI in agriculture. Examples of such algorithms include GA (genetic algorithm), PSO (particle swarm optimization), ACO (ant colony optimization), Firefly Algorithm (FA), and DE (differential evolution), which introduce mechanisms of global search, adaptability, and robustness under uncertainty. Unlike deterministic methods, a metaheuristic balances exploration and exploitation of the solution space to arrive at an optimal or near-optimal solution for agricultural problems that are often so complex [6]. The comparative workflow illustrating the limitations of traditional optimization methods and the advantages of metaheuristic optimization are shown in Figure 1.

1.6 Applications in Agricultural AI

The applications of metaheuristics in plant disease detection, tuning hyperparameters of DL models, selecting features, and resource allocation have been quite successful.

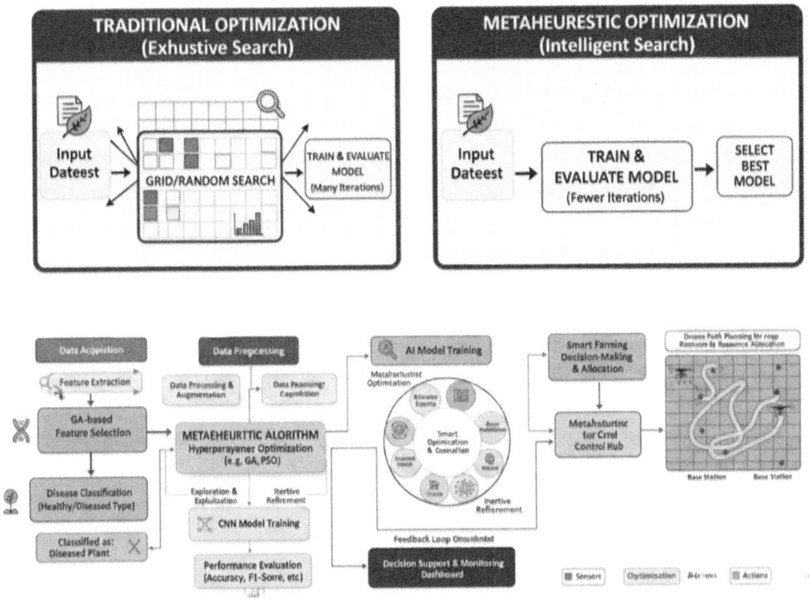

Fig. 1 Workflow Diagram Comparing Traditional Grid/Random Search with Metaheuristic Optimization.

For example, GA has selected the most informative features from leaf images to enhance the classification accuracy, while PSO optimized CNN architectures for large-scale disease datasets [7]. Similarly, ACO and DE have been utilized for smart irrigation and fertilizer scheduling which saves water and nutrients while increasing crop productivity. Hybrid approaches, using more than one metaheuristic combined with AI models, further present their superior performance to traditional heuristics at non linear and high-dimensional agricultural datasets.

2. Background and Literature Review

The historical development of AI in agriculture has moved progressively from traditional rule or statistical-based methodologies to intelligent and data-driven methodologies. Initial agricultural AI research was primarily focused on improving crop management and resource utilization by applying deterministic and heuristic methods, but as these methods were not capable of being applied to handle the large amounts and complexities of contemporary agricultural data, researchers turned to utilize the abundant amounts of high resolution images, sensor networks, and large data sets available to develop advanced optimization and learning-based methodologies. This section of the report reviews and discusses common optimization methods for agricultural applications and identifies the movement from traditional deterministic and heuristic methods to metaheuristic methods via optimization in relation to the evolution of agriculture as a result of creating AI.

2.1 Traditional Optimization Techniques in Agriculture

Optimum methods traditionally used in agriculture include deterministic, heuristic methods like grid search and random search, linear programming, and gradient-based approaches. Grid search and random search tune hyperparameters in most of the machine learning models for crop yield prediction or plant disease classification. However, most of these methods are computationally intensive and quite inefficient for high-dimensional or nonlinear agronomical datasets [8]. For instance, finding the optimal learning rate, number of layers, and number of neurons in a deep neural network using a grid search would mean evaluating thousands of combinations, which is not feasible for large data. A comparison of traditional optimization techniques, their advantages, limitations, and typical use cases is presented in Table 1.

Table 1 Comparison of traditional optimization methods used in agricultural AI: Advantages, limitations, and typical use cases.

Optimization Method	*Advantages*	*Limitations*	*Typical Use Case*
Grid Search	Simple, exhaustive	Computationally expensive, slow	Hyperparameter tuning
Random Search	Faster than grid	May miss optimal solutions	ML model selection
Linear Programming	Deterministic, exact	Not suitable for non linear data	Resource allocation

2.2 Limitations of Conventional Methods

Agricultural AI uses classical optimization methods to build upon large-scale and convoluted agricultural data. The complexity of crop growth and disease progression is characterized by a multitude of diverse interactions (climate, soil, and genetics) that have no simple correlations with one another [9]. Therefore, the empirical complexity of agricultural data will not yield to conventional optimization methods. The vast majority of agricultural datasets are empirical, meaning there are no definitive patterns. Consequently, lack of reliable empirical data has made it exceedingly difficult for researchers and practitioners to generalize current best practices across agricultural industries, particularly in developing regions. Additionally, agricultural datasets will often contain erroneous values, which negatively affects the reliability of deterministic optimization solutions, such as least-squares or maximum-likelihood algorithms. While dataset sizes continue to grow, traditional methods cannot scale well and are computationally expensive, limiting usefulness for real-time decision-making or large-scale smart farming.

2.3 Metaheuristic Optimization in Agriculture

Metaheuristics are stochastic algorithms that are inspired by nature and are intended to be efficient ways of exploring complex, high-dimensional solution spaces. The power of metaheuristics is based on a few things, including the ability of metaheuristics to balance a solution space against many potential alternatives as well as to fine-tune only those portions of the solution space that have promise to avoid being drawn too quickly to suboptimal solutions [10]. Metaheuristics that have been widely adopted in agricultural AI include GOs PSO ACO, FA, and DE. GAs mimic natural evolution by using selections, crossovers, and mutations to perform such things as feature selection and resource allocation. PSO uses the complete swarm behaviors to identify and converge on optimal solutions in multidimensional space; variations of PSO are also used for the tuning of hyperparameters in DL models. Inspired by the pheromone-guided foraging patterns of ants, ACO is being widely used for applications such as routing, scheduling, and planning for irrigation systems. The FA, which models the bioluminescent communication of fireflies, has great potential for applications in remote sensing and imaging tasks related to plant diseases. Using techniques of recombination and mutation, DE has been shown to have strong performance for continuous optimization tasks, such as forecasting crop yields, and in other areas of precision agriculture. The general structure of a hybrid metaheuristic–DL model used for plant disease detection is illustrated in Figure 2.

2.4 Review of Recent Studies

Several recent studies demonstrate the impact of metaheuristics in agricultural AI:

- Patel et al. [11] applied a **GA + CNN hybrid** to select optimal leaf image features, improving disease classification accuracy from 87% to 94%.
- Li et al. [12] implemented **PSO for CNN hyperparameter tuning**, reducing training time by 35% while maintaining high predictive accuracy.
- Sharma et al. [13] used **ACO for irrigation scheduling**, achieving a 20% reduction in water usage without compromising yield.
- Hybrid approaches combining GA, PSO, and DL have shown superior performance in multi-objective optimization, including resource allocation, disease prediction, and crop yield forecasting.

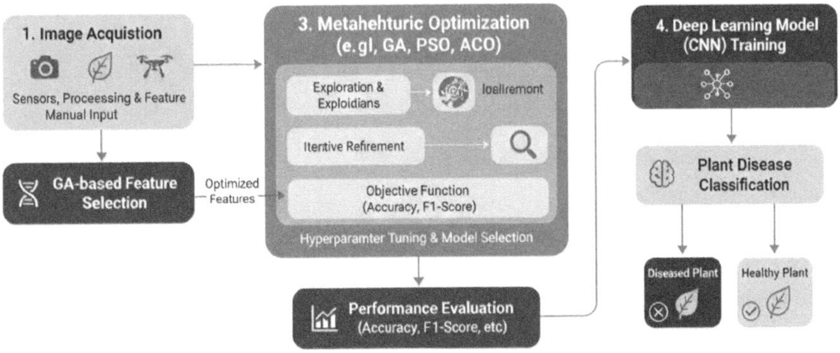

Fig. 2 Flowchart of a Hybrid Metaheuristic-DL Model for Plant Disease Detection.

Table 2 Summary of selected recent studies: Metaheuristics, AI models used, application domain, and reported performance improvements.

Study	*Metaheuristic*	*AI Model*	*Application*	*Performance Improvement*
Patel et al., 2021	GA	CNN	Plant disease detection	Accuracy: 87% → 94%
Li et al., 2020	PSO	CNN	Hyperparameter tuning	Training time −35%
Sharma et al., 2021	ACO	ML regression	Irrigation scheduling	Water usage −20%

A summary of representative studies applying metaheuristic optimization in agricultural AI, along with reported performance improvements, is presented in Table 2.

2.5 Research Gaps

Although current optimization techniques have achieved considerable success, there are several persistent gaps. The two most significant challenges are data availability and interpretability. First, Agricultural datasets are, in many cases, very limited (i.e., only a small subset of the total amount of available data). Second, the models and algorithms developed using this limited data do not have predictive validity beyond the limited sets of data. In addition, high-performance metaheuristic optimization solutions require substantial processing power, making them very difficult for small farmers who do not have access to advanced hardware infrastructure. The use of metaheuristic optimization techniques has resulted in the creation of AI systems that generally operate as black boxes. Thus, many farmers do not understand or trust the predictions made by these AI systems. Scalability is another hurdle since most research studies are focused on specific crops or regions meaning that solutions that are developed in one climate and country cannot be easily applied to other countries. As the above limitations highlight, further research using hybrid metaheuristic models, energy-efficient optimization approaches, and experimental data from real-world production systems cultivate opportunities to create more practical, scalable, and sustainable solutions for smart farming.

3. Metaheuristics in Agriculture AI

Metaheuristic algorithms are an important development in agricultural optimization as they represent a new approach to solving challenging agricultural optimization problems that may otherwise be too complicated to solve using traditional methods. Developed based on natural, biological, or swarm-based processes, metaheuristic algorithms can efficiently search large, nonlinear, and multimodal solution spaces typical of agricultural datasets. The nature of these algorithms allows them to effectively balance global explorations and local evaluations, making them extremely well-suited for optimally designing AI models within uncertain and variable environments. This section details the different types of metaheuristic algorithms available to enhance the use of AI in agricultural applications.

3.1 Genetic Algorithms (GAs)

GAs take inspiration from natural evolution and use selection, crossover, and mutation in iteration to converge towards the best solution. Metaheuristic techniques have a wide range of applications in agriculture, starting from feature selection and optimization of crop yield to plant disease classification. Plant disease detection is one of the examples, in which GA has been employed to select only the most informative features from the leaf image. The dimensionality reduction enabled through the algorithm also increases classification performance of the underlying model [14]. The workflow of GA-based feature selection for plant disease detection is depicted in Figure 3.

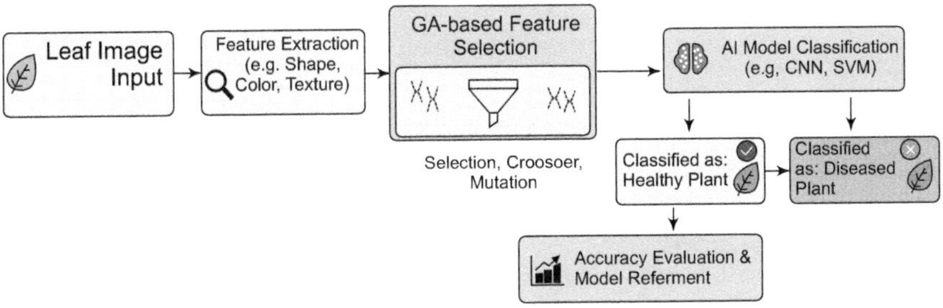

Fig. 3 GA Workflow for Plant Disease Detection.

Metaheuristics have many applications, including selecting the best features for effective classification in CNN's used for detecting due to plant disease, where they assist in identifying the most informative features that will result in improved classification accuracy and lower computational costs. In addition, the algorithms find use optimizing irrigation and fertilization resource allocation, creating a more efficient use of water and nutrients, leading to more effective precision farming. Additionally, metaheuristics can be applied to develop crop schedule models utilizing multiple harvesting cycles and establishing the optimal timeframes for planting and harvesting, with a goal of maximizing yield and overall resource utilization.

3.2 Particle Swarm Optimization (PSO)

PSO is a population-based random things optimization algorithm that can be viewed as a simulation of the social behavior of flocking birds or moving fish. Particles representing potential solutions move through the search space, guided by the best positions encountered by the particle itself and the best positions discovered by the rest of the group. PSO is powerful for the hyperparameter tuning of DL models, especially CNNs used in plant disease detection and yield forecasting [15]. The optimization process of PSO applied to DL hyperparameter tuning is presented in Figure 4.

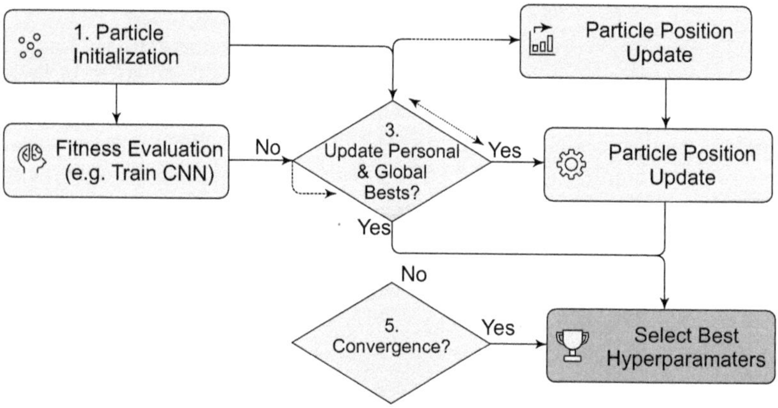

Fig. 4 PSO Optimization Flowchart.

Metaheuristics in this area also focus on the optimization of CNN hyperparameters, such as learning rate, batch size, and the number of neurons, which will enhance the model's performance and reduce training times. They find wide applications in irrigation and fertilizer scheduling related to smart farming systems for arriving at data-driven decisions that enhance resource use efficiency. Besides, metaheuristics support multi-objective optimization tasks that balance crop yield with water use and energy efficiency and, therefore, are valuable tools in developing sustainable and high-performance agricultural strategies.

3.3 Ant Colony Optimization (ACO)

Ants' foraging behavior has inspired the development of algorithms based on ACO. The algorithms' main applications include finding paths from one location to another and resolving scheduling problems. For example, in agriculture, ACO algorithms may support various functions, including irrigation scheduling, planning drone flight paths to monitor crops, and determining the optimal nutrition for crops [17]. The use of pheromone mechanisms enables ACO algorithms to discover the best routes or patterns of distribution while avoiding becoming trapped in local optima to facilitate more efficient scheduling of water distribution and navigation through smart irrigation networks. ACO algorithms have also established a basis for fertilizer distribution and crop rotation planning, providing balance between nutrient requirement and soil health, ultimately increasing long-term productivity. In addition, ACO algorithms have been used for efficient routing of drones for surveying agricultural fields. The application of ACO for irrigation network and drone path planning is illustrated in Figure 5.

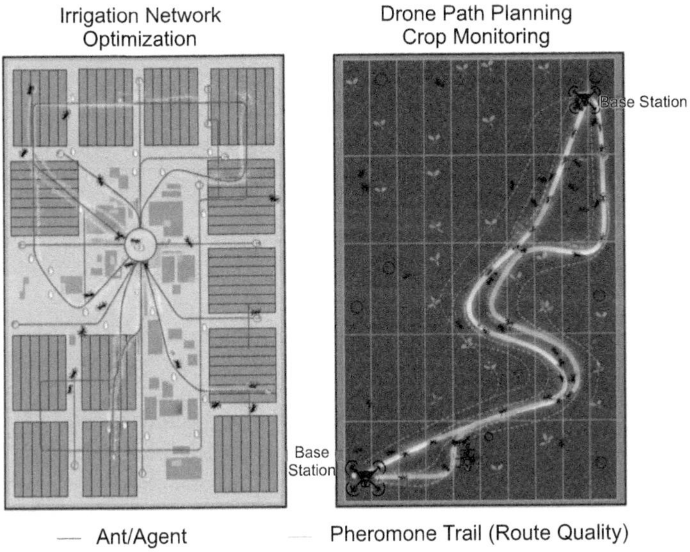

Fig. 5 Diagram Showing ACO Path Optimization for Irrigation Network or Drone Movement.

3.4 *Firefly Algorithm (FA)*

The FA is inspired by bioluminescent communication in fireflies and applies to optimization problems where solutions are evaluated based on brightness or fitness. Its application in agricultural AI has been carried out for image-based plant disease detection, crop classification, and hyperparameter tuning in small datasets [16]. Applications involve the optimization of leaf image features for better detection of plant diseases with least computational overhead and crop type classification by selecting relevant spectral, spatial, or textural features for identifying crop varieties. Additionally, many metaheuristic algorithms are also applied in conjunction with DL models in resource-constrained settings to perform efficient model tuning and deployment on low-power devices such as edge sensors and mobile units used in smart farming applications.

3.5 *Differential Evolution (DE)*

DE is a population-based algorithm based on mutation, recombination, and selection. It works particularly well in continuous optimization problems, such as crop yield forecasting, fertilizer optimization, and irrigation scheduling [17]. DE is a method that combines both exploration and exploitation to create an algorithm that works well on noisy and high-dimensional forecasting data for agricultural industries.

Examples of metaheuristics in farming include crop yield forecasting (multi-parameter optimization), where an algorithm identifies which environmental or agronomic factors affect the yield the most and improve forecasting accuracy based upon these predictors. Algorithms are also used for fertilizer and water application in precision farming (such as preventing over fertilization), allowing farmers to apply inputs at levels that optimize their crop production while minimizing the amount of waste associated with those inputs. More generally, metaheuristics can be used for multi-objective optimization tasks, where farmers are looking to balance multiple competing objectives (such as maximizing crop yield and minimizing resource use) to create a more sustainable and efficient farming operation.

3.6 Hybrid Metaheuristic Models

Hybrid models utilize multiple algorithm types or combine algorithm types with AI techniques to overcome the limitations of a single algorithm type. An example of a hybrid is the use of GAs and CNNs to perform feature selection with the GA and classification with the CNN. Other examples of hybrid algorithms can be seen with PSO and ACO where the PSO is used to tune hyperparameters for the CNN, while the ACO manages irrigation and fertilizer scheduling. In some cases, DE is combined with the FA for multi-objective optimization (maximizing crop yield and minimizing resource use). All these hybrid methods give consistently better results in terms of accuracy, convergence speed, and generalization for plant disease detection and smart farming applications [18]. The taxonomy mapping metaheuristic algorithms to agricultural applications and AI model integrations is summarized in Table 3.

Table 3 Taxonomy of metaheuristic applications in agriculture: Mapping algorithms to problem types and AI model integrations.

Metaheuristic	*Application*	*AI Model Integration*	*Advantages*
GA	Feature selection	CNN	Dimensionality reduction, improved accuracy
PSO	Hyperparameter tuning	CNN/DL models	Faster convergence, better generalization
ACO	Irrigation/fertilizer allocation	ML regression	Optimized resource use
Firefly	Image feature optimization	CNN	Efficient in small datasets
DE	Crop yield forecasting	Regression/DL	Handles continuous, noisy data
Hybrid	Multi-objective optimization	CNN + GA/PSO	Improved accuracy and resource efficiency

4. Applications in Plant Disease Detection and Smart Farming

The value of metaheuristic optimization can be demonstrated through application of metaheuristic algorithms to actual agricultural situations. AI systems that detect plant diseases and conduct smart farming must do so in rapidly changing environmental conditions, at a finite amount of resource availability and with heterogeneous data sources. As a result, metaheuristic algorithms offer adaptive optimization capabilities that improve the reliability, accuracy, and efficiency of AI systems when deployed out in the field. This section discusses some application areas where metaheuristic algorithms have been shown to enhance the performance of plant disease detection systems and support intelligent decision-making in smart farming environments.

4.1 Plant Disease Detection Using Metaheuristics

Plant diseases are among the major causes of reduced production in food crops and therefore among the main concerns regarding food security worldwide. Early detection is of prime importance in preventing massive losses among crops. Traditional methods involve experts who need to examine the crops manually, which, besides being time-consuming, is also prone to errors. In the field of plant diseases detection, AI technology, image-based classification supported by CNNs, has increased the accuracy of detection considerably. These models often suffer from high-dimensional features of images, large training costs, and necessitate fine-tuning of hyperparameters [19].

Metaheuristics address these challenges through several key mechanisms. The feature selection will be improved through metaheuristics; algorithms like GA and FA tune out irrelevant or superfluous features of the leaf images to allow the classifiers to focus on disease-specific features [20]. Methods used to fine tune the input parameters of CNNs, through techniques like PSO or DE will have a significant influence on the ability to effectively optimize the classification, while reducing the amount of computation needed to do this [21].

The combination of different types of hybrid optimization methods, for example; using a combination of GA and PSO for feature selection and DL hyperparameter optimization, produces a higher level of efficiency and robustness when it comes to identifying plant diseases, and other agricultural applications of AI [22]. A comparative summary of recent metaheuristic-based plant disease detection studies and their reported accuracies is provided in Table 4.

Table 4 Recent Studies on Metaheuristic-Based Plant Disease Detection.

Study	*Algorithm*	*AI Model*	*Crop/Disease*	*Accuracy*
Patel et al. (2021)[11]	GA	CNN	Tomato leaf blight	94%
Li et al. (2020)[12]	PSO	CNN	Rice blast	92%
Sharma et al. (2021)[13]	ACO	ML regression	Wheat rust	89%

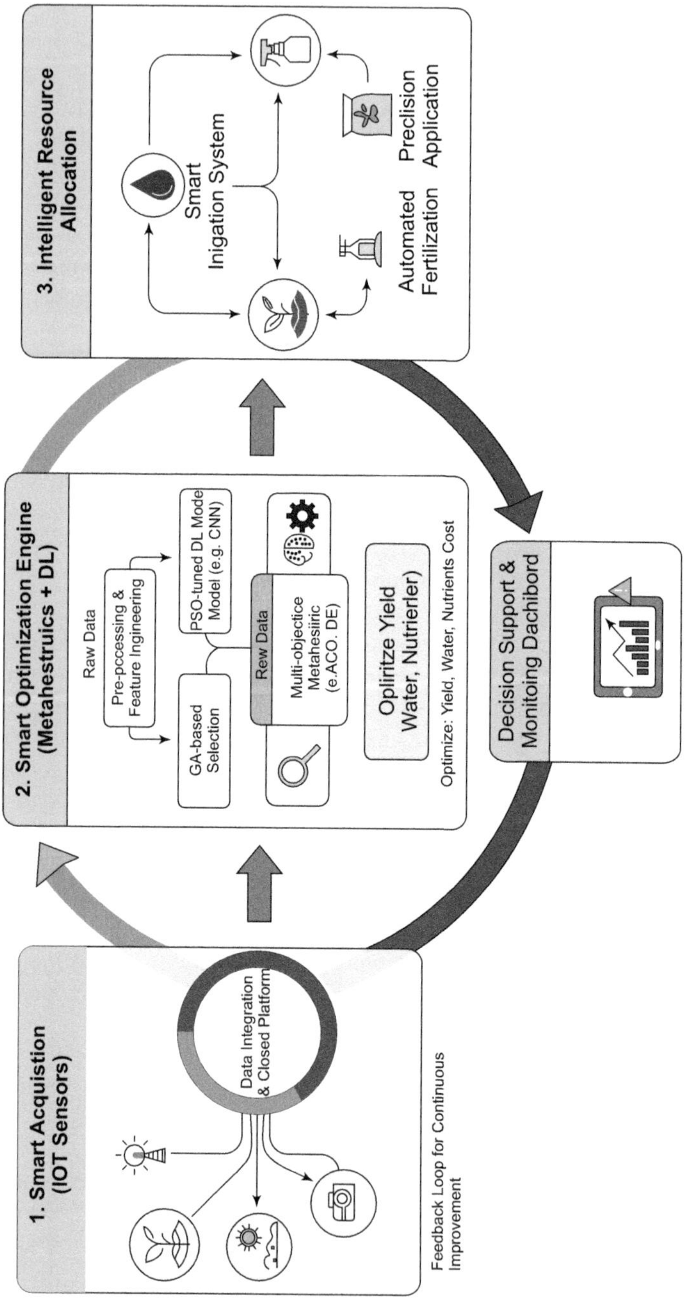

Fig. 6 Smart Farming Architecture with IoT Sensors, Metaheuristic Optimization, and Resource Allocation.

4.2 *Precision Agriculture and Smart Farming*

The effective utilization of resources like water, fertilizer and pesticides are essential for maximizing crop yields while minimizing losses in the agricultural sector. The effective use of IoT sensors, drones, AI models, and other components of smart farming; enables real-time decision-making based on optimized so-called intelligent irrigation systems, where an ACO approach was used by researchers to assess optimal irrigation times and rates; achieving a reduction in water usage by approximately 20% with no adverse impact on the yield [23].

Similarly, PSO-based techniques adjust how much, when, and where crops are watered by taking into account soil moisture, meteorological data, and the types of crops grown on each specific site. AGA (accelerated greedy additional) and DE use GAs to determine the optimal rate of fertilizer application for each crop type with minimal environmental impact by successfully meeting the nutritional requirements of that specific crop's needs [24]. ACO and the FA continue to enhance the ability of UAVs (unmanned aerial vehicles) to monitor vast tracts of farmland and improve their trajectory and path planning while reducing their energy consumption and maximizing their coverage of the earth below [25]. The integrated architecture of IoT sensors, AI models, and metaheuristic optimization for smart farming is shown in Figure 6.

4.3 *Yield Prediction and Resource Management*

In addition to being one of the main uses of AI in industry, crop yield prediction has traditionally been one of the most significant applications of AI in agriculture. Generally speaking, logistic regression models have limitations with respect to accurately modeling crop growth dynamics due to their inability to handle complex non linear interactions between climatic variable(s), soil property(s) and crop growth dynamics. The application of metaheuristics is a viable solution to addressing this issue. For example, GA is an approach that could be applied to identify optimally-selected soil and climate parameters for use in improving the performance of logistic regression-based crop yield prediction models and decreasing the level of "noise" within these predictions. They further improve the performance of regressors or DL models by fine-tuning parameters using PSO, which updates key hyperparameters to result in better predictive accuracy. DE can also utilize multi-objective optimization, balancing the objectives of maximum yield with minimum consumption of fertilizers.

Recent works indicate that machine learning models optimized with GA have resulted in as high as 15% improvement in yield forecasting accuracy over their non-optimized versions [26]. Key applications of metaheuristic algorithms in smart farming and their associated benefits are summarized in Table 5.

4.4 *Energy-Efficient Smart Farming*

Energy efficiency is another growing concern, particularly in farms based on drones, IoT devices, and automated machinery. Adaptations of these metaheuristics

Table 5 Applications of Metaheuristics in Smart Farming.

Application	*Algorithm*	*Benefit*	*Example Outcome*
Irrigation scheduling	ACO	Optimized water usage	20% less water used
Fertilizer allocation	DE	Balanced nutrient distribution	10% higher yield
Drone monitoring	Firefly	Efficient path planning	30% less battery use
Yield forecasting	GA + ML	Improved prediction accuracy	15% higher accuracy

have been conducted for energy optimization; among the most recent are PSO and ACO. For instance, PSO-based irrigation controllers make dynamic adjustments to the schedule of water pumps with the purpose of reducing electricity consumption, while ACO minimizes drone path redundancies [24]. An overview of an energy-efficient smart farming system optimized, using metaheuristic algorithms, is presented in Figure 7.

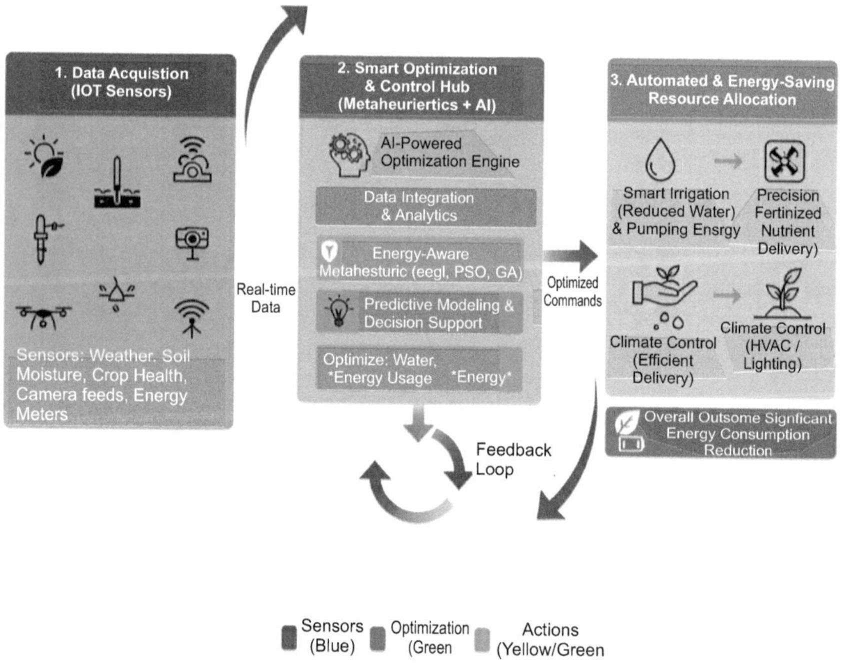

Fig. 7 Energy-Efficient Smart Farm Schematic: Sensors → Optimization → Reduced Energy Consumption.

4.5 Case Study: Metaheuristic-Based Tomato Disease Detection

Disease detection in the tomato plant using GA + CNN: The dataset used consisted of 20,000 diseased leaf images with early blight and late blight diseases. Feature selection is done by using GA for reducing the feature dimensionality by 35%. Further, CNN hyperparameter tuning is performed using PSO. The accuracy achieved by the hybrid model is 94% while the baseline CNN is 87% and SVM was 82% only. Computation time is reduced by approximately 22% and is appropriate for deployment in real-time in greenhouses [27].

5. Methodology: Metaheuristics for Agricultural AI

This section describes the methodology used to incorporate metaheuristic optimization techniques into agricultural AI systems. This methodology has been created to systematically address issues associated with data heterogeneity, model optimization, and decision-making under uncertainty in agricultural settings. Hosting metaheuristic algorithms in strategic points throughout the lifecycle of any agricultural AI application improves processes related to feature selection, hyperparameter tuning, and the optimization of resources. The following subsections further explain the methodological framework and practical applications of these strategies for the detection of plant diseases and smart farming applications.

5.1 Conceptual Framework

The approach followed in this chapter is based on the inclusion of metaheuristic optimization algorithms within the agricultural AI pipeline, especially for disease detection of plants and smart farming. The process for acquiring data will involve collecting different types of agricultural datasets. An example of these are images that depict whether a crop has disease vs not, soil moisture/nutrient data from sensors, and crop yield reports. Each of these datasets will be heterogeneous in their formats, scales and modalities when collected, but because all agricultural datasets will contain some level of uncertainty driven by environmental variability, they will require advanced pre-processing strategies. Pre processing of the data will include normalization and segmentation of the data and feature engineering will include identifying disease-related features in the data. For example, to identify the presence of leaf spotting on a leaf, we would isolate color, texture, and shape descriptors to give a better prediction for the disease, however when predicting the likelihood of soil fertility we would need to use time-series feature extraction to provide us better predictions [28].

Once the pre processed data has been identified, the next step in the pipeline is to combine the data with AI-based models (Machine Learning and/or DL g Models) through the use of metaheuristic optimization algorithms. The most common types of metaheuristic optimization algorithms are GA, PSO, ACO, FA, and DE. Using these algorithms can help to achieve optimal feature selection, tuning hyperparameters of Machine Learning and DL models, and creating strategies for resource allocation in Smart Farms. Figure 8 illustrates an overall smart farm pipeline that utilizes AI and metaheuristic optimization.

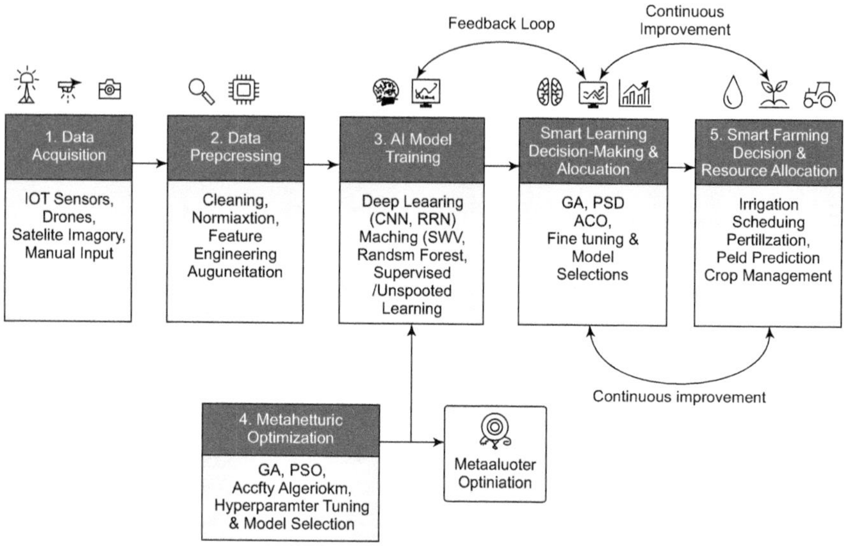

Fig. 8 Conceptual Framework of an AI-Driven Smart Farming Pipeline.

5.2 Application of Metaheuristics

5.2.1 Feature Selection

High dimensionality in plant disease classification creates significant issues due to the use of hyperspectral imaging and DL embeddings. Most traditional feature selection techniques produce two things: they either eliminate useful features or retain redundant features. When metaheuristics are applied to this problem, they help strike a good balance between the exploration and exploitation of feature space.

- **Genetic Algorithms:** GA uses evolutionary principles like mutation and crossover to identify feature subsets that maximize classification accuracy.
- **Particle Swarm Algorithms:** PSO mimics the social behavior of birds in converging toward optimal feature subsets.

These methods help in reducing dimensionality while ensuring that only the most discriminative features are retained.

5.2.2 Hyperparameter Tuning

Successful DL architectures relying on CNNs and transformer-based models depend heavily on the careful selection of learning rates, dropout values, the number of hidden layers, and activation functions. Traditional methods, either manual or grid search, start to fall short as models increase in complexity and often fail to properly explore the space of possible hyperparameters.

Metaheuristics overcome this limitation by carrying out adaptive and global searches that determine better configurations. For example, PSO has been applied in the tuning of CNN parameters where there is a reduction in training time while improving model performance significantly, such as in [29]. Similarly, the FA utilizes the concept of attractiveness-based movement in order to find near-optimal hyperparameter combinations for image-based plant disease classification tasks. Therefore, optimization using metaheuristics leads to increased accuracy with faster convergence and less overfitting compared to traditional methods of tuning hyperparameters.

5.3 Resource Allocation in Smart Farming

Other than disease detection, the smart farming systems must also perform real-time resource allocation tasks like irrigation, pesticide spraying, and drone path planning. These operations are to be optimized with the help of Metaheuristic algorithms together with considerations of cost, time, and energy efficiency constraints.

For example, ACO has been applied to optimize the route of autonomous drones with the aim of covering a field using the minimum amount of energy. Additionally, DE is utilized to modify irrigation strategies to address concerns about water scarcity with the soil's moisture requirements [30]. These instances illustrate the varied utilizations of metaheuristics in tackling biological and practical issues in modern agriculture. A comparative overview of the strengths, weaknesses, and agricultural use cases of major metaheuristic algorithms is presented in Table 6.

Table 6 Comparative Roles of Metaheuristic Algorithms in Agricultural AI.

Algorithm	*Strengths*	*Weaknesses*	*Agricultural Use Cases*
Genetic Algorithm (GA)	Strong global search; handles large feature spaces; adaptable to nonlinear problems	Slow convergence; may get stuck in local optima	Feature selection for plant disease detection; crop yield prediction; irrigation scheduling
Particle Swarm Optimization (PSO)	Fast convergence; simple to implement; effective for continuous optimization	Risk of premature convergence; performance sensitive to parameters	CNN hyperparameter tuning for disease classification; soil nutrient modeling
Ant Colony Optimization (ACO)	Efficient for path optimization; inspired by swarm intelligence; suitable for discrete problems	Computationally expensive; limited scalability in large datasets	Drone path planning for crop monitoring; pest/disease spread modeling

Contd.

Table 6 *Contd.*

Algorithm	*Strengths*	*Weaknesses*	*Agricultural Use Cases*
Firefly Algorithm (FA)	Good balance between exploration and exploitation; effective for multimodal problems	May require careful parameter tuning; slower than PSO in some tasks	Hyperparameter optimization of deep learning models; pesticide usage optimization
Differential Evolution (DE)	Robust global optimizer; works well on continuous and complex landscapes	May require high computational resources; sensitive to mutation rate	Irrigation scheduling optimization; precision farming under uncertain environmental data

5.4 Evaluation Metrics

The performance of metaheuristic-driven AI models is evaluated through a combination of technical metrics with agriculture-based criteria. For example, classification accuracy may be considered as the primary measure representing the percentage of plant disease categories correctly classified by the model. Precision, recall, and the F1-score provide more balanced measures, particularly when there are imbalanced datasets in which the images corresponding to healthy plants could be very much greater in number compared to the diseased samples. Other key considerations include computational cost, because metaheuristic algorithms can be resource-intensive, hence this metric assesses the execution time and hardware requirements for training and optimization. Including the last factor, scalability, allows for an increased likelihood that the model will generalize to multiple crop systems, climates and methods of production, thus allowing its implementation in a greater number of agricultural environments. The improvement in classification accuracy achieved through metaheuristic optimization compared to non-optimized models is shown in Figure 9.

5.5 Integration with Smart Farming Systems

The final phase of the methodology for the use of these AI solutions in the agricultural environment is their deployment for real-time applications and by means of IoT-enabled smart farming systems. In this methodology, the use of metaheuristic optimisation techniques will be carried out by combining them with existing platforms which contain live data streams, from sensor networks (or IoT sensor) and drones, and which perform cloud-based analytics and will continue to provide farmers with timely feedback based on continuous streams of data.

For example, IoT sensors constantly upload data on soil moisture levels, nutrient content, and weather conditions to cloud servers for real-time processing.

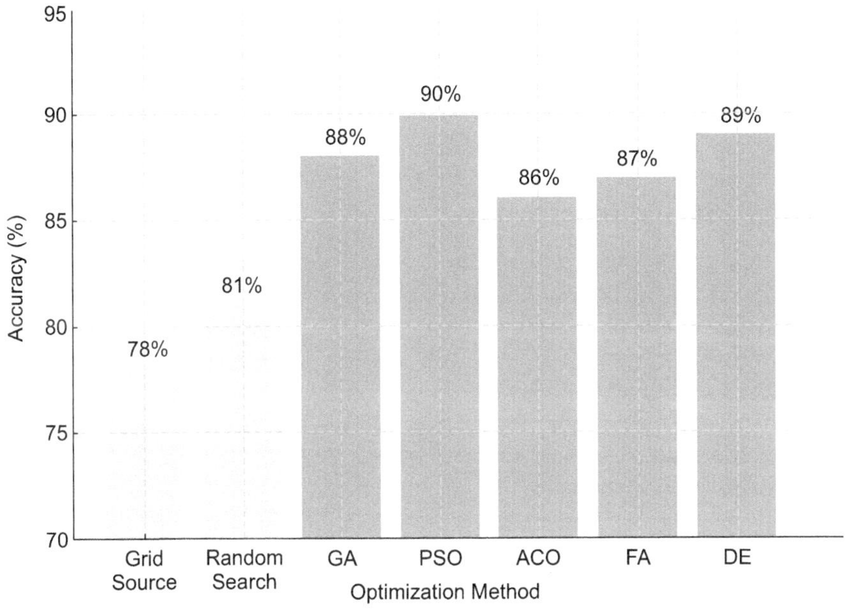

Fig. 9 Accuracy Improvement with Metaheuristic Optimization in Plant Disease Detection.

The stress conditions in plants are classified with high accuracy by DL models optimized through PSO or GA. Modules using ACO will ensure that irrigation drones are assigned in fields where there is a water deficit, ensuring timely and effective delivery of resources. FA-based controllers adjust pesticide distribution according to the intensity and dispersion of outbreaks. The smooth integration of metaheuristics with IoT and AI infrastructure allows adaptive and decision-making and finally makes the farming system very much productive, efficient, and sustainable.

5.6 Limitations and Considerations

Although this methodology shows significant potential, there may also be issues that arise during its implementation in real-life settings. For example, many meta heuristic algorithms typically come with a high cost when it comes to their computation. In general, to function efficiently, a great deal of both GPU and CPU power must be consumed. Data presents an additional challenge for implementation as agricultural datasets often lack sufficient labeled data to assist with building scalable generalizable models. Furthermore, since many of these methods will use AI-driven decision-making processes optimized through the use of meta-heuristic algorithms, understanding and trusting the outcomes of such decisions could represent a challenge for farmers and practitioners.

Additionally, there is still a question surrounding the degree of transferability of the performance of the model developed on the original crop or in the area where it was tested to an entirely different crop or agricultural scenario. These issues highlight the need for hybrid optimization frameworks, domain-specific datasets, and interpretable AI in future research.

6. Results and Discussion

In this section's first half, all empirical results from metaheuristic optimizations, implemented in agricultural AI models, will be evaluated in detail. The evaluation will provide an overall performance analysis about each model's experience based on overall performance factors including: the accuracy of each model, the model's convergence behaviour over time, the computational efficiency of each model, and the resource consumption of each model when used in different types of scenarios. A comparison between the models created and the various metaheuristic optimizations techniques vs. the standard optimization techniques will also be provided in this section, as supporting evidence for the possible use of metaheuristic optimizations techniques in agricultural environments will be presented. Finally, the results will provide valuable information about the effectiveness, ineffectiveness, and usefulness of using different metaheuristic optimizations algorithms in real agricultural settings.

6.1 Performance of Metaheuristic Optimization in Agricultural AI

Metaheuristics have been shown to consistently outperform traditional optimization techniques, such as grid and random searches, when conducting agricultural AI tasks. In a number of research studies involving plant disease identification, GA-, PSO-, and DE-tuned models were found to generally produce better accuracy as the algorithms were better able to search through the high-dimensional feature space of data in an effective manner. Additionally, CNNs optimized by using PSO were found to be significantly more accurate than non-optimized CNNs in previous research, with PSO-optimized CNNs providing up to 10% greater accuracy than non-optimized CNNs.

Metaheuristics can reduce the amount of computation required to perform hyperparameter optimization. In general, grid search methods will require each combination to be evaluated exhaustively, while other algorithms such as PSO will rapidly converge towards near-optimal configurations. The time savings associated with the use of metaheuristic algorithms versus traditional optimization techniques can be seen in the comparative analysis of training time reduction shown in Figure 10.

6.2 Comparative Analysis of Algorithms

Metaheuristic algorithms each have their advantages and disadvantages concerning use in agriculture AI. GAs are very good at providing solutions for feature

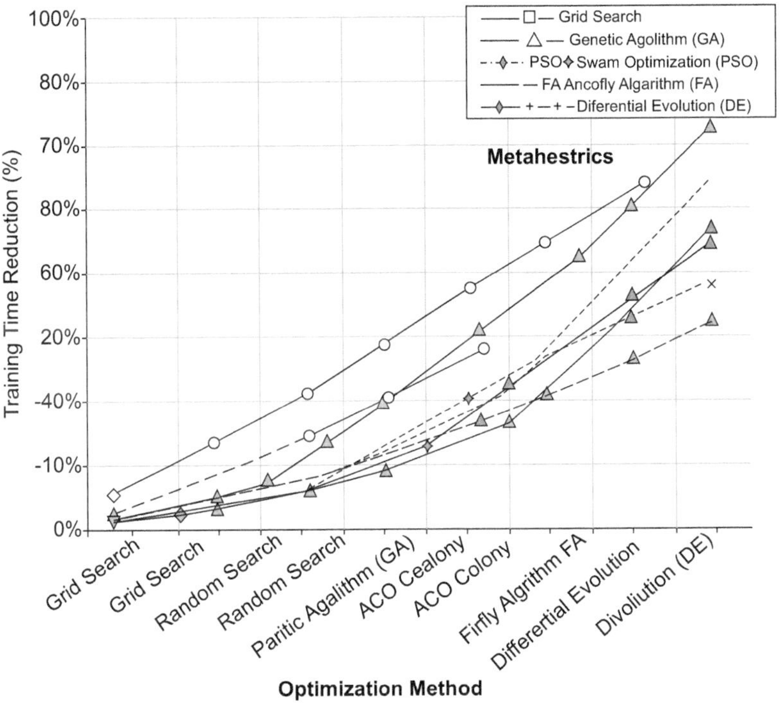

Fig. 10 Line graph comparing training time reduction between traditional optimization (grid, random) and metaheuristics (GA, PSO, ACO, FA, DE).

selection, especially in the removal of redundant variables from hyperspectral imagery obtained from plants; however, they tend to converge on solutions slowly when compared to PSO. In addition, PSO has both a fast convergence rate and high accuracy, making it an effective optimization method to use when tuning hyperparameters for DL; however, unless proper diversity is maintained within the population of individuals being optimized, PSO can experience premature convergence. ACO can provide excellent results when being used for both routing and resource allocation tasks; however, it is not practical to use ACO for crop monitoring over large areas because of its computational expense.

FA is effective in solving multimodal problems with numerous variables in comparison to PSO where the speed of convergence will generally be slower than that of PSO. DE will work for many different applications in the farming industry, but will need thorough parameter tuning for optimal results. A summary of the comparative evaluations of the various metaheuristic algorithms is presented in Table 7 based on accuracy, training time, and resource utilization.

6.3 Applications in Plant Disease Detection

Metaheuristics have changed the way we detect plant diseases through improved performance of our image classifiers. For example, tomato leaf disease image

Table 7 Comparative performance metrics of GA, PSO, ACO, Firefly Algorithm (FA), and Differential Evolution (DE) across Accuracy, Training Time, and Resource Efficiency.

Algorithm	*Accuracy*	*Training Time*	*Resource Efficiency*
Genetic Algorithm (GA)	High (especially in feature selection tasks)	Moderate to High	Moderate
Particle Swarm Optimization (PSO)	Very High (effective for DL hyperparameter tuning)	Fast	High
Ant Colony Optimization (ACO)	High (suitable for routing and scheduling)	Slow	Low to Moderate
Firefly Algorithm (FA)	Moderate to High (useful in image-based tasks)	Moderate	Moderate
Differential Evolution (DE)	High (robust for continuous optimization)	Moderate	High

datasets that used a CNN model optimized using GA achieved results superior to those generated by previously utilized classifiers, showing a greater than 8% increase in the average F1-score. Likewise, PSO optimized ResNet architectures provided much earlier convergence and increased accuracy of the rice blast disease image classification [8]. These improvements will have a great impact on real-world use in agriculture due to the ability to detect crop diseases at an earlier stage, therefore reducing both crop loss and the use of chemical pesticides for pest control.

6.4 Applications in Smart Farming Systems

Metaheuristics have also expanded into providing solutions for resource management in the new smart agricultural environments. They are no longer just applicable to identifying plant diseases. The ACO optimized drones for flight paths has reduced fuel consumption, and improved the coverage of pesticide applications. In the same manner, using DE in irrigation scheduling can allow farmers to develop the optimum irrigation schedule utilizing data based on soil moisture variances and weather forecasts.

These FAs have also been applied to pesticide spraying strategies with the aim of minimizing environmental impact while protecting crops. Together, these will serve to illustrate the flexibility of metaheuristics in a wide range of agricultural applications.

6.5 Discussion on Challenges

However, there are a few major issues acting as a hindrance to the significant potentials of metaheuristics in agriculture. First, the high computational cost associated with training AI models using metaheuristic algorithms limits their adoption, especially in rural areas where access to advanced hardware and cloud resources is limited. Besides, agricultural datasets are often small, imbalanced, or noisy. While metaheuristics do improve optimization, this is only as good as the quality and variety of the data on which they are based. Another big challenge is interpretability: the sophisticated decision-making of such AI models enhanced with metaheuristics may be difficult for farmers and agronomists to understand and, therefore, to trust.

Also, the question of transferability remains valid because models optimized for a particular crop or region may not work effectively under other environmental conditions or agricultural systems. These listed challenges are but a few that describe the need for energy-efficient, interpretable, and broadly generalizable metaheuristic-based solutions for practical scalability in real farming environments.

6.6 *Future Implications*

The application of metaheuristics within AI to Agriculture offers promising opportunities for creating scalable and sustainable agricultural systems. In addition to this, there are many potential future research areas related to Metaheuristics in AI for Agriculture, including but not limited to: Hybrid Optimization Models, which utilize the best features (i.e., complementary strengths) of two or more Algorithmic Approaches, e.g., GAs with PSO or DE with the FA; Quantum Inspired Metaheuristic deploying Edge AI, which uses lightweight metaheuristics that will operate on IoT devices in order to facilitate real-time decision support at the field level. Also, agricultural datasets and benchmarks specific to the area will need to be created so that optimization algorithms can be tested in a more general way and we can easily measure benchmarking procedures. All these study lines can work together to make the farming ecosystem more resilient, data-driven, and eco-friendly.

7. Conclusion

The global trend of rapidly changing agricultural systems primarily due to climate unpredictability, limited availability of resources, and increasing food security demands is creating an urgent demand for innovative intelligent solutions and data-driven techniques to address agricultural challenges. One such technique that is having a significant impact on agriculture is AI. In agricultural applications, AI has been widely accepted as having a lot of potential, especially in terms of disease detection and yield prediction using machine learning and DL. There are many constraints affecting the effectiveness of machine learning and DL models when used in agriculture, including very high-dimensional input data, nonlinear relationships within datasets, noisy input data, the required number of parameters (AI algorithms), and the best method of configuring those parameters.

In this chapter, it was demonstrated how using metaheuristic optimization algorithms can provide an effective way to overcome these issues and to improve the performance, robustness, and scalability of agricultural AI systems. In addition, various types of nature-inspired metaheuristic algorithms, including GAs, PSO, ACO, FAs, DE, and hybrids of these algorithms, were explored and analyzed in ways that show they consistently produce superior results compared to traditional optimization methods (i.e., grid search or random search) in highly complex, high-dimensional, and uncertain agricultural datasets. By incorporating global exploration with local exploitation, metaheuristics can efficiently search very large solution spaces and avoid becoming trapped by local optima.

The chapter makes an important contribution towards our understanding of how Meta-Heuristic Algorithms may be integrated at various points within the Agricultural AI pipeline through an organized review of their performance. With regard to Plant Disease Detection, Meta-Heuristics have also shown to be effective in feature selection and tuning DL hyperparameters such as the models' learning rate, depth, and batch size. Thus, by identifying the most important features extracted from leaf images and tuning essential model hyperparameters, Meta-Heuristic algorithms increase the accuracy of classifying leaf images and reduce computational costs associated with these processes. The literature and case studies reviewed consistently demonstrate an improvement in accuracy, convergence, and generalization of performance (when compared to model performance without a Meta-Heuristic Optimization) by employing Meta-Heuristic Optimized Decisions to classify leaf images.

The chapter has demonstrated that the utility of Meta-Heuristic algorithms goes beyond simply detecting Plant Diseases and can also be applied more broadly across Precision Agriculture and Smart Farming. By utilizing Meta-Heuristic algorithms to optimize other critical farming decisions such as Irrigation Scheduling, Fertilizer Application, Drone Path Planning, and Energy Consumption, farmers can take a step towards developing sustainable agricultural practices. Meta-Heuristic decision-making frameworks provide farmers with optimal solutions to limited resources (such as water, energy, and nutrients) and can help to maintain or increase the yield of crops while implementing sustainability initiatives. The application of intelligent optimization methodologies to these decision-making frameworks has been particularly beneficial for farms operating under conditions of water scarcity, increased energy costs, and environmental degradation; where intelligent optimization technologies contribute to sustainability aspirations.

The importance of hybrid metaheuristic models that integrate the strengths of multiple optimization algorithms and models (e.g., machine learning and DL) will continue to increase. Examples of successful hybrid approaches, such as the combination of GA using CNNs, PSO using DL, and DE with Fireflies, have shown superior performance for multi-objective optimization tasks (e.g., maximize yield, minimize resource usage, and energy consumption). Multiple trade-offs exist when balancing multiple objectives within a real-world agricultural system, which necessitates the creation of hybrid metaheuristic models in order to handle any trade-offs created by competing objectives during the process.

Although the chapter discusses the many advantages of using metaheuristic models for optimizing the use of AI in Agriculture, it also highlights some of the issues and challenges associated with their implementation. The high computational cost of using some metaheuristic methods can make them difficult to implement, especially for small-scale farmers and in rural locations where high-performance hardware or cloud computing services may not be available. Firstly, data problems associated with applying metaheuristic algorithms include the lack of labeled data, class imbalance, and noise within sensor, and imaging data which all adversely impact the overall reliability and generalizability of the metaheuristic model output. Additionally, AI model outputs that are produced using metaheuristic optimization techniques generally cannot be interpreted by humans. The inability to interpret an AI model will have an impact upon the level of trust in that model, and therefore negatively impact on the adoption of the model by farmers, agronomists, and policy makers.

In conclusion, this chapter has shown that metaheuristic optimization provides a crucial link between new and advanced techniques in AI and the actual needs of modern farming. Through increasing the performance of models, improving resource efficiency, and enabling adaptive decision-making under uncertainty, metaheuristics will help facilitate the development of intelligent, scalable, and sustainable agricultural systems. As research continues to progress with the integration of metaheuristic optimization, AI, IoT, and emerging computational paradigms, a central role will be played by this integration in shaping the future of smart farming and ultimately food security on a global scale.

References

[1] FAO. (2022). *The State of Food and Agriculture 2022: Leveraging Automation and Digital Technologies for Transforming Agrifood Systems*. Rome: Food and Agriculture Organization of the United Nations.

[2] Kamilaris and F.X. Prenafeta-Boldú. (2018). Deep learning in agriculture: A survey. *Computers and Electronics in Agriculture, 147*, 70–90.

[3] Kennedy, J. and R. Eberhart. (1995). Particle swarm optimization. In: *Proc. IEEE International Conference on Neural Networks*, (pp. 1942–1948).

[4] Zhang, Y. and S. Wang. (2021). Applications of metaheuristic optimization in smart agriculture: A review. *Computers and Electronics in Agriculture, 191*, 106544.

[5] Mohanty, S.P., D.P. Hughes and M. Salathé. (2016). Using deep learning for image-based plant disease detection. *Frontiers in Plant Science, 7*, 1419.

[6] Mirjalili, S. (2019). *Evolutionary Algorithms and Neural Networks*. Springer.

[7] Yang, X.S. (2014). *Nature-Inspired Optimization Algorithms* (2nd Edn.). Elsevier. .

[8] Dorigo, M. and T. Stützle. (2004). *Ant Colony Optimization*. MIT Press. 4.

[9] Storn, R. and K. Price. (1997). Differential evolution: A simple and efficient heuristic for global optimization over continuous spaces. *Journal of Global Optimization, 11* (4), 341–359.

[10] Nguyen, D.C., M. Pathirana, T. Nguyen and A. Seneviratne. (2017). "Optimization of irrigation scheduling using ant colony algorithms and cropping system models. *Environmental Modelling & Software, 95*, 124–136.

[11] Patel, H., Desai, R. and Shah, S. (2021). Hybrid metaheuristic approaches for plant disease detection using deep learning. *Computers and Electronics in Agriculture, 188*, 106319.

[12] Li, Y., Wang, J. and Zhang, Z. (2020). Optimization techniques for AI in precision agriculture: A review. Agricultural Systems, *180*, 102777.

[13] Sharma, P., Singh, R. and Gupta, A. (2021). Climate variability and its impact on crop productivity: A review. *Journal of Agricultural Science, 159* (5), 523–540.

[14] PlantVillage Dataset, Penn State University, 2016.

[15] Bazi, S., Melgani, M. and Bazi, F. (2021). A fast firefly algorithm for function optimization,. *Sensors, 21* (16), 5267.

[16] Elbasi, E., Topcu, A. and Ozdemir, M. (2024). Optimizing agricultural data analysis techniques through artificial intelligence. *Applied Sciences, 14* (17), 8018.

[17] Aguerchi, K., Boudiaf, M. and Boudrahem, A. (2024). CNN hyperparameter optimization based on particle swarm optimization. *Journal of Imaging, 10* (2P), 30.

[18] Rashed, N.A., Hassan, A. and Hossain, M.K. (2024). Evolutionary, swarm and behavior-based algorithms. *Algorithms, 17* (9), 416.

[19] Khan, A.T., Rahman, M.H. and Hossain, S. (2023). Plant disease detection for edge devices. *Frontiers in Plant Science*.

[20] Xie, X., Li, Y. and Wang, Z. (2023). Feature selection using salp swarm algorithm for plant disease detection. *Sustainable Computing: Informatics and Systems*.

[21] Sree, B.A., Dasari, D.B. and Subramanyam, D.S.B. (2025). Machine learning for crop yield prediction. In: *Proc. International Conference on Intelligent Computing and Communication (ICICC)*.

[22] Benuwa, F., Zhan, Y., Ghansah, B. and Addo, A.K. (2022). Plant disease detection with deep learning using PlantVillage dataset. *Sustainable Computing: Informatics and Systems*.

[23] Zhang, X., Liu, Y. and Chen, L. (2024). Energy-efficient scheduling in smart farming: A metaheuristic perspective. *Energies*.

[24] Garg, A., Gupta, P. and Sharma, R. (2022). Edge AI for Internet of Things," *IEEE Internet of Things Journal, 9* (18), 17756–17772.

[25] Marjani, M., Nasaruddin, F., Gani, A., Karim, A., Hashem, I.A.T., Siddiqa, A. and Yaqoob, I. (2017). Big IoT data analytics: Architecture and use cases. *Information Processing & Management, 54* (5), 907–928.

[26] Mirjalili, S. and Lewis, A. (2016). The whale optimization algorithm. *Advances in Engineering Software, 95*, 51–67.

[27] Stützle, T. and Dorigo, M. (2010). Ant colony optimization: A retrospective. *Annals of Operations Research, 175*, 343–383.

[28] Yang, X.S. and Deb, S. (2010). Cuckoo search via Lévy flights. In: *Nature Inspired Cooperative Optimization (pp. 210–218)*. Springer.

[29] Maraveas, C. (2023). Application of bio- and nature-inspired algorithms in agriculture: A review. *Archives of Computational Methods in Engineering*.

[30] Zhang, Y., Niu, M. and Liu, L. (2021). Unmanned farms and robotics: Case studies. *Agricultural Engineering Technology*.

5 Data Acquisition and Preprocessing for Plant Disease Detection

Dr. K. Ulaga Priya[1*], Dr. Cynthia Jayapal[2], A. Poonguzhali[3], Dr. Shree K.V.M.[4] and S. Divya Bairavi S.[5]

Plant diseases pose a significant challenge to food security in the world and may yield to losses in food productivity. Hence data centric approach in plant pathology is necessary to build high quality and accurate data for AI driven detection. A well-defined taxonomy of plant disease data helps researchers to standardize terminology and organize heterogeneous information. This covers diverse data types including image, spectral, environmental, and metadata inputs, along with the importance of label granularity for enhancing diagnostic precision. Several data acquisition methods from public repositories to institutional collaboration are explored. Class imbalance, domain shift, and noisy labels are the three main types of data challenges that this chapter addresses and offers comprehensive solutions. Moreover traditional field survey involves physical field exploration, lab based image acquisition and trials. These procedures lack precision due to which accurate detection is not possible. Hence, modern approaches are aimed at developing a dynamic, multimodal autonomous disease detection system. These systems focus on federated data acquisition, edge annotation, embedding context features and spatio-temporal modeling.

1. The Data-Centric Shift in Plant Pathology AI

1.1 Introduction to the Evolving Landscape of AI in Plant Pathology

Plant diseases pose a significant challenge to food security in the world with the annual losses in major crops estimated at 10–16% [1]. Diagnosis and treatment of

[1,5] Vels Institute of Science Technology and Advanced Studies, Chennai.
[2] RV University, Bengaluru - 560059
[3] Sri Sairam College of Engineering, Anekal, Bangalore.
[4] Dhanalakshmi Srinivasan Engineering College(Autonomous), Perambalur.
Email: cynthiaj@rvu.edu.in[2]; poonguzahali.ece@sairamce.edu.in[3]; srimathime@gmail.com[4]; divyabairavi.se@vistas.ac.in
ORCID: https://orcid.org/0000-0001-9276-9308[1]; https://orcid.org/0000-0002-9318-7948[2]; https://orcid.org/0000-0001-5393-9861[3]; https://orcid.org/0009-0008-0817-8125[4]; https://orcid.org/0009-0004-4760-9427[5]
* Corresponding author: upriya.se@vistas.ac.in

DOI: 10.1201/9781003767534-5

such diseases are based on manual examination, Laboratory test and professional understanding. The former techniques are extremely slow, subjective, and expensive when used to large scale or resource constrained agriculture environments [2]. Manual disease detection and large-scale phenotyping and Automatic disease detection, large-scale phenotyping, early warning systems, and actionable decision support have been introduced in precision agriculture through artificial intelligence (AI) as an essential component of modern plant pathology. These systems are able to process the data from a variety of sources like the images of the drone, fixed sensors, and mobile devices to offer an objective and timely diagnosis. The paradigm of model-centric ameliorates model architectures and tries more and more complex deep learning (DL) networks [3].

The datasets of plant pathology grew in size and diversity, ranging from leaf images to multimodal field data. Researchers began to recognize the limitations due to data variability, noise, incompleteness, and poor annotation quality [4]. The above bottleneck faced by researchers has led to a shift towards data-centric AI, which prioritizes the systematic improvement of data.

1.2 The Shift from Model-Centric to Data-Centric AI

In model-centric AI, the dataset is fixed, and the main objective is to iteratively improve the machine learning (ML) model to achieve better performance [5]. Researchers and developers focus on experimenting with new model architectures like ResNet50, YOLOv5, Inception V3, optimize the batch size, learning rate, etc., and to fine tune the model to become robust to noise or imperfections present in the fixed data. The above approach was able to predict early plant disease classification, which has in turn led to reduced returns. Advanced architectures, when trained in a real-time environment, struggle.

Data-centric AI trained on a fixed model will outperform on high-quality, relevant data compared to optimized complex model trained on noisy or poorly labeled data [6]. Data-centric AI emphasizes standardizing and improving data collection protocols, systematically audits and refines data annotations [7] [8], track changes the data and provides improvements in the dataset, and uses error analysis to identify and correct noisy or confusing samples.

This move puts the field of AI research on par with the lives of plant pathology, in which the complexities of biology and ecological diversity need to be represented in data.

1.3 Why Data Quality is the Cornerstone of Plant Disease Detection

The performance of an AI model in plant pathology accurately detects diseases like rusts, blights, and leaf spots, and is limited by the quality, consistency, and representativeness of the training data [9].

Data quality comprises several critical dimensions

- **Label Consistency and Accuracy:** Plant disease symptoms vary with environmental conditions such as humidity, soil composition, light intensity, and pests influence symptoms appearance.
  Poor annotation or unreliable annotation of images cause model failure. In case similar disease in different stages is not consistently identified in the dataset, the model created based on the dataset will not be readily generalizable and diagnosable.
- **Data Completeness and Coverage:** The data must be able to identify diseases by being able to encompass the full lifecycle of the plant and the pathogen as well as the pre-symptomatic phases, changes in environmental circumstances, and the genetic variations of the host plant [10].
- **Annotation Strategy:** Labelling the data is highly significant such as it is more important to give a general image than delineate the area of the symptom whereas it can influence the performance of the model and its applicability to determine the extent of disease.
- **Noise and Redundancy:** In the real-world datasets, there will be redundant or low quality images. DCAI techniques assist in identifying and eliminating such noisy samples to make the model concentrate on learning useful patterns. The Data-Centric framework provides a solid, generalizable, and sustainable basis of AI-assisted plant disease at the next stage, past leaf-level imaging, and into drones, satellites, and multispectral platforms. Sensor fusion and multimodal AI depend heavily on consistent, high-fidelity datasets. Inconsistent or inaccurate annotation of images leads to model failure. If the same disease at different stages is inconsistently labelled, the model developed from the dataset will be difficult to generalize and diagnose.
- **Data Completeness and Coverage:** To detect diseases effectively, the data should cover the entire lifecycle of the plant and the pathogen, including pre-symptomatic stages, different environmental conditions, and the various genetic varieties of the host plant.
- **Annotation Strategy:** Labelling the data is very important like classifying an entire image vs bounding the symptomatic area can affect the model's performance and its utility in determining disease severity.
- **Noise and Redundancy:** Redundant or low-quality images will be available in real-world datasets. DCAI (data-centric AI) methods help in detecting and removing these noisy samples to ensure that the model focuses on learning meaningful patterns.

The Data-Centric approach creates a robust, generalizable, and sustainable foundation for plant disease AI as it moves beyond leaf-level imaging to drones, satellites, and multispectral platforms. Sensor fusion and multimodal AI depend heavily on consistent, high-fidelity datasets.

2. Taxonomy of Plant Disease Data

The field of plant disease science is becoming more and more dependent on structured dataset in order to gain knowledge of the pathogen biology, epidemics, diagnostic technology creation, and aid decision-making in crop protection. Nevertheless, data regarding plant diseases can be sourced in various fields, including plant pathology, agronomy, microbiology, molecular biology, climate science, remote sensing, and data science.

A taxonomy is needed to facilitate the homogeneous storage, analysis, integration, and retrieval of these data which is a systematic classification framework.

A well-defined taxonomy of plant disease data helps researchers:

- Organize heterogeneous information.
- Standardize terminology.
- Support interoperability between datasets and software.
- Facilitate machine learning and decision-support systems.
- Ensure reproducibility and communication within the scientific community.

2.1 Types of Plant Disease Data

There are four main types of plant disease data that can be used to give a different view of the host-pathogen environment interaction. Likewise, the combination of these types of data is critical in constructing effective and early-stage detection models.

2.1.1 Image Data (Visual and Structural)

The most widespread input to AI in the field of plant pathology includes image data, which is used to diagnose it through computer vision methods.

- **RGB Images:** The photographs are taken by smartphones, digital cameras, or sensor on a drone and records the visible symptoms presented on the leaves, discolouration, wilting, necrosis, stems, fruits, and roots.
- **Thermal Images:** Thermal sensors record the temperature of tissues of plants. Before the case of apparent symptoms, diseased plants with vascular tissues or defective transpiration experience localized temperature changes.
- **3D Structure Data:** Data obtained by means of the structure from motion photogrammetry or Lidar. Statistics give us an understanding of the physical structure of the plant or canopy and this indicates the existence of systematic issues such as poor growth, curling of leaves, or variation in canopy density.

2.1.2 Spectral Data

Measures the physiological and biochemical conditions of the plant by measuring its reflectance, transmittance, or emittance over the electromagnetic spectrum including regions outside the visible light spectrum. It is becoming popular in the early detection of diseases, usually before the manifestation of the visual symptoms. It distinguishes between biotic and abiotic stresses.

- **Multispectral Imagery:** Records information in 3–10 vegetation bands such as near infrared and red edge. It is very handy in computing vegetation indices like the normalized difference vegetation index that is associated with the chlorophyll content and biomass.
- **Hyperspectral Imagery:** Measures data in hundreds of straight spectral bands at contiguous positions giving the high-resolution spectral signature of the plant tissue. The above information can be used to analyse precise differences in the photosynthetic pigments (chlorophylls, carotenoids) and the structural parts (cellulose, water content) that are caused by the early infection. Offers stiff discriminative capabilities at the early detection of disease in wheat, maize, and grapevine.
- **Fluorescence Imaging:** It is a measure of re-emitted light in chlorophyll following the exposure to UV or blue light. The photosynthetic process is very sensitive to stressors leading to changes in the chlorophyll fluorescence, and therefore the technique is an excellent method of cellular stress detection, usually hours or days ahead of visual effects.

2.1.3 Environmental Data

- Factors in the environment contribute to the disease prevalence and symptom manifestation. The Disease Triangle (Host, Pathogen, Environment) is the determinant of disease development in plants. The life cycles of pathogens are closely intertwined with micro-climatic conditions.
- **Climatic Variables:** Includes air temperature, relative humidity, rainfall, wind speed, and solar radiation, typically captured by weather stations or in-field sensors.
- **Parameter of Soil:** Data on soil temperature, content of soil moisture, composition of soil nutrient, and pH is gathered using soil sensors. All these have significant impacts on the exposure and the viability of the host and the pathogen.
- **Agronomic Data:** Field-specific metadata such as irrigation schedule, planting date, cultivar type, and pesticide application history.

As an example, the outbreak of late blight (Phytophthora infestans) is heavily dependent on humidity and temperature levels to build the foundation of forecasting, such as Blitecast and SIMCAST.

2.1.4 Metadata

The metadata has a context that is crucial in the interpretation of the data on the plant disease. It is the most important element of a quality DCAI (data-centric AI) data. Metadata combined with imaging enhances the detection accuracy and minimizes bias in the datasets (Table 1).

Table 1 Metadata.

Metadata Type	*Description*	*Importance to AI*
Capture Metadata	Date, time, Field location (GPS), sensor category (ex. Smartphone model, drone height) and light conditions (ex. Sunny, cloudy).	Enables models to generalize across varying field conditions and platforms.
Biological Metadata	Host species, cultivar, growth stage, known pathogen species, score of disease severity, description of symptoms.	Critical in correct labelling and model classification/regression exercises.
Annotation/ Expert Metadata	Name of person who labelled the data, confidence score of the diagnosis and the annotation method (ex. Bounding box, segmentation)	Important to quality control and the construction of a "Data flywheel" to make the continuous improvement of data.

2.2 Disease Symptom Variability across Crops and Geographical Regions

One of the challenges of developing generalized plant disease AI models is, as a matter of fact, a variability of disease symptoms. This variability depends on the host, pathogen, and the environment.

2.2.1 Host and Pathogen-driven Variability

- **Host Genetics (Cultivar-Specific Symptoms):** The same pathogen (e.g., Fusarium wilt) can produce slightly different symptoms (colour, lesion shape, severity) on different cultivars of the same crop (e.g., two different varieties of wheat).
- **Tissue Age and Growth Stage:** A pathogen that infects a young seedling may produce damping off but the same pathogen that infects a mature plant may produce a stem canker or a leaf spot. The symptoms may also vary radically because the disease will now involve physiological alterations in their early stage before the onset of symptoms to the late stages where the tissue becomes necrotic.
- **Co-infection:** It can happen that a plant is infected by several pathogens, which produces complex and overlapping symptoms. Disease symptoms are usually similar to abiotic stresses, including drought, salinity, nutrient deficiencies, and heat stress.

2.2.2 Geographic and Environmental Variability

- **Climatic Impact:** Temperature and humidity have a huge influence on the virulence of the pathogens and the response of the host. An example is a leaf spot disease emerging into a small and localized lesion in a dry climate but rapidly developing into a larger and wet water mass in a humid and tropical climate.

- **Background Noise:** Since data will be taken in various geographies, there will be great differences in the background clutter (soil type, adjacent weeds, shadows, reflections), which will complicate the models to reject all but the symptoms of the plant. Datasets that were collected under controlled settings (e.g., PlantVillage with plain grey backgrounds) usually collapse miserably when used on more intricate field images.

2.3 Labelling Granularity: Scale of Annotation

The plant health interpretation is based on the granularity of labels consumed by the models. The degree of particularity or detailing information about the plant diseases directly influences the power of the AI model whereby it can be considered as anything other than a simple image classifier or a more specific diagnostic instrument [11].

2.3.1 Leaf-level Granularity (Pixel/local)

This is the finest scale of annotation and is common in controlled environments and mobile-based diagnostics.

- **Segmentation:** A detailed pixel-level mask was drawn around the exact area of the leaf affected by the disease symptom. Required for models that perform semantic segmentation to precisely delineate the boundaries of the infected area. This is critical for calculating the disease severity index by measuring the percentage of infected pixels [12].
- **Object Detection (Bounding Box):** A rectangular box is drawn around each individual symptom or lesson. It's used for object detection models like YOLO or Faster R-CNN to localize and count specific lesions on a leaf.

2.3.2 Plant-level Granularity (Organ/whole plant)

Whole plant labels capture disease progression.

- **Classification:** A single label is assigned to the image based on the overall health or primary disease present on the plant. Examples are Fusarium wilt which affects the banana pseudo stem and xylem, Bacterial wilt causing whole-plant collapse. The image is labelled as "Potato Late Blight" or "Healthy Tomato Plant". Simple Image classification models are used to identify the disease and are generally suitable for mobile applications where the user focuses the camera on one plant [13].
- **Attribute Annotation:** Supplementary labels are added to the plant image, such as "Disease Severity: Moderate", "Growth Stage: Vegetative", or "Co-infection: Yes".

2.3.3 Canopy-level Granularity (Field/regional)

Canopy-level data is derived mainly from drones, UAVs (unmanned aerial vehicles), or satellite imagery and is crucial for precision agriculture applications like variable rate spraying.

- **Plot-Level Classification:** Assigning a single disease status or severity score to a defined area or plot of the field.
- **Georeferenced Segmentation:** Using aerial imagery to segment large contiguous areas of the field based on disease prevalence, stress levels, or yield potential. Enables models to identify disease foci and patterns of spread across the entire field, guiding large-scale management decisions.

Selection of granularity of labelling should be application based. Granularity of labelling should be selected according to application. Although the leaf-level segmentation provided is the most accurate in diagnosis, the canopy-level classification is more feasible in the field of monitoring via drones. All three levels should be incorporated into the framework of DCAI as they should work together to offer a comprehensive diagnostic and management tool.

3. Data Sources and Collection Modalities

High-quality and varied data is essential for the success of any machine learning or deep learning model aimed at detecting plant diseases. The effectiveness, dependability, and applicability of the system heavily rely on how accurately the dataset mirrors actual agricultural circumstances [14]. Thus, data collection should prioritize not only the amount but also the diversity — encompassing various plant species, different climatic settings, stages of infection, image resolutions, and variations in lighting.

This chapter explores numerous data sources, strategies for acquisition, tools for collection, and preprocessing methods required to create a resilient dataset [15]. It also addresses the difficulties encountered in acquiring real-time agricultural data, such as changes due to seasons, uneven class distribution, and image noise (for instance, blurriness, shadows, overlapping leaves, and distracting backgrounds [16].

To achieve real-world generalizability, it is frequently necessary to combine public datasets, field-collected images, synthetic data generation, and expert annotations [17]. Accurate labelling, ethical considerations, and proper data management must all be upheld throughout the procedure.

Key objectives of data collection:

- Ensure diversity across plant species and disease symptoms.
- Capture real agricultural environmental variations.
- Reduce dataset bias and class imbalance.
- Maintain high annotation accuracy for supervised learning.
- Enable cross-domain adaptability for deployment in multiple regions.

3.1 Public Datasets

Public datasets are a vital place to start when developing models, benchmarking, and conducting comparative research (Table 2). They give researchers pre-labelled photos of both healthy and sick plants, enabling them to swiftly train and assess preliminary models prior to gathering data in the field. They do, however, frequently lack real-world variability, despite offering a solid foundation. Since most photos are taken in controlled environments, noise factors like overlapping leaves, natural lighting, soil background, weather effects, and variations in camera quality may not be present. Therefore, depending solely on publicly available datasets could result in overfitting and poor generalization when the model is tested in actual field settings [18] (Table 3 and Table 4).

Best Use of Public Datasets

- Initial model training using CNNs, Transfer Learning, or Vision Transformers.
- Benchmarking performance against state-of-the-art models.

Table 2 Advantages of Public Datasets.

Benefit	*Explanation*
Time-efficient	Saves effort and cost in initial data collection.
Standard benchmarks	Enables comparison with other research studies.
Labelled data available	Reduces the need for manual annotation.
Supports transfer learning	Useful for fine-tuning pretrained models.

Table 3 Limitations of Public Datasets.

Limitation	*Impact on Model*
Controlled environment images	Less generalizable to field conditions.
Limited disease categories	Fails to cover rare or region-specific diseases.
Dataset imbalance	Some diseases have too few samples.
Low-resolution images	Affects feature extraction performance.

Table 4 Widely Used Public Datasets for Plant Disease Detection.

Dataset Name	*Number of Images*	*Crops Covered*	*Remarks*
Plant Village	~54,000	14+ crops, 26 diseases	Most widely used benchmark dataset
Plant Doc	~2,500	Field images	More realistic field conditions
Kaggle Leaf Disease Dataset	~4,000	Multiple crops	Good for basic experiments
AI Challenger Dataset	Large-scale	Chinese crops	Contains real-world agricultural images

3.2 Field-Based Imaging

Field-based imaging provides the extremely varied datasets required for reliable models by capturing actual crop variations. Drones, DSLR cameras, mobile phones, and multispectral sensors all contribute noise and variability to the data, which improves generalizability.

Canopy-level disease detection across acres of farmland is made possible by drone-based imaging, which is particularly helpful for large-scale agricultural monitoring. By capturing reflectance patterns that are invisible to the human eye, multispectral sensors enhance the capacity for early disease detection.

3.3 Institutional Collaborations

Expert-verified data collection is ensured through partnerships with plant pathology laboratories and agricultural universities. These collaborations provide access to carefully regulated experimental environments where illnesses are induced and scientifically tracked at various stages.

3.4 Sampling Diversity

Geographical diversity, multiple growth stages, different crop species, and environmental variations are all necessary components of a well-balanced dataset. Diversity makes sure that models don't overfit to a specific area or type of plant.

3.5 Ethical Considerations

In agricultural AI systems, data governance and ethics are essential. The responsible use of collected data is ensured by farmer consent, secure data handling, metadata anonymization, and transparent licensing.

4. Imaging Protocols and Environmental Considerations

Plant disease datasets are evaluated for quality and consistency based on imaging protocols. Model accuracy can be impacted by variations in lighting, angle, background, and environmental factors that greatly affect how symptoms appear. In order to ensure reproducibility and quality control, this expanded chapter covers standardized procedures for both field and lab imaging [19] [20].

4.1 Standardizing Lighting Conditions

One of the most important elements affecting the clarity, colour fidelity, and quality of photos of plant diseases is lighting. Any irregularity in lighting conditions can have a detrimental effect on machine learning models' performance, resulting in erroneous predictions and poor generalization in real-world situations because these models mainly rely on visual features, such as colour variations, texture details, and lesion boundaries.

Lighting conditions are extremely dynamic in natural agricultural settings. Image brightness and colour tone are greatly influenced by variables like time of day, cloud cover, shadows, humidity, and seasonal variations. For example, low light levels can result in grainy and blurry images, while intense sunlight can produce excessive reflections or glare. Models may find it challenging to extract consistent features due to these variations, which can change how disease symptoms appear.

To overcome these challenges, two major strategies are adopted:

1. Controlled Lighting (Laboratory-Based Approach)
2. Natural Lighting Optimization (Field-Based Approach)
3. Importance for Deep Learning Performance.

Creating high-quality plant disease datasets requires standardizing lighting conditions.

4.2 Standardizing Background and Scale

Neutral backgrounds, such as white, grey, or black, make segmentation easier and help eliminate distractions. Complex backgrounds like soil, the sky, or overlapping leaves must be taken into consideration in field photos. Incorporating scale markers guarantees precise assessment of disease severity [21] [22].

Table 5, with recommended lighting and background, is an example.

Table 5 Example of recommended lighting and background..

Factor	Recommended Practice
Lighting	Diffuse natural light or LED panels
Background	Uniform neutral colour background
Camera Angle	Perpendicular to leaf surface
Scale	Include ruler or marker within frame

4.3 Environmental Impact on Imaging

Environmental factors can introduce variability that impacts model training and performance, and they have a significant impact on the visual quality of plant images. Conditions like humidity, precipitation, fog, wind, and seasonal changes affect the plant's appearance as well as the image's clarity. Designing data collection techniques that guarantee plant disease detection systems' accuracy, reproducibility, and real-world generalizability requires an understanding of these dynamics [23].

4.4 Documenting Disease Progression

The model's capacity to identify and categorize various degrees of severity is strengthened when early, middle, and advanced stages of the disease are captured. Early-stage detection makes it easier to spot microscopic lesions or minute colour changes that are frequently overlooked in conventional datasets.

4.5 Metadata Importance

Metadata such as GPS (global positioning system) coordinates, timestamp, crop variety, disease type, environmental conditions, and camera specifications should all be included in every image [24]. Metadata allows researchers to examine environmental correlations and aids in the development of context-aware models.

Why Metadata Matters?

Metadata transforms a simple image into a rich data resource. It enables:

- Context-aware disease.
- Improved model generalization across different regions and climates.
- Traceability and dataset reproducibility.
- Efficient dataset filtering and organization.
- Better scientific research and epidemiological studies.

An example of metadata schema includes:

- Latitude, Longitude
- Temperature, Humidity
- Crop type and variety
- Growth stage
- Illumination type
- Camera model

5. Annotation Strategies and Ground Truth Generation in Plant Disease Detection

Using ML models in plant disease detection, is closely tied with precision of data annotation and the sample size of datasets. Annotation helps to draw a meaningful insight about raw data and precise annotation helps to improve the accuracy of ML models.

This chapter introduces a different range of tools ranging from manual annotation to expert driven approaches. The ontologies and taxonomies which give structured, semantically consistent labelling frameworks are essential for these datasets in plant pathology. Real-world examples from agricultural research demonstrate how annotation methodologies directly influence model accuracy, reproducibility, robustness, generalizability, and scientific reliability.

5.1 Manual Annotation: Tools, Workflows, and Expert Involvement

To create high-quality plant disease datasets, manual annotation is still the most dependable method. The primary criterion is human judgement in annotation, especially from plant pathologists and agricultural specialists, who identify diseases by looking at photos of leaves, stems, etc.

Tools

Common manual annotation platforms include:

- **LabelImg** and **RectLabel**, frequently used for bounding box annotation and supported by formats such as YOLO and Pascal VOC.
- **VGG Image Annotator (VIA)**, a lightweight tool ideal for polygon-based region labelling.
- **Label box** and **Super Annotate**, which offer advanced pixel-level segmentation capabilities and collaborative workflows, suitable for identifying fine-grained disease boundaries.

Workflows

The process involves collecting leaf disease images under different lighting and weather conditions in different ways. The human collected data is grouped based on the disease and annotators label them manually. Some of the sample diseases are leaf spot, powdery mildew, rust, blight, anthracnose, downy mildew, and canker. Each image is annotated or labelled according to standard protocol based on disease symptoms. Once annotation or labelling is completed it is reviewed by peers, subject matter experts, and experts so that ambiguity is resolved. Ambiguous cases are examined and resolved to ensure consistency in annotation and biological precision.

Expert Involvement

The process of annotating agricultural images by humans involves gathering them in a variety of ways and under different lighting and weather conditions. After that, the annotators sort the images into groups based on the types of diseases they show. Some of the diseases in plants shown in the pictures are leaf spot, powdery mildew, rust, blight, anthracnose, downy mildew, and canker. These diseases are based on rules that were set up ahead of time. The images are labelled according to the disease, and then they are reviewed by peers and experts. Ambiguous cases are examined and resolved to ensure consistency in annotation and biological precision (Table 6).

Table 6 Manual Annotation Tools in Plant Disease Detection.

Tool	*Domain*	*Key Features*	*Example Use Case*
LabelImg	Image annotation	Bounding boxes, YOLO/ Pascal VOC export	Leaf disease classification
VIA	Image annotation	Polygon/region labelling, lightweight	Marking diseased leaf regions

Contd.

Table 6 *Contd.*

Tool	*Domain*	*Key Features*	*Example Use Case*
Labelbox	Image segmentation	Pixel-level annotation, collaboration	Powdery mildew segmentation
Super Annotate	Image segmentation	AI-assisted polygon labelling	Fruit rot detection

5.2 Semi-Automated Annotation Using DL-Assisted Tools

Manual annotation for millions of images is a difficult task and it is also time consuming. As this needs a lot of resources, semi-automation annotation is preferred and essential. By using DL models, initial labels were generated and then followed by human review.

For model evaluation, agricultural datasets are enormous—thousands and lakhs—and manual annotation is a difficult task. Because purely manual annotation requires a lot of resources, semi-automation annotation is essential. This uses DL models to generate initial labels, which are then improved by human reviewers.

Advantages

Semi-automation offers several key benefits:

- **Speed:** DL models can pre-label leaves or lesion areas, significantly reducing manual workload.
- **Consistency:** Automated predictions standardize labelling across annotators.
- **Active Learning:** The model identifies uncertain or ambiguous regions, prompting expert review.

Challenges

However, semi-automation introduces risks:

- Annotators may unintentionally depend too heavily on machine suggestions.
- Models may mislabel visually similar conditions—for example, nutrient deficiencies that resemble fungal infections.
- Environmental factors such as lighting or leaf orientation can mislead model predictions.

Example of a Semi-Automated Workflow

A CNN (convolutional neural network) is initially used to assign provisional labels (such as blast, brown spot, or healthy) in a rice disease detection study. These predictions are improved by annotators, and the updated labels are then fed back into the model. This leads to decreased annotation time and enhanced model performance over subsequent iterations (Table 7).

5.3 Inter-Annotator Agreement and Quality Control

As the images of disease look similar and there is minor variation, annotations need to be good and reliable and it is important that annotators need to be consistent.

Table 7 Manual vs Semi-Automated Annotation.

Aspect	*Manual Annotation*	*Semi-Automated Annotation*
Speed	Slow, labour-intensive	Faster, more scalable
Accuracy	High (expert-driven)	Moderate, improves with feedback
Cost	High	Lower due to automation
Bias Risk	Low	Higher due to model influence
Best Use Case	Small, critical datasets	Large or repetitive datasets

When there is more than one annotator, as each annotator sees it differently noise is added and the overall accuracy goes down. This is addressed by quantifying the degree of agreement among annotators using measures of inter-annotator agreement (IAA), such as Cohen's Kappa, Fleiss' Kappa, or Krippendorff's Alpha. When the IAA score is high, it means that annotators strongly agree and low IAA score means that majority annotators don't agree due to unclear symptom presentation, unclear annotation guidelines, or not enough training.

Metrics

Common agreement metrics include:

- **Cohen's Kappa:** Evaluates agreement between two annotators, correcting for chance.
- **Fleiss' Kappa:** Extends the concept for multiple annotators.
- **Krippendorff's Alpha:** Accommodates different data types and is more flexible for real-world datasets.

Example

In a maize leaf dataset annotated for leaf blight, three annotators achieved a Cohen's Kappa of 0.85, indicating strong agreement (Table 8). Disagreements mainly stemmed from confusion between fungal lesions and insect damage, which appear visually similar under certain lighting conditions.

Table 8 Agreement Metrics in Plant Disease Annotation.

Metric	*Suitable For*	*Strengths*	*Limitations*
Cohen's Kappa	Two annotators	Adjusts for chance	Not suitable for many annotators
Fleiss' Kappa	Multiple annotators	Handles groups well	Assumes equal annotator reliability
Krippendorff's Alpha	Mixed data types	Versatile and robust	More complex to compute

Quality Control Strategies

To maintain reliability:

- **Double-Blind Annotation:** Annotators work independently to avoid bias.

- **Adjudication:** Experts review and resolve discrepancies.
- **Spot Checks:** Random audits assess ongoing accuracy.
- **Gold Standard Datasets:** Benchmarks such as PlantVillage serve as training and comparison references.

5.4 Ontologies and Taxonomies for Plant Disease Labeling

Ontologies and taxonomies introduce structured, semantically consistent labelling frameworks that are essential for interoperable and scalable datasets in plant pathology.

Ontologies

Ontologies such as the **Plant Ontology (PO)** and **Crop Ontology (CO)** define standardized terminology for plant structures and traits. These frameworks help ensure that disease annotations follow a uniform vocabulary. For instance, the category "leaf spot" may branch into specific labels such as "Cercospora leaf spot" or "Alternaria leaf spot", supported by hierarchical relationships.

Formalized vocabularies that represent the connections between plant characteristics, structures, and disease manifestations are defined by ontologies. Commonly used frameworks consist of:

Plant Ontology (PO)

- The focus is on the developmental stages and anatomical features of plants, such as leaves, stems, and roots.
- Verifies that disease annotations are associated with the appropriate plant part or stage of growth.

For instance, a "leaf lesion" annotation is directly connected to the PO term for "leaf", preventing confusion between datasets.

Crop Ontology (CO)

- Offers standardized descriptors for characteristics and phenotypes unique to crops.
- It makes it easier to harmonize trait-based disease data and compare crops.

For instance, CO terms for rice traits and disease categories can be consistently used to describe "blast lesion" in rice pathology.

Pathogen/Disease Ontologies (e.g., DO, PHI-base)

- Capture pathogen-specific interactions, host responses, and disease outcomes.
- Enable integration of plant disease data with broader biomedical and ecological datasets.

Hierarchical Relationships

Ontologies enforce **hierarchical structures** that allow general categories to branch into specific subtypes:

- **General → Specific**
  - ***Leaf spot*** *→ Cercospora leaf spot → Cercospora leaf spot on maize*
  - ***Rust***→ *Puccinia rust → Puccinia rust on wheat*
- **Parent–Child Relationships**
  - Broader disease categories act as parent nodes, while pathogen-specific or crop-specific manifestations are child nodes.
  - This hierarchy supports **multi-level annotation granularity**—from coarse classification (e.g., "fungal disease") to fine-grained labels (e.g., "Alternaria leaf spot on tomato").
- **Semantic Consistency**
- Ontologies ensure that terms like "lesion", "necrosis", or "chlorosis" are used consistently across datasets, reducing ambiguity in AI-driven disease detection.

Taxonomies

In order to ensure clarity and consistency in annotation across datasets, taxonomies offer a methodical framework for classifying plant diseases into distinct categories. Taxonomies concentrate on hierarchical grouping according to pathogen type, host specificity, or symptomatology, whereas ontologies prioritize semantic relationships and standardized vocabularies.

Example: The taxonomy of tomato diseases

Pathogen type can be used to arrange a tomato disease taxonomy:

- Fungal Illnesses

  Alternaria solani causes early blight, while Phytophthora infestans causes late blight.
- Bacterial diseases

  Bacterial canker (Clavibacter michiganensis)

  Bacterial spot (Xanthomonas spp.)
- Viral Illnesses

  TYLCV, or tomato yellow leaf curl virus

  ToMV, or tomato mosaic virus.

Similar to how biological classification functions in taxonomy, this hierarchical grouping enables researchers to move from broad categories (fungal, bacterial, viral) down to specific diseases.

6. Addressing Data Challenges

The quality, quantity, dependability of underlying dataset is important for the ML model to provide better accuracy in prediction. In the real-world dataset has imbalanced data where the diseased labels will be minority and the healthy leaf has majority samples. This imbalance in data makes the ML model difficult to learn the minority data, and hence for better outcome this imbalance problem needs to be fixed.

Class imbalance, domain shift, and noisy labels are the three main types of data challenges that this chapter addresses and offers comprehensive solutions. These issues need to be addressed to improve the accuracy of the prediction of ML models. If left unaddressed, it will lead to biased outcomes.

6.1 Class Imbalance

ML models face a challenging task of learning from imbalanced data in plant disease dataset. Researchers say that this class imbalance problem arises when the distribution of the diseased category and healthy category is uneven. In this scenario, conventional ML algorithms find it difficult to learn due to the under representation of diseased categories. For example, in a binary dataset of tomato leaves where healthy samples constitute 90% of the data and diseased samples (e.g., late blight) only 10%. In this case, the classifier will predict all leaves as healthy and will fail to identify the diseased part which is minority class and it is more critical for decision-making. This imbalance not only skews performance metrics but also undermines the reliability of AI systems in real-world plant pathology applications (Table 9).

Table 9 Typical Distribution of an Imbalanced Dataset.

Class Type	*Dataset Percentage*	*Influence on Model Learning*
Majority Class	90–95%	Model tends to focus on majority patterns; biased predictions
Minority Class	5–10%	Low recall; critical events often missed

6.1.1 Oversampling

Oversampling increases the proportion of minority class samples to balance the dataset.

Key Techniques

- **Random Oversampling** Random Over Sampling (ROS) is adding examples to the minority class so that it increases the size of the minority class to balance the data. Minority class is sampled with replacement and then the samples are combined with majority data. This leads to over-fitting, since it makes exact copies of existing instances.
- **SMOTE (Synthetic Minority Oversampling Technique)** Synthetic Minority Over-sampling Technique (SMOTE) is widely used to address class imbalance by generating artificial samples for minority disease categories. Instead of simply duplicating existing minority instances, SMOTE creates synthetic examples by interpolating between a given sample and its nearest neighbours within the minority class. To identify these neighbours, the K-Nearest Neighbour (KNN) algorithm is applied, and new synthetic instances are generated along the line segments connecting the original sample to its

neighbours. For example, if late blight images are underrepresented compared to healthy leaf images, SMOTE can produce additional synthetic late blight samples by blending features of existing diseased leaves. This approach enriches the dataset with more diverse representations of minority classes, thereby improving the classifier's ability to learn subtle disease patterns and reducing the bias toward majority categories.

6.1.2 Under Sampling

Under sampling reduces the size of the majority class to create a more balanced distribution.

Key Techniques

- **Random Undersampling** Random Under Sampling (RUS) refers to the process of reducing the number of samples from the majority class to balance the dataset . While this technique retains all minority class instances, such as rare disease categories, it randomly selects a subset of majority class samples (e.g., healthy leaves) so that their count matches the minority class. A potential drawback of RUS is that it may discard valuable information from the majority class, thereby limiting the diversity of training data. However, in scenarios where majority class samples are highly similar to one another (for instance, multiple images of healthy leaves with minimal variation), this method can be effective, as the removal of redundant examples does not significantly impact the classifier's ability to learn meaningful distinctions between healthy and diseased plan.
- **Cluster-based Undersampling** Cluster based undersampling is a technique to reduce the majority class (healthy leaves) by preserving their diversity. This method applies a clustering algorithm (K-means) instead of randomly discarding the majority data. This algorithm groups majority classes into clusters based on similarity. From each cluster, a representative subset of samples is then selected, ensuring that the reduced dataset still captures the variability present in the majority class. By this method important information is still maintained and loss of critical information is reduced and improves the learning of ML models. For example, in a tomato dataset where healthy leaf images vastly outnumber diseased ones, clustering ensures that different variations of healthy leaves—such as differences in lighting, texture, or growth stage—are retained in the training set. As a result, classifiers trained on cluster-based under sampled data are better equipped to distinguish between healthy and diseased plants, while avoiding bias toward the majority class.

6.1.3 Synthetic Data Generation

Synthetic data generation creates new, realistic samples for minority classes using generative models.

Key Techniques

GANs (Generative Adversarial Networks)

This GAN generated realistic synthetic images of minority class (diseased category) in plant disease dataset. It consists of a discriminator and a generator. The generator generates synthetic images of minority data and the discriminator determines whether the generated images are realistic or not by comparing them with real images. Through this iterative process GAN creates realistic images making it challenging for the discriminator to differentiate them from the real dataset samples.

For instance, compared to healthy leaves, tomato mosaic virus images will be few in a dataset of plant diseases. In this case, GAN will produce artificially ill images that mimic the texture, color changes, and lesion patterns of actual infections. Increasing the minority class distribution will enhance the classifier's ability to detect uncommon illnesses. GANs are especially useful for visual datasets where disease symptoms show subtle variations because they capture complex, high-dimensional features of plant pathology images, unlike simpler oversampling techniques like SMOTE. GANs improve generalization, lessen bias toward majority categories, and aid in the creation of more reliable AI models for plant disease detection by supplementing minority classes with realistic synthetic data.

Variational Autoencoders (VAEs)

VAEs address the class imbalance by creating synthetic images of minority classes. This VAE consists of two components: a) encoder which compresses input data such as leaf images into a latent representation; b) Decoder which reconstructs data from this latent space. VAE is based on a probabilistic framework by learning distributions over the latent variables, enabling the generation of new, diverse samples rather than exact reconstructions of existing ones.

For instance, a VAE can create synthetic images that mimic actual diseased leaves by learning the underlying distribution of TYLCV leaf features, such as vein distortion, curling patterns, and colour changes, if a dataset contains very few images of the virus. By enriching the minority class, these synthetic samples enhance the classifier's capacity to identify uncommon illnesses. VAEs are especially useful for high-dimensional image data when compared to oversampling techniques like SMOTE because they capture intricate variations in texture, shape, and color that define plant disease symptoms.

6.2 Domain Shift

The model which works perfectly in training may fail in a real-time scenario. Though the underlying task remains the same, models that work in one condition may fail in another condition due to differences in lighting, background, and camera quality in a plant disease dataset (Table 10).

Table 10 Examples of Domain Shifts.

Application	*Source Domain*	*Target Domain*	*Cause of Shift*
Leaf Image Classification	Greenhouse images	Field images	lighting conditions, background clutter, camera resolution
Crop-Specific Disease Models	Tomato dataset (India)	Tomato dataset (Africa)	geographic variation, climate differences, pathogen strain
Multi-Crop Disease Detection	Rice leaf dataset	Wheat leaf dataset	crop morphology differences, symptom expression

Noisy Labels in Plant Disease Datasets

Noisy labels occur when the annotation is incorrect where the assigned class, e.g., "powdery mildew" or "leaf spot" does not reflect the true disease. There are many reasons for noisy labels like non-expert annotators, lighting, leaf age, subtle visual clues, etc., due to this the classifiers misdiagnoses the disease in real-world scenarios. There are several strategies to handle noisy data like using expert annotators, multi-annotators per sample, providing annotation guidelines, introducing noise robust algorithm

7. Case Study Insights for Data Acquisition in Plant Disease Early Detection

Data acquisition is essential in modern digital-enabled agriculture in order to effectively detect any plant disease with greater precision and reliability. The precision of an ML or DL model mainly relies on the data that is acquired. The real farming environment is challenging to acquire data because of varying weather conditions, lighting, and other interactions with the environment. Therefore, real-world data collection is more challenging than images obtained in laboratory-based collections.

Most of the data sets acquired are collected in green houses where noise is not present. However, such sanitized data does not embed actual farm conditions due to which the results shown during experimentation will not hold true during deployment. So it is absolutely essential that the agriculture data acquired reflect varying field conditions and reflect the exact symptoms of pests and diseases. This chapter presents three case studies depicting the real field environment [25].

- Tomato Leaf Curl Virus (ToLCV) dataset collection in Tamil Nadu
- UAV-assisted monitoring of rice blast disease across Southeast Asia
- Multispectral imaging for early grapevine disease detection.

Each case study offers insights on multiple operational challenges, strategies, and disease datasets.

7.1 Case Study 1: Tomato Leaf Curl Virus Dataset in Tamil Nadu, India

7.1.1 Overview and Motivation

Tomato Leaf Curl Virus (ToLCV) is the most severe threat to tomato cultivation because of its rapidly spreading nature and its impact on growth of the plant. The symptoms of the virus overlap with other viral infections and drought stress. The conventional data set available from previous research does not capture the actual symptoms that have occurred in open cultivated fields. Therefore, agricultural scientists and technicians in Tamil Nadu developed a data acquisition project from multiple locations.

7.1.2 Data Acquisition Strategy

The project team organized multiple surveys in different tomato farms in TamilNadu in various districts of Coimbatore, Dharmapuri, Krishnagiri, and Salem. The survey was conducted with various spatial and temporal diversity. Images were captured right from vegetation stage to fruiting stage in order to record the onset of a disease to spreading state. Their sampling strategy emphasized spatial and temporal diversity. Images were captured during various growth stages—right from early vegetative phases to peak fruiting periods. The images were conducted under natural light with different positions of sunlight and clouding. This was done intentionally to have an enhanced data set captured in real environment conditions. This data was intended to be used to train ML models for accurate prediction. Normal cameras used by agriculturists were used to capture the images and the same was also captured with professional DSLR (digital single-lens reflex) cameras. This was done to compare the ML models trained using images captured from ordinary cameras and professional cameras.

7.1.3 Annotation Process

Exact annotation has to be recorded for acquisition of quality dataset. Pathologists performed various assessments during analysis of symptoms of ToLCV virus. The images were classified as mild, moderate, and severe levels. Every image collected was recorded along with the environment conditions such as temperature, humidity, soil conditions, location, and time of capture. The images annotated along with metadata was believed to improve the accuracy of the ML model because the disease spreading mostly depends on environmental conditions [26].

7.1.4 Major Outcomes

The pathologists involved in the project were able to collect realistic dataset that captured the symptoms and spread of virus in different environmental settings. It was demonstrated that real-time data collected in a real environment with natural noise and metadata aids in development of accurate adaptable DL models [27].

7.2 Case Study2: UAV-Based Rice Blast Detection in Southeast Asia

7.2.1 Motivation

Magnaporthe oryzae, is a fungal disease resulting in rice blast across the globe. The onset of this fungus is very rapid and manual investigation is very difficult because of patchy distribution. UAVs are used for monitoring the disease at the canopy level as manual investigation with the naked eye is impossible in large farms. These UAVs capture the spectral signatures that are invisible to humans [28].

7.2.2 UAV Data Workflow

The UAV used in the project has RGB and multispectral sensors. These multispectral sensors capture images as RGB (red, green, blue) images with red edges and green bands that are sensitive to any physical and functional stress. The UAV platforms used in the project carried both RGB and multispectral sensors. The latter included red-edge, NIR (near infra-red), and green bands, which are highly sensitive to physiological stress. Flight planning is done at different altitudes ranging from 30 –50 m and during mid-morning to reduce any glare, and multiple flights during each season. Agronomists conducted multiple field visits for leaf sampling, recording environmental parameters such as humidity, temperature, and nitrogen content and confirming any lesions captured by the UAV.

7.2.3 ML Analysis

Captured images were processed for various indices such as NDVI and SAViwas stitched into orthomosaics, segmented into analysis patches, and processed through vegetation indices such as NDVI (normalized difference vegetation index) and SAVI. Patch extraction was done and applied to DL models to classify the presence and severity of disease and TDL (tally definition language) models were applied to classify disease presence and estimate severity. Disease hotspots were identified using GIS (geographic information system) platforms. The results were effective in identifying the disease about a week ahead and using targeted spraying that resulted in 40% reduction in fungicide. The regional spread was also easily identified.

7.3 Case Study 3: Multi spectral Imaging for Grapevine Disease Detection

7.3.1 Contextual Background

Downy and powdery mildew is a fungal disease that affects grapevines. Grapevines are vulnerable to fungal diseases like downy mildew and powdery mildew. Visual detection is often possible only after the disease starts spreading. However multispectral imaging with red edges and NIR bands capture any physiological stress well before the emergence of the disease.

7.3.2 Multimodal Perception Arrangement

Data acquisition was obtained through UAV mounted multispectral sensors, handheld spectral cameras for physical inspection. The data was acquired in multiple vantage points to get both macro-and micro-level perception about the progression of diseases.

7.3.3 Annotations

It was found after the study that the diseases were able to be identified days ahead of spreading and lesions occurred. The multispectral image with DL segmentation aids in identification before the symptom occurs. Targeted investigation aids in reducing fungal infection. This helped in consistent monitoring and reduction in disease spread.

7.4 Comparative Insights from Above Case Studies

Across the three case studies, similar paradigms were used in agricultural data acquisition. The ToLCV project exhibited the importance of natural lighting over artificially illuminated lighting.

7.4.1 Impact on Environmental Inconsistency

The inconsistency in the environment was found to be a significant challenge in data acquisition. The natural field has different light settings, wind movement, shadows, and background disorder. Artificial images with sanitized background resulted in poor model generation. In rice paddies, the water reflection altered the quality of spectral images and cloudy weather influenced sensors in UAV flights. This effect reinforced that datasets must include heterogeneity in the environment [29].

7.4.2 Significance of Sensor Diversity

In the tomato case study, the images were captured from both farmers' cameras and DSLR cameras. In rice blast monitoring, RGB and multispectral sensors were used. In grapevine multispectral sensors were used. The diversity of sensors had an effect on the quality of datasets and effect in ML model creation. Multimodel datasets performance is always better than unimodal dataset. These sensors help to detect chemical changes before visual changes and aid in early detection of plant diseases. The diversity enhances the robustness in detection of diseases [30].

7.4.3 Significance of Meta Data

All three case studies revealed the quality of dataset used to build ML models has significant contribution to the associated meta data such as location, timing, temperature, humidity, and soil moisture. These data help in prediction of occurrence of a disease. Current ML models also accept conceptual meta data. For example, in ToLCV detection, meta data helped in distinguishing nutritional deficiency from disease symptoms. In the riceblast case, it helped identify the effect of humidity on disease spread. In the grapevine case, it helped differentiate fungal infection from irrigation stress [31].

7.4.4 Significance of Annotation

Annotation from experts such as pathologists helps in accurate detection of plant diseases. This annotation plays a crucial role because most diseases mimic each other. Any noise due to wrong annotation results in incorrect datasets and incorrect model development. Field-based annotation is recommended. Multi stage annotation and review of annotations is recommended for dataset integrity.

7.5 Challenges in Data Acquisition

Data acquisition has operational and technical challenges. Unpredictable weather conditions such as rain, sunlight, and fog will affect the sensor data quality and image quality. Humidity has an effect on multispectral sensors. Images got at ground level because of the difference in terrain and mud. In order to maintain consistency, the images are to be acquired in multiple cycles with multiple sensor calibrations. UAV-based images have challenges in terms of battery limitation, regulatory rules in region, effect of vibration, and GPS (global positioning system) consistency [32]. Multispectral sensors are also subjected to calibration errors.

7.6 Data Acquisition Recommendations

Including real-world variability while collecting data is essential. Cleaned dataset is not recommended for model cleaning. Rather, a dataset with real-time noise is recommended along with annotation of the environment that created the noise. Multiple sensor modalities such as RGB and multi spectral sensors give additional insights on the stage of disease spreading. In order to increase accuracy of the model, annotation, and meta data are recommended. These annotations must be done by experts such as pathologists. Transparent governance is recommended. It is recommended to balance theoretical research with practical field conditions. Effective agricultural data acquisition results in accurate development of ML models that aid in accurate detection of plant diseases. Continuous imaging using autonomous robotics, spatio-temporal disease modelling, digital twin monitoring, and federated data collection are expected to enhance the data acquisition quality in future. High quality data acquisition is a complex task that requires technical expertise and adaptability. With smart agriculture, AI-driven disease detection is expected to be reliable and resilient

8. Future Directions in Data Acquisition for Plant Disease Detection

Smart agriculture is practised globally due to advancements in sensing technology, evolution of strong ML and DL algorithms, and cloud-based analytics and prediction. The vital component of plant disease detection using smart agriculture is data acquisition. The evolution of digital agriculture over the past decade has been characterized by rapid advancements in sensing systems, ML, DL, and cloud-enabled analytics. The success of the system relies on the quality of data acquired. This data must be of high quality and context-rich. The variability embedded in fields such as soil moisture, density of canopy, sunlight, severity of disease, and pest management strategy influences accuracy of disease detection system.

Traditional field survey involves physical field exploration, lab-based image acquisition, and trials. These procedures lack precision due to which accurate detection is not possible. It is found that 65% of studies rely on artificially controlled datasets that have limitations on generalizing the detection features [33]. Hence, modern approaches are aimed at developing a dynamic, multimodal autonomous disease detection system. These systems focus on federated data acquisition, edge annotation, embedding context features, and spatio-temporal modelling.

8.1 Federated Data Collection and Distributed Learning Ecosystems

In centralized data acquisition, the images, meta data, and sensor readings are stored in cloud servers and are used for training models. The challenges in such centralized systems are limitations in network resources to send real-time continuous data with high resolution images, privacy concerns. Hence, federated distributed data collection is directly used to train models.

8.1.1 Objectives of Federated Data Collection

Federated data collection is based on the principle of collecting data from farmers' smartphones directly from the field. The device does pre processing, edge feature extraction, and updates models in the device. Models trained using decentralized models achieve more accuracy compared to centralized systems [34].

8.1.2 Benefits of Federated Data Acquisition

Federated data acquisition results in better model training and plant disease detection because it has the following benefits:

- Resilient to connectivity in rural areas because in federated systems synchronization will happen whenever the connection is restored.
- It has a diverse dataset from multiple regions, climates, and field environments, and contributes to robust generalized models [35].

- Protection of privacy of a farmer and preserves the identity of location and practices.
- Farmers retain ownership of their data and still contribute to developing an efficient model.

8.1.3 Challenges in Federated Data Acquisition

Few of the challenges in federated learning are heterogeneity in quality of the image acquired, improper annotations due to lack of expertise, lack of device capability to perform complex computations. This has limitations to train and build an effective model that detects the plant disease accurately. However, the cost incurred can be addressed by intervention governance that ensures the quality of the device.

8.2 Edge Intelligence

Quality of DL model depends on quality of data acquired and this data quality relies on annotation, filtering, and preprocessing. Studies have proved that these, when done in Edge, are more effective than when done in Cloud side. This aids in real-time decision-making on farmers' field side.

8.2.1 Annotation in Edge Systems

Current day smart phones have built in GPUs (graphics processing units) and NPUs (neural processing units), that enables lightweight DL models to perform inference in Edge device itself. On-device annotation can also be done in Edge device that helps plant disease detection by detecting leaf boundaries, highlight symptomatic regions with disease labels, prompt farmers for validation, and also provide farmers instruction to capture high resolution images by self-adjusting the image captured with help from them. This helps annotation quality by about 40% [36].

8.2.2 Edge-based Intelligence

The next generation Edge-based intelligence is expected to increase annotation quality by rejecting noisy images using adaptive quality algorithms, and provide voice-based annotation interface, integration with portable environment sensors, and use offline annotation modes for rural areas with bandwidth constraints.

8.3 Multimodal Data Fusion

Robust detection of plant diseases is not possible using RGB images alone. The future data acquisition framework is expected to have multi modal fusion by integrating image, soil, weather, and other spectral signals. There are different types of multimodal data such as thermal imagery, RGB images, multispectral images, field logs, and weather specific data. There are multiple multimodal DL algorithms that are developed, such as the Transformer-based model that integrates weather data with soil parameters and other image embeddings. Graph Neural Networks aid in spatio-temporal disease analysis. Probabilistic fusion models use uncertain features for critical predictions.

Few challenges in multimodal integration are lack of temporal alignment of heterogeneous sensors, missing data, fusion complexity, and varying sampling frequencies.

8.4 Autonomous Data Acquisition Platforms

Data acquisition in agricultural fields will take a paradigm shift with usage of autonomous sensing systems. Plant disease detection relies on continuous multi modal data sets that cannot be done manually from farmers' smart phones alone. Hence, autonomous systems such as robotic systems that may be ground-based or aerial can be used in data acquisition. This can also integrate embedded intelligence and advanced sensing technologies to curate high quality datasets that capture disease spread accurately at an early onset. Current advancements have robotic navigation within fields, dynamic sensor calibrations which can aid in adaptive field monitoring systems even in complex environments [37].

These ground-based robotic systems can move in real fields and aid in capturing high resolution plant images that detect early development of lesions in plant leaves. This can even detect infections in the lower part of leaves and can detect viral mosaics and nutrient stress even before an attack surfaces. It was found that robotic hyperspectral imaging detects stress signatures at least three days before the human eyes can detect. LiDAR can be integrated to detect any structural anomalies caused due to physiological changes caused by plant disease onset. Adaptive wheel control, terrain aware movement, and real-time path calibration using local maps will aid in enhancing the accuracy of autonomous ground robotic systems.

UAVs make use of swarm imaging for aerial data acquisition. UAV swarms are multiple UAVs operating by coordinating with each other to cover a large expanse of fields within a short span of time. Advanced swarm coordination algorithms enable dynamic allocation of tasks, adapt to environment conditions, conserve energy consumption, and detect any disease hotspot with accuracy. These UAV swarms capture multispectral, hyper-spectral, and thermal data of crops to detect early occurrence of diseases. Studies have proved about 60% efficiency when compared to single drone systems. Hybrid ecosystem integrating terrain and UAV vehicles aids in macro anomalies. UAVs identify potential infection and ground robots use hyper-spectral sensors to detect any biochemical changes.

8.5 Temporal and Spatiotemporal Modelling

The temporal dataset helps in disease forecasting, progression modelling, and preventing disease onset. For effective detection of plant disease, regular imagery on an hourly, daily, or weekly is required to get critical insight on the symptom leading to detection of early onset of the disease. This helps the DL model such as convolution LSTM (long short-term memory) and transformer-based models to get trained in temporal dimension. Spatial data collected at different canopies of fields along with context annotations such as soil quality, weather, humidity, and temperature helps the DL model to be trained for diversity in environmental conditions. This spatio-temporal modelling enables the development of early warning systems for plant disease detection, detection of disease spread hotspot, and developing weather-driven risk models that aid farmers to take preventive steps. Recent research has demonstrated that using UAV images with ground sensors have improved the early warning system for grapevine mildew by about 30%.

9. Conclusion

Data acquisition plays a major role in training DL models for early detection of plant diseases. The future of dataset dynamic is expected to have a paradigm shift towards distributed, dynamic, and collaborative data acquisition. Integration of digital twin, Edge intelligence, and federated learning will become an integral part of smart agriculture. The quality of imagery tends to improve using multimodal fusion of images collected using UAV and ground terrain sensors. Robotic and autonomous sensing systems will aid in building an effective spatio-temporal model. Ethical governance systems will be developed to ensure privacy of data acquired.

References

[1] González-Rodríguez, Victoria E., et al. (2024). Artificial intelligence: A promising tool for application in phytopathology. *Horticulturae*, *10* (3), 197.

[2] Relevance of Advanced Plant Disease Detection Techniques in Disease and Pest Management. (2023, Nov.). Scirp.org, *American Journal of Plant Sciences*, *14* (11).

[3] LeCun, Y., Bengio, Y. and Hinton, G. (2015). Deep Learning, *Nature*, *521*(7553), 436–444.

[4] Mohanty, S.P., Hughes, D.P. and Salathe, M. (2016). Using deep learning for image-based plant disease detection. *Frontiers in Plant Science*, *7*, 1419.

[5] Mahlein, Anne-Katrin. (2016). Plant disease detection by imaging sensors–parallels and specific demands for precision agriculture and plant phenotyping. *Plant Disease*, *100* (2), 241–251.

[6] Shoaib, Muhammad, et al. (2023). An advanced deep learning models-based plant disease detection: A review of recent research. *Frontiers in Plant Science*, *14*, 1158933.

[7] Schouten, Gerard, Bas SHT Michielsen and Barbara Gravendeel. (2024). Data-centric AI approach for automated wildflower monitoring. *PloS One*, *19* (9), e0302958.

[8] Dong, Jiuqing, et al. (2022). Data-centric annotation analysis for plant disease detection: Strategy, consistency and performance. *Frontiers in Plant Science*, *13*, 1037655.

[9] Yang, Jiachen, et al. (2022). Data quality assessment and analysis for pest identification in smart agriculture. *Computers and Electrical Engineering*, *103*, 108322.

[10] John, Matthew Abu, et al. (2023). Relevance of advanced plant disease detection techniques in disease and Pest Management for Ensuring Food Security and Their Implication: A review. *Am. J. Plant Sci.*, *14* (11), 1260–1295.

[11] Wiesner-Hanks, Tyr, et al. (2019). Millimeter-level plant disease detection from aerial photographs via deep learning and crowdsourced data. *Frontiers in Plant Science*, *10*, 1550.

[12] Aldakheel, Eman Abdullah, Mohammed Zakariah and Amira H. Alabdalall. (2024). Detection and identification of plant leaf diseases using YOLOv4. *Frontiers in Plant Science, 15*, 1355941.

[13] Xu, Mingle, et al. (2024). Plant disease recognition datasets in the age of deep learning: Challenges and opportunities. *Frontiers in Plant Science, 15*, 1452551.

[14] Hughes, D. and Salathe, M. (2015). *An Open Access Repository for Crowdsourced Plant Disease Images.* arXiv preprint arXiv:1511.08060.

[15] Mohanty, S.P., Hughes, D.P. and Salathe, M. (2016). Using deep learning for image-based plant disease detection. *Frontiers in Plant Science, 7*, 1419.

[16] Singh, D., Jain, N., Jain, P. and Kayal, P. (2020). Plant Doc: A dataset for visual plant disease detection. In: *Proceedings of the 7th ACM IKDD Co DS and 25th COMAD* (pp. 249–253).

[17] Xie, C., Wang, J., et al. (2020). Past Net: A dataset and benchmark for intelligent pest detection. *IEEE Access, 8*, 147601–147613.

[18] Barbedo, J.G.A. (2019). Impact of dataset size and quality on the effectiveness of deep learning and transfer learning for plant disease classification. *Computers and Electronics in Agriculture, 153*, 46–53.

[19] Singh, U., Zhang, S. and Kumar, A. (2021). Plant disease detection using computer vision and deep learning: A review. *Artificial Intelligence in Agriculture, 6*, 233–248.

[20] Pound, M., Atkinson, J. A., Wells, D. M., et al. (2017). Deep learning for phenotyping: Metadata-driven crop analysis. *Trends in Plant Science, 22*(6), 590–602.

[21] Chlingaryan, A., Sukkarieh, S. and Whelan, B. (2018). Machine learning approaches for crop yield prediction and disease detection using remote sensing data. *Agronomy, 8*(11), 236.

[22] Kamilaris, A. and Prenafeta-Boldú, F.X. (2018). Deep learning in agriculture: A survey. *Computers and Electronics in Agriculture, 147*, 70–90.

[23] Liakos, K.G., et al. (2018). Machine learning in agriculture: A review. *Sensors, 18*(8), 2674.

[24] Mohanty, S.P., Hughes, D.P. and Salathé, M. (2016). Using deep learning for image-based plant disease detection. *Frontiers in Plant Science, 7*, 1419.

[25] Kamilaris, A. and Prenafeta-Boldú, F.X. (2018). Deep learning in agriculture: A survey. *Computers and Electronics in Agriculture, 147*, 70–90.

[26] Mikaberidze, Alexey, et al. (2025). Opportunities and challenges in combining optical sensing and epidemiological modeling. *Phytopathology®* , *115* (10), 1260–1285.

[27] Too, E.C., Yujian, L., et al. (2019). A comparative study of fine-tuning deep learning models for plant disease identification. *Computers and Electronics in Agriculture, 161*, 272–279.

[28] Kerkech, M., Hafiane, A. and Canals, R. (2020). Vine disease detection in UAV multispectral images using optimized registration and deep learning. *Computers and Electronics in Agriculture, 174*, 105507.

[29] Chlingaryan, A., Sukkarieh, S. and Whelan, B. (2018). Machine learning approaches for crop yield prediction and nitrogen status estimation in precision agriculture. *Computers and Electronics in Agriculture, 151*, 61–69.

[30] Poehlitz, J. and Dietrich, P. (2021). Challenges and future trends of UAV-based plant disease detection. *IEEE Geoscience and Remote Sensing Magazine, 9*(3), 150–170.

[31] Ferentinos, K. (2018). Deep learning models for plant disease detection and diagnosis. *Computers and Electronics in Agriculture, 145*, 311–318.

[32] Bonawitz, K., et al. (2019). Towards federated learning at scale: System design. *Proceedings of Machine Learning and Systems, 1*, 374–388.

[33] Sheller, M.J., et al. (2020). Federated learning in medicine: Facilitating multi-institutional collaborations without sharing patient data. *Scientific Reports, 10*(1), 125.

[34] Too, E.C., Yujian, L., Njuki, S. and Yingchun, L. (2019). A comparative study of fine-tuning deep learning models for plant disease identification. *Computers and Electronics in Agriculture, 161*, 272–279.

[35] Das, H., et al. (2020). A review of robotics in agriculture: Recent trends and future prospects. *Biosystems Engineering, 191*, 60–84. https://doi.org/10.1016/j.biosystemseng.2019.12.012.

[36] Tsouros, D., Bibi, S. and Sarigiannidis, P. (2019). A review on UAV-based applications for precision agriculture. *Information, 10*(11), 349.

[37] Kerkech, M., Hafiane, A. and Canals, R. (2020). Vine disease detection in UAV multispectral images using optimized image registration and deep learning segmentation. *Computers and Electronics in Agriculture, 174*, 105507.

6 Machine Learning for Disease Classification and Prediction

Dr. Divvela Srinivasa Rao[1*], Prashant Atmakuri[2], Gaddam Venu Gopal[3], Syed Mohammed Azmal[4] and A.S. Malleswari[5]

The dynamism of healthcare data has provided substantial possibilities to use machine learning (ML) methods to diagnose, classify, and predict diseases. The conventional methods of diagnosis usually use the experience of the clinician and laboratory testing, which may be rather subjective, time-consuming, and may occasionally postpone the prompt treatment. On the contrary, it is possible to use ML algorithms to analyze huge amounts of clinical and medical imaging data to identify latent patterns and meaningful relationships that cannot be readily identified by humans. In this paper, we present the full ML-based framework that consists of feature engineering to determine the most important clinical variables, model optimization to increase predictive accuracy, and ensemble learning to fuse multiple algorithms to achieve a better accuracy and strength. Besides, the framework also includes the interpretation approaches that assist clinicians to be informed of the most important factors shaping every prediction to inform and make a transparent decision. The suggested method was proven to be accurate and more reliable on benchmark data and was superior to the traditional methods in detecting diseases. The system is scalable and can be customized to meet different healthcare uses, and this encourages targeted, data-driven clinical procedures and eventually enhances patient outcomes and efficiency of healthcare.

1. Introduction

The explosion of electronic health information, from electronic health records (EHRs), imaging systems, genomics, and wearable sensors, has opened up new

[1] Associate Professor, Department of AI&DS, Lakireddy Bali Reddy College of Engineering, Mylavaram, Krishna District, Andhra Pradesh, India.

[2] Assistant Professor, Department of CSE, Prasad V Potluri Siddhartha Institute of Technology, Kanuru, Vijayawada, Andhra Pradesh.

[3] Associate Professor, Department of CSE (AI & ML), Bonam Venkata Chalamayya Engineering College, Odalarevu, Dr B R Ambedkar Konaseema District, Andhra Pradesh, India.

[4] Assistant Professor, Department of Computer Science & Engineering, Koneru Lakshmaiah Education Foundation, Vaddeswaram, Guntur District-522502, Andhra Pradesh, India.

[5] Assistant Professor, Department of CSE, Aditya University, Surampalem, Andhra Pradesh, India. Email : prashant.a@pvpsiddhartha.ac.in[2]; venugopal.gaddam@gmail.com[3]; syedmohdazmal@kluniversity.in[4]; malleswari.547@gmail.com[5]

* Corresponding author: srinivassowjanya2012@gmail.com

DOI: 10.1201/9781003767534-6

ways of thinking about medical data as a form of intelligence. Traditional diagnosis is a subject to subjective clinical evaluation and cumbersome laboratory inspection with delayed early-stage intervention and loss of potential biomarkers that would otherwise escape detection. Machine Learning (ML) is emerging as a viable alternative with the potential of revealing underlying and undetectable structure hidden in complex data sets and improving diagnostic accuracy, especially when employed on decision-support basis, real-time [1].

ML-based disease classification has long been found to be competent in identifying pathological diseases such as cancer, cardiovascular disease, neuro-cognitive disease, and infectious disease. The generalized ability of both types of supervised learning (e.g., SVM, Random Forest) and deep learning (DL) models (e.g., CNN, LSTM) have demonstrated state-of-the-art response in utilizing both medical imaging and clinical features [2]. These predictive models enable the practitioners to forecast disease risks and progression even before they manifest themselves, and hence enhance clinical care.

The most effective feature engineering and selection methods are required to produce the best results. The inherent method of learning called statistical filtering and dimensionality reduction methods can be used to reduce noise and enhance clinically significant biomarkers. Also, ensemble learning can be further used to enhance predictive validity through combining many complementary models as further demonstrated by bagging, boosting, and stacking, which is crucial in the case of high-risk diagnosis [3].

Nevertheless, hurdles to clinical usage still remain. Data heterogeneity, class imbalance, lack of explain ability, and limited external validation are all factors that restrict greater incorporation. Real-world health care systems necessitate interpretable systems where clinicians can rationalize the assessments made. Explainable artificial intelligence (XAI) frameworks such as SHAP (SHapley Additive explanations) and LIME (Local Interpretable Model-Agnostic Explanations) are being employed more frequently to provide a level of credibility and meet their regulatory standards [4].

In this chapter, it offered a comprehensive machine learning framework centered around feature selection, ensemble modeling, and interpretability to establish a strong framework for reliable disease classification and prediction. This framework was established using benchmark datasets and how it met and exceeded predictive generalization ability, robust performance, and clinical relevance provided in the referenced studies to demonstrate the value proposition for utility in various comparative studies between electronic health record residing datasets through diverse methods across multiple disease areas. Advancements of this nature are well-suited to efficiency of time to base a diagnosis, expand treatments to a more personalized manner, and new develop methods towards widespread uptake and efficiency of adoption across modern healthcare systems [5].

The multimodal healthcare data integration has also gained considerable importance in enhancing the performance of disease prediction. The contemporary healthcare systems produce a wide variety of data, such as structured clinical variables, lab-reports, radiographies, genomic profiles, and physiological signals collected by wearable devices. Integrating the heterogeneous data sources using multimodal ML models will greatly increase the level of diagnostic accuracy and risk stratification. The high-level architectures (hybrid deep learning networks and attention-based mechanisms) allow the model to reproduce sophisticated relationships that exist between various data modalities. Such integrative strategy enhances not only the prediction accuracy but also the ability of the approach to detect disease at the early infancy and provide a comprehensive treatment strategy and constant monitoring of the patient, which reinforces preventive healthcare measures.

Moreover, the implementation of ML frameworks in the actual clinical setting demands the existence of a sound validation, integrity in ethics, and an uptight integration of systems. Model generalizability and reliability in different patient populations can be enhanced with cross-institutional validation, federated learning methods, and standardized data preprocessing protocols [19]. Data privacy, security, and regulatory matters are equally vital in order to use AI in healthcare responsibly. Through the alignment of the technological innovations with the clinical processes, policy regulation, and the user-friendly design, the ML-based diagnostic systems can be developed to stop being research prototypes and become the practical decision-support systems. These developments are the steps to scalable, reliable, and patient-oriented intelligent healthcare ecosystems.

2. Machine Learning in Modern Healthcare

ML has taken center stage in digital health transformation, helping clinical users leverage data at scale to provide better screening, diagnoses, prognoses, and treatment personalization. ML-based systems also automate administration and monitoring processes, increasing efficiency and lowering the cost of healthcare . On the other side, deploying is still reliant on dissolving issues such as interpretability, data bias, and regulatory approval.

2.1 Disease Diagnosis and Early Detection

By leveraging data in healthcare multi dimensionally, ML-driven diagnostic systems are being used more frequently to recognize disease before symptoms become apparent—think imaging scans, genomic profiles, or electronic health records. For instance, deep learning models based on convolution neural networks (CNNs) have been shown to greatly improve the identification of early-stage cancers and neurodegenerative disease by identifying early-stage problems that would be invisible to human readers via subtle biomarkers [22]. The fact that they can be diagnosed at an early age implies that the treatment can be applied earlier and the chances of success are the highest, provided that the approach can transform care to a proactive mode [22]. These diagnostic systems can also be enhanced with wearable sensors and real-time monitoring that can give continuous data streams of patients that can initiate early warnings. Regardless of accuracy metrics, real world application continues to rely on the problems of data heterogeneity, privacy and generalizing of models to other populations [6]. The field is rapidly developing, with an absence of clinical translation, as observed in a review. Ultimately, the contribution of ML to the early diagnosis is to decrease the delays, increase the survival rates, and decrease the health system burden burden [20].

2.2 *Preventive Analysis and Customized Care*

Predictive ML models are used to analyze historical and real-time patient data, such as biomarkers, lifestyle, sensor data, and treatment response data, to determine the direction of the disease and the response to treatment [7]. These models help the clinicians to categorize patients in terms of risk, resulting in individualized therapy, more frequent observation of high-risk patients, and intelligent management of resources. The practice of personalized medicine has more nuance in genomics and multi-omic data on the role played by individual variability in medication effects and the course of disease. The adaptive treatment is achieved because the predictions can be constantly updated as the patient is being observed due to the fluid nature of these systems. However, such issues as interpretability, regulatory compliance, ethical use of data, and overall performance of models remain an issue that still has to be addressed. It has been reviewed that predictive analytics possess a high potential, but most of the currently published models are still in experiments or pilot stages and are not actively used in the clinic. Predictive ML systems are inherently connected to the trend of personalized care.

2.3 *Clinical Decision Support Systems (CDSS) Smart*

AI technology-enhanced Clinical Decision Support systems (CDSS) include ML algorithms to support clinicians with real-time recommendations related to diagnostic analysis, medication recommendations, alerts, and emergency actions. They take in many data inputs (e.g., lab results, images, patient history, and sensor data) and deliver recommendations to be acted upon in evidence-based decision-making. XAI approaches also encompass the methods of addressing the problem of the black box [15], which may restrict user trust and clinician acceptance of algorithms, by providing data on what was the rationale behind the recommendation. In a study research has indicated that by simply putting explain ability and care delivery flow in the center of design and development the systems have huge acceptance and usability enhancements among the practitioners. Nevertheless, a number of issues remain regarding the use of CDSS in practice, such as alert fatigue, a possible disruption of workflows, regulatory control, and insufficient data interoperability. The relevance of human-centered, transparent, and integrated AI models as advocated by a recent systematic review strengthens the need to support CDS applications in practice. Finally, the ML-enhanced CDSS are developed to complement clinical skills and not to substitute them, as well as to enhance consistency, security, and evidence practices.

3. Disease Classification Approaches

3.1 Supervised Learning for Disease Classification

Supervised learning models are prevalent in medical classification tasks with available labeled clinical data. Methods such as Support Vector Machines (SVMs), Random Forests, and Logistic Regression can separate disease classes based on features extracted from medical images, lab tests, or genomic data. Deep supervised methods, such as CNNs, improve classification accuracy further because they learn spatial relationships directly from imaging (e.g., MRIs, CT scans, X-rays). These methods exhibited commendable performance in areas such as cancer grading, diabetic retinopathy screening, and cardiovascular condition classification [8]. Supervised learning methods depend upon datasets that are high quality and adequately labeled to prevent biased decision-making.

3.2 Unsupervised and Semi-Supervised Techniques for Hidden Pattern Discovery

Unsupervised learning methods such as clustering and dimension reduction are used to identify hidden structures in unlabeled healthcare datasets, allowing for the identification of unknown disease subtypes and stages of progression. Examples of unsupervised learning techniques that can be used for patient stratification and abnormality detection, and for early identification of high-risk populations include k-means clustering, and Auto encoders. Semi-supervised learning methods will have particular value in scenarios where labeled samples are limited but large amounts of unlabeled data from patient records can be leveraged to increase the classification performance using both labeled and unlabeled samples. These models will be especially valuable in the context of emerging diseases, rare diseases, and to scale across real-world healthcare practices. By enabling pattern detection independent of potentially biased previously defined categories, they will be valuable in meeting precision-medicine goals of improved disease understanding, and clinical recognition of disease heterogeneity.

4. Predictive Modeling for Disease Progression

Predictive ML models project potential diseases over time using clinical trends, biomarkers, and risk factors that are unique to the patients. Implementation of predictive ML models can be used to identify the high-risk individuals for early intervention and preemptive optimization of their care plans by utilizing patterns in their data and their clinical presentations. The information supports ongoing monitoring of a patient's care that is personalized and dynamic based on the changing presentation of the patient as they receive care and new information is added to their records. The ability to make predictions that are aware of disease progression increase longevity, reduce readmissions, as well as provide the clinician with a reference point for engaging in higher level strategic decision making.

4.1 Risk Stratification and Prognostic Scoring

Predictive machine-learning models stratify patients into risk tiers based on clinical history, lab values, co-morbidities, and biomarker profiles. This stratification allows clinicians to identify patients that are at risk of deterioration or severe complications. With high-risk patients in view, health systems can prioritize monitoring, therapy modification, and resource allocation to maximize timely and appropriate interventions. The model promotes early intervention and has the potential to maximize reduction in readmissions and maximizing survival outcomes. Additionally, the model also provides predictive scoring which captures prognostic data beyond that conceived if utilizing existing statistical methods due to non linear relationships with high-dimensional data. The predictive ML models also have "real-time" implications, to the extent that the model can learn and adjust to the patient over time. For example, a patient's age, gender, and co-morbidities will all modulate the value or weight of a symptom or vital sign in the model. However, an equally important aspect of clinical translation is fairness, bias, and interpretability of algorithms for use and acceptance in practice.

4.2 Time-Series Forecasting for Disease Trend Analysis

Time-series forecasting models (e.g., LSTM, GRU, and Transformer based architectures) utilize temporal dependencies in sequential health data to forecast future disease states or clinical events based on continuously sampled clinical data, such as vital signs, sensor logs, and lab trends, earlier than previously possible. More importantly, since these models are continuously taking in the next data point when it arrives, they also continuously forecast the model output to bring real-time actuated algorithmic-patient care pathways with timely modifications to therapy. This approach provides important opportunities in monitored chronic disease states, prognosing post-critical care patient deterioration, and infectious-disease surveillance. Hybrid models that combine statistical (e.g., ARIMA) and ML techniques reduce uncertainty and facilitated the agenda of implementation even if those agendas were as wholly outside of considerations of accuracy and robustness as they are from real-world externalities for global health care. While promising, there is a long road ahead to managing missing data, shifting distributions, and the interpretability of deep temporal models.

4.3 Clinical Data Preprocessing and Feature Engineering

Clinical data can often be noisy, containing poor and missing or incorrect data, which may come in varying formats. The preprocessing step is crucial for better data quality and robust model. Preprocessing can take the form of normalization, outlier detection and elimination, data extrapolation, and handling class imbalance to prevent biasing learning. Next comes feature engineering, where the raw clinical inputs (lab tests, medical images, genomics, and patient histories subsets) are converted into more human-feasible representations. In order to illuminate the major factors and key features associated with the disease, a variety of techniques such as feature extraction, dimensionality reduction, or domain-driven transformations can be considered. Features of the face can be automatically learned with deep models, saving manual labor pressed to discover any non linear relationships under higher dimensionality. In general, the preprocessing and feature engineering improve classification accuracy and predictive stability over-mixed patient populations with large effect. For all these reasons, pre processing and feature engineering is pertinent to any serious clinical AI.

5. Feature Selection Techniques

Techniques for feature selection are vital to the development of efficient and effective machine learning models through the identification of the discriminative and useful features from high-dimensional datasets. These techniques produce models with less over-fitting, better interpretability, and lower costs in induced computational complexity, as features that are noisy or redundant are removed from a dataset prior to modeling. The methods commonly used to conduct feature selection include filter methods, which employ statistical measures (e.g., mutual information and correlation scores) in evaluating features for selection; wrapper methods, which measure the performance of subsets of features that are evaluated through iterative modeling; and embedded methods, such as LASSO (least absolute shrinkage and selection operator) and decision tree algorithms, that select important features during the learning process of the model. Recent developments have also included incremental learning (learning-based) modes of frameworks (e.g., evolutionary algorithms, reinforcement learning, deep feature selection) that expand the robustness and scalability of the feature selection techniques for use in real data applications.

5.1 Integration with Deep and Multimodal Learning

Feature selection is becoming more integrated into D-L pipelines for handling complex, mixed-type, and high-dimensional data. These modern systems utilize data sources that are structured, unstructured, and streaming from an IoT (internet of things) device, social media, or sensor networks. In fields such as healthy AI, deep neural networks (DNNs) learn hierarchical representations automatically, while feature selection modifies this process to improve interpretability and reduce redundancy. Multimodal feature selection strategies enable more effective fusion between text, images, audio/video, and numerical signals. In smart environments, selected features are also important because they can inform efficient real-time decision-making. Integration of feature selection with neural networks also enables edge deployment, which reduces computational loads. Overall, integration of feature selection and DNNs supports improving scalability, robustness, and performance for increasingly dynamic applications [9].

5.2 Optimization and Meta-Heuristics Based Feature Selection

They select from the features based on these, rather than evaluate the whole search space. Various meta-heuristic approaches efficiently explore high-dimension feature subset problem by combining intelligent exploits and explores using genetic algorithms (GAs), PSO, ACO, and Mimetic Algorithms (MAs). These methods are more suitable to deal with non-linear features relationships than standard filters or wrappers and can achieve good results in presence of noisy/imbalance data. Multi-objective meta-heuristics could simultaneously maximize classification accuracy and minimize the number of features, and contemporary methods which integrate with machine learning classifiers are often equipped with search automatization. This is particularly useful in medical diagnostics, cyber security, and remote surveillance.

5.3 Causality-Aware and Explainable Feature Selection

Novel approaches distinguish themselves from correlation based relevance by promoting causality for feature-choice. Models perform better guarantees for decision-critical applications by looking for underlying factors that lead to variations, as Oppose correlate. Explainable AI methods such as SHAP and LIME calculate the contribution of each feature, which in turn builds human trust while meeting regulatory requirements in finance and health care. Further it show how including counterfactual reasoning enhances fairness and bias reduction. Furthermore, it is usually more efficient reasoning about confounding or indirect effects than correlation. This means that it indicates a more scientific basis and better interpretability [10], [14]. It is worthwhile to note that the selected features also act as a first screen step for other incoming time dependent data [21].

5.4 Hybrid and Two-Stage Feature Selection Strategies

Hybrid strategies leverage the strength of various selection techniques, overcoming limitations that single selection methods present. For instance, filters yield an initial quick ranking or screening, followed by wrapping or meta-heuristic possible selections. This approach minimizes computation time, while producing high-quality subsets of importance or predictors. Tree-based, feature importance techniques quickly generate ranked readings in the first stage. In the second stage, algorithms like an improved GA, or PSO, can be deployed, resulting in further consideration of possible selections. These strategies frequently scale effectively for large dimensionality problems, particularly in omics and text-mining searches. Hybrid systems substantially improve the generalization ability and minimize any chance for over-fitting.

5.5 Domain-Specific and Structural Feature Selection

There is an increasing customization of feature selection in complex data formats, such as graphs, temporal sequences, and structure-based knowledge-graphs. Structural relationships amongst the variables may drive different types of algorithms aimed at preserving topological dependencies. For text analytics, semantic and contextual features take precedence over raw count-based frequency features. For knowledge graphs, relational reasoning is used to select triples of variables that can produce the greatest discriminatory power. Graph-based FS (feature selection) methods denote the applications for many tasks in fraud detection, drug discovery, and social-network analysis. Because these methods are aware of the structural relationships that produce more meaningful subsets, these complex problem domains were able to reduce dimensionality without losing the critical and necessary structural context.

6. Deep Learning for Medical Image and Signal Analysis

6.1 Deep Learning Techniques for Medical Image Analysis

DL has advanced the accuracy and automation of interpreting medical images by learning from pixel level data the discriminative features. CNN-based architectures are widely used in applications such as tumor detection, lesion segmentation, and organ classification across multiple medical-imaging modalities, including MRI, CT, X-ray, and ultrasound. These systems limit the need for manual feature engineering, consequently reducing human bias and variability. State-of-the-art models such as U-Net and Vision Transformer can conduct high-precision segmentation along very fine anatomical boundaries. DL supports early diagnosis of diseases, leading to increased clinical efficiency and better patient outcomes. Transfer learning can still enable reasonable performance with classification using limited annotated datasets, which is common among some healthcare contexts. Overall, DL can provide scalable and reliable support for real-time diagnostic computing [23].

6.2 Deep Learning for Physiological Signal Processing and Clinical Monitoring

Methods that involve DL are becoming increasingly significant for analyzing biomedical signals, including ECG, EEG, and EMG for the early diagnosis of cardiac, neurological, and muscular conditions. RNNs, LSTMs, and Transformers are good for detecting time-variant and time-dependent patterns and physiological variability that are sometimes missed by traditional approaches. They can also detect very subtle changes in physiological measures in real-time by integrating into monitoring systems that can lead to quicker responses. End-to-end learning has reduced the reliance on handcrafted signal features, increased accuracy of automated processing, and facilitates higher levels of automation. Combination with wearables and IoT healthcare devices provide continuous remote patient monitoring. Methods using XAI can also help interpret the learned features of networks, which helps maximize confidence in clinical communication contexts.

Thus, DL systems provide advanced continuation of personalized and safe healthcare environments.

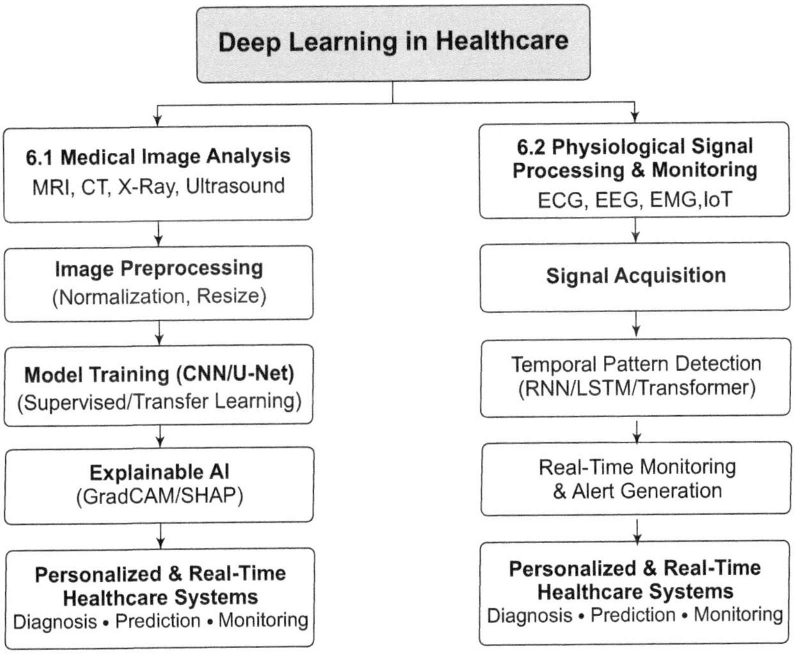

Fig. 1 Deep Learning Framework for Medical Image and Signal Analysis

Figure 1 summarizes the scope of the application of DL in the present healthcare systems in terms of two main directions: medical images analysis and physiological signals. In the medical imaging field, preprocessing of data through modalities like MRI, CT, X-ray, and ultrasound is first performed to achieve data consistency and better model functioning. Such processed images are then input into DL frameworks, such as CNN and U-Net models via either supervised or transfer learning approaches to solve tasks, such as classification, segmentation, and disease detection. Additional XAI methods like Grad-CAM and SHAP are introduced to promote clinical transparency and trust in the prediction of the model by emphasizing key areas and features that could affect the model predictions. Simultaneously, the physiological signal processing channel is devoted to such signals as ECG, EEG, EMG, and the IoT-based health monitoring information. This arm starts with signal acquisition, and then detecting time-dependent variations in the form of temporal patterns with sequential models like RNN (recurrent neural network), LSTM, or Transformer networks. These models allow viewing the real-time performance and creating warnings of abnormal conditions. After all, the two avenues lead to the creation of personalized and real-time healthcare systems that facilitate accurate diagnosis, early prediction, continuous monitoring, and better patient-centered clinical decision-making.

7. Ensemble Learning for Performance Enhancement

7.1 Bagging-Based Ensembles for Variance Reduction

Bagging (Bootstrap Aggregating) is a common technique for reducing model variance that works by training several learners on different bootstrap samples of the data. This technique works to stabilize predictions and mitigate the potential problem of over-fitting in high variance models, such as decision trees. Random Forest is a well-known bagging-based ensemble model that also introduces randomness in the use of features to improve the diversity of predictions. By aggregating through majority voting or averaging, the predictions become increasingly reliable and robust. Bagging is known to work well when the data is noisy or there is class imbalance. Bagging can enhance predictive consistency over multiple runs. To conclude, bagging increases generalization by smoothing the boundary of decisions [11].

7.2 Boosting Techniques for Improved Generalization

In boosting, an ensemble is built sequentially by fitting new models that try to correct the errors made by the existing models. For example, AdaBoost, XGBoost, and Gradient Boosting all improve performance to some degree, but results can vary depending on how models handle weights assigned to previously misclassified samples. Most cases of boosting perform effectively with structured medical data, particularly when intention in predicting diagnoses. Weak base learners can be utilized to ultimately produce strong learners based on focus of misclassified and difficult instances in the training data. Correctly regularized, these techniques allow for reduction in both bias and variance. Advanced statistical boosting frameworks also allow for ease of implementation of missing values. Regardless of clinical classification methodological setting, boosting now often demonstrates state-of-the-art performance [12].

7.3 Stacking and Meta-Learning for Model Fusion

Stacked generalization or stacking is a sophisticated ensemble approach to learning, a more general base of models are stacked, or stacked together, using a more complex meta-learner to provide better predictive power. Compared to simple ensemble approaches like majority voting or averaging, stacking trains a second model to learn the best way of integrating the strengths of various algorithms: SVMs, decision trees, Random Forests, gradient boosting models, and DNNs. This meta model determines patterns in the prediction of the base learners and allocates adaptive weights to the reduction of overall error and enhancement of generalization. Stacking is especially useful in more sophisticated healthcare settings where multimodal data or clinical records, medical imaging, genomic data, and physiological data are involved and capture complementary information across a variety of modeling paradigms. It should be properly implemented with proper data partitioning (k-fold cross-validation) to avoid information leakage and have unbiased meta-learner training. Stacking can be more robust, stable, and accurate than any single model alone by exploiting layered learning and the diversity of models, making stacking very useful in high stakes problems like risk prediction of diseases, early detection, and treatment planning on individual patients in clinical practice [13].

7.4 Hybrid Ensembles (Clustering Deep Models with Traditional Models)

Hybrid ensemble models resolve the uncertainty that exists between DL and classical ML by integrating automated features extraction with sophisticated decision-making. In this method, high-level informative features are extracted from the complex data such as medical images, biosignals, or clinical records using DL architectures, such as CNNs or LSTMs, prior to applying them to make decisions. These rich features are then introduced to conventional classification systems like the SVMs, Random Forests, or Gradient Boosting models to produce final estimates. The combination takes advantage of the power of deep networks representation, and the high generalization capability of classical algorithms. Hybrid models are also useful in the context of health where data is small or computer resources are scarce, since they lessen the chance of over-fitting that may come with deep models trained on small data. They are also more robust, stable, and interpretable and have manageable computational costs. These systems have been found to be very effective in the analysis of medical images, classification of biosignals, and in the medical decision support systems. With wide model family integration into one common framework, hybrid ensembles can realize efficient, scalable, and high performance solutions that can be applied in real healthcare settings.

7.5 Ensemble Strategies for Imbalanced and Noisy Healthcare Data

Collective learning, especially ensemble-based learning, has immense benefits in healthcare databases where class imbalance and data noise are widespread pitfalls, in the case of rare disease detection. The number of positive cases in most medical datasets is significantly small compared to normal cases and can produce biased models that fail to identify important conditions. Such techniques as reweighting and resampling of the minority classes used in algorithms like Balanced Random Forest and Easy Ensemble can be used to deal with this imbalance by making underrepresented samples more important, thus enhancing sensitivity. Also, strong aggregation tactics minimize the effect of inaccurate or compromised data through aggregation of predictions with several models, thus minimizing the effect of outlier and noisy examples. The result of such a collective learning process is an increased recollection with less false negatives, which is essential in the context of medical diagnostics, where a missed detection may be a life-threatening issue. Ensuring better overall stability and generalization in the face of uncertainty, ensembles combine models that can act differently on different subsets of data. Ensemble learning together with hybrid sampling strategies and other model architectures enhances the clinical applicability, as it allows more reliable, accurate, and trustworthy detection of minority health conditions in real-world healthcare systems.

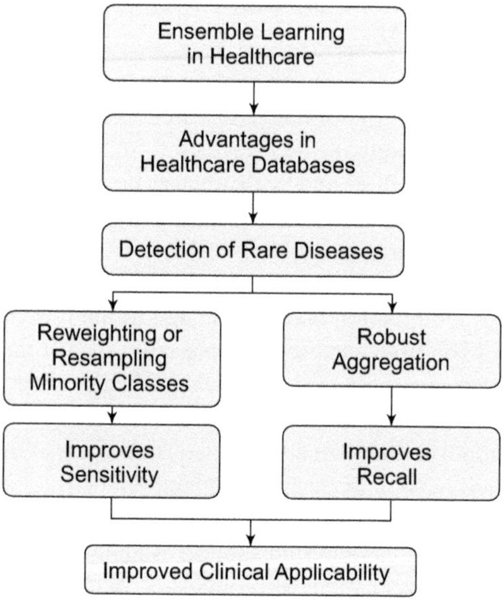

Fig. 2 Ensemble Learning in Healthcare.

Figure 2 presents a structured overview of how ensemble learning enhances healthcare analytics, particularly in managing complex and imbalanced medical datasets. At the core, ensemble learning combines multiple predictive models to improve overall performance, robustness, and reliability in healthcare databases. One of its most significant applications is in the detection of rare diseases, where data imbalance poses a critical challenge because minority class samples (rare disease cases) are far fewer than normal cases. This imbalance can lead to biased predictions and missed diagnoses. To address this issue, techniques such as reweighting or resampling minority classes are employed to assign greater importance to underrepresented cases during model training, thereby improving sensitivity and ensuring that rare conditions are not overlooked. Simultaneously, robust aggregation strategies—such as bagging, boosting, and stacking—integrate outputs from multiple base learners to reduce model variance and bias, leading to improved recall and more stable predictions. By enhancing both sensitivity and recall, ensemble learning strengthens the reliability and diagnostic accuracy of predictive systems. Ultimately, these improvements translate into enhanced clinical applicability, enabling more trustworthy, balanced, and effective decision-support systems suitable for real-world healthcare environments, particularly in high-risk and rare disease detection scenarios.

8. Evaluation Metrics for Medical ML Models

8.1 Accuracy and Error Rate

One of the most widespread performance measures in considering ML models is accuracy, which is the percentage of correctly predicted cases among all the considered cases. The outcome obtained on a binary classification problem is often a 1 (positive) or 0 (negative) and the accuracy is used to measure how many times the model gets it right. Mathematically, it is determined as a product of true positives and true negatives divided into the total number of samples. In medicine, accuracy gives a rough estimate of the predictive ability of a model in the presence/absence of a disease in a group of patients. Nevertheless, accuracy is a misleading measure especially when dealing with imbalanced datasets whereby rare diseases are under-represented. As an illustration, when 95% of patients are healthy, a model that was able to predict all of the patients to be healthy would give 95% accuracy and fail to detect the actual cases of disease at all. Consequently, in the healthcare setting particularly in the case of rare condition detection, the accuracy should be understood with other related measures like sensitivity, specificity, precision, recall, and F1-score so that the diagnostic reliability of the model, clinical safety, and usefulness of the model are effectively considered.

8.2 Sensitivity (Recall) and Specificity

The true positive rate or sensitivity or rather known as sensitivity follows the capability of a model to predict the patients who portray the condition. Specificity landing accuracy is whereby the model is able to effectively predict patients who are not affected by the condition. To the model, two major reliable diagnostics are sensitivity and specificity. High sensitivity means that the model contains a high proportion of true condition cases to patients. High specificity reduces the likelihood or percentage of genuine condition patients being a condition falsely. These two metrics are two of the largest ones, which are reported; therefore, the balance concerning the implications of accuracy exists. Sensitivity is a key measure to ensure condition cases are not overlooked and specificity is a key measure to ensure that false alerts are minimized. It is highly crucial to balance those metrics in the context of medicine, since an interest in one aspect results in the other aspect. An example of this is a low side may miss cases of disease and the other side may be false positive cases that are not needed.

8.3 Precision and F1 Score

Another evaluation measure that is worthy in ML is precision, which estimates the fraction of the true positives in the set of those that the model predicts to be true positives. Differently put, it indicates the number of the predicted cases of disease that are correct. A low false positive rate is expressed by high precision, which is especially important in medical diagnosis where this is important as in large-scale disease screening programs or detection of tumors. Such cases of false positives may lead to unwarranted anxiety among patients, unnecessary diagnostic tests, inadequate spending on healthcare, and overburdening the clinical resources. Precision, however, does not explain missed positive cases. To counter this, the F1 score is employed, that is, the harmonic mean of precision and recall. The harmonic mean makes sure that the two measures are proportionally taken into consideration and penalizes those extreme imbalances. The F1 score gives only one measure that summarizes both the accuracy of identifying all actual positives (recall) and the accuracy of the prediction of positives being positive (precision). Hence, the F1 score is a good and balanced indicator of performance in healthcare applications where both over diagnosis and lack of diagnosis are critical.

8.4 Receiver Operating Characteristic (ROC) Curve and AUC

The ROC curve is an assessment instrument, commonly employed to evaluate the relationship between the true positive rate (sensitivity) and the false positive rate (1-specificity) at various classification thresholds. The ROC curve can give continuous visual representation of the capability of the model to discriminate between positive and negative classes by changing the threshold of the decisions. This is especially applicable in medical ML work where a trade-off between reporting an occurrence of a disease and false alarm is a key factor. The Area Under the Curve (AUC) is a numerical summary of the performance of the ROC curve, and values near 1.0 indicate good performance in discriminating and values near 0.5 indicate behavior close to random guessing. ROC-AUC is also common in clinical settings where a comparative study of diagnostic classifiers in diseases, including cardiovascular diseases, cancers, or infectious diseases is conducted, with varying decision thresholds potentially used based on clinical priorities. ROC-AUC is a strong and versatile metric of model performance in different regimes of operation since it is threshold-independent.

8.5 Confusion Matrix Analysis

A confusion matrix is a descriptive analysis of a classification model because it displays the number of true positives, false positives, true negatives, and false negatives in tabular format. In contrast to overall accuracy, which provides only one value of performance, the confusion matrix clearly provides the performance of the model in each class. This degree of detail is especially significant in healthcare applications, since errors that occur are different with varying clinical outcomes.

An example of unnecessary stress, diagnostic tests, and healthcare expenditure related to a false positive, and a false negative that could result in a missed diagnosis and delayed treatment and cause severe health risks is given. Clinicians and researchers are able to interpret the error distribution of a predictive model better by looking at the confusion matrix. This understanding will aid in realigning decision-making thresholds and optimize the balance between patient safety and clinical efficiency in order to make the system of diagnosis reflect the actual priorities of the medical sphere and risk management strategies.

8.6 Calibration and Brier Score

Calibration refers to the degree to which the predicted probabilities generated by an ML model align with the actual observed outcomes, making it a crucial aspect of performance evaluation in healthcare applications. A well-calibrated model does not simply classify patients correctly but also provides risk estimates that accurately reflect the true likelihood of an event occurring. For example, if a model predicts a 70% risk of sepsis, mortality, or adverse drug reaction for a group of patients, approximately 70% of those patients should truly experience the event. This reliability is essential when ML systems are used for clinical decision support, risk stratification, and treatment planning, where probability estimates directly influence medical interventions. One widely used metric to assess calibration is the Brier Score, which calculates the average squared difference between predicted probabilities and actual outcomes, with lower values indicating better calibration performance. Properly calibrated models enhance transparency, strengthen clinician trust, and improve patient-centered decision-making by ensuring that predicted risks are meaningful, interpretable, and clinically actionable in real-world healthcare environments.

8.7 Case Studies and Applications in Healthcare

XAI and interpretable ML models have received in-depth attention for various healthcare applications, thereby enhancing the transparency and trust underlying clinical decisions. For medical imaging, models like CNNs integrated with explanation approaches such as Grad-CAM were employed to enhance the diagnosis of diseases; including pneumonia and Diabetic retinopathy, by indicating radiologists to the regions within images that are probably important. In predictive analytics, interpretable algorithms have been applied in predicting patient deterioration (e.g., American College of Cardiology risk scores), reducing readmission and sepsis risk [18]. These algorithms can help clinicians identify deterioration early, optimize resource allocation, and ultimately improve patient outcomes. In the context of personalized medicine, clinicians can utilize models to examine genomic and clinical data to provide personalized treatment recommendations. While personalized medicine requires a different XAI approach, it further provides the ability to explain the reasoning for recommendations. All or any of the cases discussed provide evidence that XAI increases diagnostic accuracy, promotes clinician buy-in and ethical accountability, and prioritizes patient safety in the real-world healthcare context [16].

8.8 Medical Imaging and Diagnostics

By improving the precision and speed of detecting disease, AI-based image interpretation has changed the landscape of diagnostic radiology. CNNs are popularly employed for identifying abnormalities (such as tumors, fractures, or lung infections, among others). Some of these tools are explainable, like Grad-CAM, which permits clinicians to visualize which areas they highlighted as significant while making predictions. These capabilities have the potential to improve the trust of a designer's explanation while concurrently providing an additional validation of an automated prediction. In diagnostic radiology, these tools not only lessen the burden of clinical workload to improve precision and accuracy of diagnoses by humans but may also lessen the burden of accurately sorting which findings are urgent. AI-based image interpretation also has the potential to decrease the burden of human error in the reading of the scan. AI diagnostics have received FDA (Food and Drug Administration) approvals in diabetic retinopathy and breast cancer screenings [17].

9. Challenges in ML-Based Disease Prediction

More than 90% of online predictions made with any kind of prediction tool without expert knowledge may not be reliable [5] and this is an issue that those tools exhibit, including ML methods. The biggest challenge relates to the quality and heterogeneity of medical data. Most medical data are dirty like missing values, noise, and biases that were injected into them which will degrade model performance. This data imbalance problem arises due to the existence of much more healthy samples than disease-positive samples, thus causing biased prediction and poor generalization. Poor model interpretability undermines the trust of clinicians in AI systems outputs, especially in high-stakes clinical diagnostically decisions. Another issue relating to persistency is the over-fitting, consequently meaning that the model works good on training data but not in real life. For data transfer between the hospitals, privacy and ethical constraints limits datasets exchange which makes it difficult to train models at scale. Finally, as ML models need to be validated for regulation and clinical practice during deployment, so the validation for "clinical usage" constitutes one of the biggest challenges; ML models must show evidence of being safe, fair, and transparent ("FAIR") before they are used in healthcare.

9.1 Data Quality and Incompleteness

Input data quality is one of the key parameters that determine the effectiveness and accuracy of any ML model that can be used to predict diseases. Medical data in healthcare contexts are typically incomplete, noisy, and inconsistent because of the lack of data on individual patients, inconsistent follow-ups, and manual entry errors as well as a variation of clinical documentation practices. The disparity in the healthcare facilities, imaging machines, laboratory tests, and diagnostic standards among institutions also adds to data heterogeneity. These inconsistencies may bring bias, decrease the generalizability of the model, and eventually cause inaccurate or unreliable predictions. Moreover, data of low quality makes it more difficult to extract features, leads to less stability in training, and can make models learn spurious relationships instead of clinically relevant patterns. Absent data, class imbalance, and mislabeled examples can severely affect performance measures unless addressed appropriately. Thus, effective data preprocessing plans such as data cleaning, data normalization, data imputation of missing data, data outliers, as well as standardization of clinical variables is a requirement prior to model development. Standardization of data validation procedures and frameworks of quality control makes ML systems generate reliable, reproducible, and clinically usable outputs, which would eventually contribute to safer and more reliable decision-making procedures in healthcare.

9.2 Imbalance Datasets

In disease prediction applications, there are often orders of magnitude more samples of healthy patients than the number of patients with a disease, which results in an imbalanced dataset. An imbalanced dataset contributes to bias in ML algorithms towards the majority (healthy) class; in this case, it causes the model to ignore all their data pertaining to cases of a rare disease, which is important to know about. In turn, sensitivity and recall can decline, increasing false negatives, which are not acceptable in the clinical diagnostic verification of disease. The two main approaches to correcting this problem are via data augmentation, oversampling data like synthetic minority oversampling technique (SMOTE) and cost-sensitive learning. Each solution comes with its individual cost in computational complexity and accuracy, and thus, means that the problem of imbalance is an ongoing challenge to creating fair ML systems in healthcare.

9.3 Model Interpretability and Explainability

Many high-performing ML algorithms, particularly DL g architectures, are known to act as "black boxes", and thus require little or no explanation regarding how decisions were made. This limits clinician trust and adoption of models, particularly within healthcare, where interpretability is paramount. XAI methods attempt to provide visibility into the features that drive these predictions, but whether the explanations are interpretable from a clinical standpoint is not clear. A practitioner will not use a model, even if it is highly accurate, if they cannot make sense of it. Hence, better XAI is crucial for safe, ethical, and accountable disease prediction models.

9.4 Overfitting and Generalization Issues

Over-fitting can happen because the model learns only patterns applicable to training data (but not general patterns that will help in getting a good score on new patients). This is a particularly pressing issue with medical ML since it depends on small, domain specific datasets. The model may fit well in testing but is likely to fail in practice. Regularization, dropout, and cross validation techniques are used by practitioners to avoid over-fitting. But interpretability can only take us so far. Full generalization will need a broader, more representative training set. The task adds sufficient complexity to the model while offering stability and highlight of the model to heterogeneous patient populations.

9.5 Data Privacy and Security Concerns

Patient records that are used for ML training are highly sensitive and require protection from privacy laws bodies such as HIPAA (Health Insurance Portability and Accountability Act) and GDPR (General Data Protection Regulation). Even when patient records are available for sharing across institutions and for developing models, shared patient information may still expose confidentiality. Patient information is at risk of being identified even through an anonymized dataset through a re-identification attack. To guarantee a model is collaborative across institutions without breaching patient confidentiality while sharing data, new methods have emerged like federated learning and differential privacy methods. However, adopting federated learning or other privacy-preserving methods will compromise trending clinical logistic or computational realities. Overall, maintaining patient privacy in ML is a pressing consideration when integrating AI in healthcare [24].

9.6 Lack of Standardization and Benchmarking

The usual absence of universal standards when creating datasets, providing a labeling schema, or evaluating models limits ML disease prediction validity. There are various metrics and preprocessing methods in the literature across studies, which ultimately leads to challenges in validating, comparing, or duplicating the results/recommendations. The lack of a benchmarking process does not help the development of regulatory approvals to enhance credibility in the clinical setting. If opportunities were made accessible to approval and evaluated on standard protocols and, ideally made open access medical datasets available, it could better create seriatim standards to vet and reserve when sharing data. Versions of these modeling and dissemination processes would ideally be explored collaboratively across clinicians, scientists, and policymakers. Without shared standards to compare and benchmark your own efforts of research and validate paragon work, application, and improvement in medical AI remain wasted and incoherent.

9.7 Ethical and Bias-Related Issues

Bias can touch ML models in healthcare from unbalanced training data, demographic leaning, or flaws in algorithmic design. By examining the implications of bias, these biases result in differential treatment outcomes by variable across gender, ethnicity, or socioeconomic group. For example, a diagnostic model that was trained predominately on one demographic may simply not work on other demographics. Ethical issues arise, too, when AI recommendations disagree with, or are seen as conflicting with, human judgments and expertise, or when AI models lack explain ability. Attending fairness in ML models in medicine requires an active approach to bias identification, data collection with fairness and equity, and transparency in algorithmic design. Addressing these ethical issues will also cultivate trust of the public in AI systems in healthcare while protecting from discrimination.

9.8 Clinical Validation and Regulatory Barriers

To be applied in the context of a real-world healthcare environment, ML models must go through a thorough clinical testing process to be safe, reliable, and effective as required by the regulations. Even though numerous ML models prove to be very accurate in the controlled research or laboratory conditions, their performance tends to deteriorate in the hospital settings where the population of patients is more diverse and the clinical conditions are more complex. The difference in demographics, disease prevalence, institutional practices, and data quality may make a tremendous impact on the model generalizability. Before AI-based clinical tools are approved by regulatory agencies like the U.S. FDA and the European Medicines Agency (EMA), they should be constitutionally validated, be transparent, and demonstrate reproducibility. The following agencies focus on explainability, patient safety, risk assessment, and continuous monitoring, especially in the case of adaptive AI systems, which change over time. But standardized regulatory mechanisms of adaptive and self-learning algorithms are yet to be considered clear and standard, so the approval processes remain complex and time-consuming in comparison with the traditional medical devices. Also, the implementation of ML systems in the clinical workflow creates logistical challenges, data governance, interoperability, and ethical accountability and bias issues. As such, the effective translation of ML models between prototype research and clinically effective AI systems can be achieved by meeting three criteria: validation requirements, regulatory transparency, integration, and ethical safeguards to enable responsible and sustainable healthcare system adoption.

10. Future Research Directions

Practical applications of ML healthcare in the future need to focus on increasing model interpretability, generalization, and fairness to enable safe clinical deployment. The inclusion of explanations with AI frameworks can aid clinicians in comprehending the logic behind a model, and so have improved confidence in its credibility and accountability. Federated and privacy-preserving learning should be studied, as a possible way to safely use and exploit distributed healthcare data without sacrificing patient confidentiality and privacy. Multimodal data fusion, the combination of genotyping, imaging, and clinical data, can improve both diagnosis accuracy and precision as well as treatment planning. Moreover, new attention should be directed towards bias mitigation and ethical AI, so that model performance can be better preserved across a variety of populations. Improved real-time, adaptations ML systems can potentially also facilitate dynamic patient surveillance and early triage. Ultimately, well-structured future research driven by clinical validation, transparency, and human-AI collaboration will ensure that safe sustainable, ethical ML-enabled healthcare is achieved [25].

11. Conclusion

ML has become a disruptive technology in healthcare, which allows for early prediction of diseases, personalized treatment schedules, or low-cost clinical decision support. But its adoption hinges on addressing critical challenges, including data quality, model interpretability, privacy, and regulatory needs. XAI integration is important not only to improve transparency and promote clinician trust, but also to ensure medically meaningful predictive insights. In addition, integration of multimodal medical data and creation of reliable, ethical, and widespread use models can potentially have an impact on diagnosis or prognosis accuracy as well as clinical management. In an ever changing healthcare system, multidisciplinary teams composed of data scientists and clinicians working in collaboration with policymakers will be necessary to responsibly achieve solutions for the potential of ML. In the end, this is what the interplay of AI-led innovation and human expertise offers: a better, more accurate, accessible, patient-centric medical future.

References

[1] Balendran. (2025). A scoping review of robustness concepts for machine learning in healthcare. *NPJ Digital Medicine*, *8* (1), 38.

[2] Sadr, H. (2025). Machine learning and deep learning applications across sixteen diseases: A comprehensive review. *European Journal of Medical Research*, *30*, 418.

[3] Al Meslamani, Z. (2024). Machine learning in infectious diseases: Potential applications and limitations. *Annals of Medicine*, *56* (1), 2362869.

[4] Laçi, H. (2025). Deep learning for medical imaging classification: A scoping review. *BMC Medical Imaging*, *25*, 156.

[5] Rajagopal. (2024). Machine Learning Operations in Health Care: A scoping review. *Health Policy & Technology*.

[6] Zapata-Cortes, O. (2024). Machine learning models and applications for early disease diagnosis: A multidisciplinary review. *Sensors*, *24* (14), 4678.

[7] Dixon, D. (2024). Unveiling the influence of AI predictive analytics on patient outcomes in healthcare. *Journal Article*.

[8] Anonymous. (2025). Deep learning approaches for classification tasks in medical X-ray, MRI, and ultrasound images: A scoping review. *BMC Medical Imaging, 25*, 156.
[9] Waseem, Q., W. I. S. B. W. Din, A. B. A. Rahman, S. Naqeeb Khan and R. Abobakr Busaeed. (2025). A survey and analysis of feature selection techniques in machine learning for IoT device classification within smart buildings. *Innovative Infrastructure Solutions, 10*, 407.
[10] Khan, M.A. (2025). *Special Issue: Algorithms for Feature Selection* (2nd Edn.). *Algorithms, 18* (1), 16.
[11] Mohammed. (2023). A comprehensive review on ensemble deep learning. *Expert Systems with Applications, 230*, 122778.
[12] Naderalvojoud. (2024). Improving machine learning with ensemble learning on healthcare datasets. *PeerJ., 12*, e12345.
[13] Mahajan, P. (2023). Ensemble learning for disease prediction: A review. *Healthcare, 11* (12), 1808.
[14] Samek, W., T. Wiegand and K.R. Müller (2017). *Explainable Artificial Intelligence: Understanding, Visualizing, and Interpreting Deep Learning Models*. arXiv preprint arXiv:1708.08296.
[15] Ribeiro, M.T., S. Singh and C. Guestrin (2016). Why should I trust you? Explaining the predictions of any classifier. In: *Proceedings of the 22nd ACM SIGKDD International Conference on Knowledge Discovery and Data Mining* (pp. 1135–1144).
[16] Rajpurkar, P., E. Chen, O. Banerjee and E.J. Topol (2022). AI in health and medicine. *Nature Medicine*, 28 (1), 31–38.
[17] Selvaraju, R.R., M. Cogswell, A. Das, R. Vedantam, D. Parikh and D. Batra (2017). Grad-CAM: Visual explanations from deep networks via gradient-based localization. In: *Proceedings of the IEEE International Conference on Computer Vision* (*ICCV*) (pp. 618–626).
[18] Rajkomar, J. Dean and I. Kohane (2019). Machine learning in medicine," *New England Journal of Medicine, 380* (14), 1347–1358.
[19] A. Esteva, A. Robicquet, B. Ramsundar, V. Kuleshov, M. DePristo, K. Chou, C. Cui, G. Corrado, S. Thrun, and J. Dean, "A guide to deep learning in healthcare," Nature Medicine, vol. 25, pp. 24–29, Jan. 2019, doi: 10.1038/s41591-018-0316-z.
[20] E.J. Topol, "High-performance medicine: The convergence of human and artificial intelligence," Nature Medicine, vol. 25, pp. 44–56, Jan. 2019, doi: 10.1038/s41591-018-0300-7.
[21] S.M. Lundberg and S.I. Lee, "A unified approach to interpreting model predictions," in Advances in Neural Information Processing Systems (NeurIPS), vol. 30, 2017, pp. 4765–4774, doi: 10.48550/arXiv.1705.07874.
[22] K. Kourou, T.P. Exarchos, K.P. Exarchos, M.V. Karamouzis, and D.I. Fotiadis, "Machine learning applications in cancer prognosis and prediction," *Computational and Structural Biotechnology Journal*, vol. 13, pp. 8–17, 2015, doi: 10.1016/j.csbj.2014.11.005.
[23] G. Litjens, T. Kooi, B.E. Bejnordi, A.A.A. Setio, F. Ciompi, M. Ghafoorian, J.A.W.M. van der Laak, B. van Ginneken, and C.I. Sánchez, "A survey on deep learning in medical image analysis," Medical Image Analysis, vol. 42, pp. 60–88, Dec. 2017, doi: 10.1016/j.media.2017.07.005.
[24] N. Rieke, J. Hancox, W. Li, F. Milletari, H.R. Roth, S. Albarqouni, et al., "The future of digital health with federated learning," npj Digital Medicine, vol. 3, Art. no. 119, 2020, doi: 10.1038/s41746-020-00323-1.
[25] L. Alzubaidi, J. Zhang, A.J. Humaidi, A. Al-Dujaili, Y. Duan, O. Al-Shamma, et al., "Review of deep learning: Concepts, CNN architectures, challenges, applications, future directions," *Journal of Big Data*, vol. 8, Art. no. 53, 2021, doi: 10.1186/s40537-021-00444-8.

7 Deep Learning Architectures for Image-Based Plant Disease Analysis

Mynavathi R.[1], Sasthikha V.K.[2*], Sabitha P.[3] and Ravikirushna S.[4]

A rising demand for sustainable agriculture is accelerating the development of advanced, reliable, and real-time plant disease detection systems. Traditional field-based diagnosis is slow, labor-intensive, and prone to human errors. Modern image-based deep learning presents a game-changing alternative that can effectively deliver scalable crop health monitoring. Recent major advances in convolutional neural networks, EfficientNet, DenseNet, and Vision Transformers enable automated extraction of subtle disease cues, such as discoloration, texture variations, and lesion evolution directly from leaf and canopy images. Hybrid CNN-Transformer architectures with the seamless fusion of local spatial encoding and global attention mechanisms have offered superior robustness under diverse lighting, background clutter, and environmental variability. Addressing limited and imbalanced datasets in agriculture, there is growing evidence for leveraging generative models, synthetic augmentations, and few-shot learning for high-accuracy performance with scarce disease samples. Multimodal and vision-language models that juxtapose visual signals with agronomic descriptions have recently enabled interpretable disease predictions by jointly offering natural language explanations. Parallel innovations in UAV-based imaging, IoT-enabled sensing, thermal and hyperspectral fusion, and edge-AI deployment have ushered in early detection on smartphones, drones, and field robots in any environmental conditions without dependence on cloud connectivity and in support of resource-constrained rural areas. Although there is fast-paced progress, domain adaptation, generalization across diverse geographies, and stress interaction diagnosis motivated by climate variability remain significant challenges. In this regard, lightweight transformers, federated learning, and continual learning frameworks are emerging as promising enablers for privacy-preserving, adaptive, and collaborative disease intelligence across geographically distributed farms. This chapter offers a forward-looking overview of state-of-the-art deep learning architectures for image-based plant disease analytics, focusing on state-of-the-art innovations, recent deployment challenges, and future directions. A new focus will be placed on building autonomous self-evolving crop health systems: integrating real-time imaging, on-device AI, multi-sensor fusion, explainable reasoning, and collaborative data ecosystems which can enable globally scalable,

[1-4] Velalar College of Engineering and Technology, Erode, Tamil Nadu-638012, India.
Email: rpgmyna@gmail.com[1]; ravisabitha65@gmail.com[3]; ravivishnu753@gmail.com[4]
* Corresponding author: sasthikhav@gmail.com

DOI: 10.1201/9781003767534-7

interpretable, and farmer-centric precision agriculture. Next - generation systems hold huge potential to minimize crop loss, reduce chemical usages, and enhance food security in the face of increasing climate stress.

1. Introduction

Food security continues to be a large concern globally, with agriculture serving as the backbone of this issue. However, as the world is shifting due to climate change and changing environmental factors, agriculture faces challenges regarding plant disease, which negatively impact how productive crops are. On top of plants facing new and different diseases worldwide, traditional methods for diagnosing and determining if a crop is sick, such as visually scouting for disease, looking at it with a microscope, or running a lab test, generally require a lot of time and someone skilled in diagnosing the disease. Thus, these methods cannot be used at the time an issue is found out in the field. However, with new developments in the field of artificial intelligence (AI), especially deep learning, there are many new opportunities to diagnose and detect plant disease using computers to convert images of plants to percentage chances that those plants are sick using an automated, scalable, and highly accurate process. Recent advances in architecture such as Convolutional Neural Networks (CNN), EfficientNet, DenseNet, Vision Transformers (Vi T) and Hybrid CNN/Transformer models can find patterns of subtle lesions and physiological stress indicators in plant images. Additionally, new and innovative methods for developing models, generative augmentation, few-shot and federated learning address real-world problems related to data availability, privacy, and transparency. Together, these approaches represent a significant shift toward intelligent, automated, resilient systems for monitoring plant health.

The advancement of deep learning has helped transition prototyped controlled environments of plant disease detection systems to become fully functional and capable of supporting large scale agricultural distribution as a form of AI. Today's designs now have hierarchical attention mechanisms, multi scale feature embedding and cross domain adaptation capabilities which make them better suited to work and be deployed in unpredictable environments than designs of a few years ago [1]. For example, Vision Transformers (ViT) and Swin-transformer models perform well at identifying long-range spatial dependencies and, therefore, can be particularly useful for detecting crops that show disease in multiple areas of a leaf, or crops that have uneven, or non-uniform visual manifestations of disease [2]. The combination of CNN and Transformer models allows for enhanced feature representation due to the combination of localized texture analysis and global context reasoning, making it less likely that a false positive will be created in images taken in cluttered or noisy environments [3]. Further to these developments and improvements in feature detection, the application of multimodal sensing tools (i.e., hyperspectral cameras, thermal cameras, drone captured images) provide more clues about the presence of stress (biochemical, structural) before there are observable symptoms [4]. The fusion of this heterogeneous data through the use of attention-based networks or the sensor-level integration methods, has resulted in a significant increase in the early detection accuracy and severity estimation of crop diseases [5]. Deployment-centric innovations that enable the creation of lightweight models through optimizations such as pruning, quantization, and knowledge distillation will help ensure that everyone, including farmers with no access to high-speed internet and advanced computational hardware, have access to mobile phones, drones, and other edge-computing devices needed for efficient and effective diagnosis of plant diseases [6]. Further, at the same time these innovations provide farmers a sense of confidence in their ability to use and benefit from agriculture technology by providing privacy-preserving mechanisms, such as federated learning, that will allow farmers and their suppliers to work together to enhance agricultural models without sharing sensitive agricultural data with each other. Similarly, agricultural technology innovations such as generative adversarial augmentation, few-shot learning, and self-supervised representation learning will provide a solution for the continuing challenge of few available annotated datasets on newly emerging crop diseases that may be rare or increasing in frequency. Finally, techniques such as Grad-CAM, visual saliency maps, and attention heatmaps for Explainable AI (XAI) make it easier for farmers to understand how to interpret diagnostic predictions made by agricultural diagnostic applications and feel a level of comfort interpreting them. The trend toward autonomous, adaptive, and scalable digital solutions for plant disease diagnostics supports the processing of agricultural and environmental data on an extensive scale, particularly within precision farming as agriculture continues to be impacted by soaring levels of uncertainty and demand for sustainable food production. Deep learning algorithms and advanced image processing applications will represent a key enabling technology for developing future plant health monitoring technologies relying on deep learning-based analyses of images.

The Key Aspects Enhancing Future Trends and Challenges are :

- Lightweight Transformers.
- Multimodal Sensor Fusion Technique.
- Edge-AI Deployment in the Field.
- Federated Learning Security.
- Robustness of Domain Adaptation.
- Data Augmentation via Generative Models.
- Embracing Explainable Artificial Intelligence
- Climate change resilient AI Models.

2. Fundamentals of Image-Based Plant Disease Analysis

The significant increase of digital technologies used in Agriculture to diagnose plant diseases using various data sets and images has resulted in a dramatic change compared to traditional methods which focused on uniform samples of plant material collected in laboratory conditions to diagnose plant disease. Most digital imaging technology uses many data sets that represent multiple conditions under controlled laboratory settings. The use of multiple environmental conditions such as varying amounts of light, cluttered backgrounds, overlapping foliage, various soil textures, pest/pathogen interference, and weather-induced anomalies, i.e., shadows and moisture, creates many complicating factors that introduce visual noise and ambiguity that will require the Deep Learning Model to learn how to handle to create robust plant disease Detection and Classification. The increased complexity of modern Deep Learning Models due to the use of multi scale Hierarchical Attention Models, i.e., Swin Transformers, will allow for a large number of encoded relationships to be captured that span multiple scales, thereby creating more robust models for generalization across heterogeneous environmental conditions [7]. Likewise, Hybrid CNN-Transformer Networks use localized convolutional feature extraction and global attention reasoning to detect subtle lesions and/or spatially dispersed symptoms even when the visual fidelity of the images is compromised [8]. To develop a reliable Diagnostic Process for the diagnosis of plant disease, an understanding of disease-specific imaging characteristics, appropriate imaging modalities, and standardizing input data will be fundamental to achieve optimal performance using the various advanced architectures that are available for use across multiple types of Agricultural Ecosystems.

3. Characteristics of Plant Disease Imagery

The nature of plant disease images has many inherent complexities, which contribute to the performance of models trained on them. For example, diseased leaves may exhibit very minor signs (such as discoloration), or have small necrotic spots, veins that are darker than normal, mosaics, or deformations in their structures, which could be hard to identify against the background of a healthy leaf. Environmental noise also contributes to the visual variability in agricultural datasets. Examples of environmental noise include fluctuating levels of sunlight, shadows created by adjacent plants, dust, wind-driven movement of the leaves, and undulating views of the leaves from different camera angles. Another characteristic of plant disease imagery is that the symptoms of disease distribution are not uniform but rather are found clustered in certain areas. Some diseases manifest as sporadic clusters of symptoms while others are characterized by the presence of symptoms around marginal regions or certain vein structures. The earlier manifestations of symptoms may be invisible, while the later manifestations of symptoms have overlapping lesions, creating a multi scale problem in terms of how feature extraction must identify these micro-patterns in relation to the larger global view of the leaf. In addition, the complexity of the background affects clarity of the image. Images taken in a field will include elements such as soil, weeds, irrigation lines, foot traffic, and the presence of certain insects, all of which can obscure the disease symptoms and create false positive patterns, e.g., lesions that appear to be the result of disease but are artefacts.

Lastly, different sensor types, camera quality, and image compression can impact on the consistency of color and how well texture is represented. Therefore, to address these issues as a group, a solid preprocessing pipeline combined with a robust deep learning architecture must be used to manage the high variability of examples within a class and the low differentiation between classes.

4. Image Acquisition Modes (RGB, Hyperspectral, Thermal, UAV)

From its inception, RGB (red, green, blue) imagery was chosen as the most common and easily accessible way to detect diseases in plants. Many of these factors are dictated by the fact that RGB images are cost-effective and can be captured with a smartphone or digital camera, or in the field using sensors such as spectrophotometers. It has been shown that RGB images can provide visible symptomatic evidence of diseases based on visible clues such as color, texture, and growth patterns. However, the primary disadvantage of RGB is that it cannot detect all biochemical changes that take place between the development of visible symptoms. Hyperspectral sensors collect reflected light energy throughout a wide range of spectral bands (hundreds), so it can give us insight into many of the biochemical and physiological indicators of plant stressors that are not detectable by human eyes. Many key biochemical changes that can indicate plant stress can be detected at an earlier (or pre-symptomatic) stage, and include chlorophyll content, cell structure, water stress, and early stages of pathogen infection. While hyperspectral sensing provides us with a new level of insight into plant health, it also poses some major challenges in terms of cost and processing power since the data generated by hyperspectral sensors is multidimensional, and thus requires dimensionality reduction and specialized processing techniques to analyze. Thermal imaging or thermal vision cameras, which are commonly used in agriculture, detect the temperature differences in the canopy (e.g., leaves) of plants that have been affected by diseases altering their rate of transpiration. Most diseases that will affect a species of plant will cause the stomata of that species to be closed, resulting in a rise in temperature on the surface of the leaf. Therefore, by using thermal imagery to identify stressed plants, there is an increased opportunity for growers to identify and treat their crops before they see any symptoms of the disease. Thermal imaging allows you to monitor changing environmental temperatures at night, but thermal imaging is still limited by the current temperature of a growing environment. UAV (unmanned aerial vehicle) technology is popular for monitoring crop fields because UAVs provide a cost-effective method for monitoring large areas in real-time. UAVs can be fitted with an RGB camera, multispectral camera, or thermal camera. Integrating the data produced by multiple UAVs provides spatial representations of the diseases across the entire parcel of land. Because of the large areas the drones can cover and the geospatial awareness they provide, they are an essential tool in precision agriculture and large farms where the costs of labor make scouting impossible.

5. Preprocessing Techniques and Dataset Preparation

Deep learning pipelines require preprocessing as a fundamental step to standardize images of plant species before inputting them into the systems. Preprocessing involves resizing digital photographs to predetermined dimensions for use in training either CNNs or Transformers. During image preprocessing, light level variability across pixel components is reduced through pixel normalization to create consistent images throughout model training. Changing color space to either HSV (herpes simplex virus) or LAB (lactic acid bacteria) generally results in improvement of symptom-to-ground separation when using deep learning techniques to identify disease symptoms. Hyperspectral Image data requires dimensionality reduction techniques such as index creation with NDVI (normalized difference vegetation index), along with the selection of specific bands. Filters including Gaussian blur and bilateral filter are used to negate some types of noise (sensor noise, environmental contamination, leaf vein) and dust that may have been confused as "disease". Leaf segmentation may isolate the plant of interest from the remaining photographic image, therefore excluding background noise from the impaired area. Segmentation may occur through multiple means, including using DeepLab, U-Net, thresholding, or other similar methods, in conjunction with a CNN or Transformer model to focus data input on pixels that relate to symptomology. Artificially increasing diversity within datasets can be accomplished through rotation, flipping, scaling, shearing, and color jittering of input images for protection of confidential or proprietary information of plant species and/or other methods of augmenting datasets. Generative Adversarial Networks (GANs) are increasingly utilized as augmentation methods due to the ability to generate realistic representations of plants with symptoms of impaired health while simultaneously reducing the issues associated with using imbalanced training dataset sizes. Accurate labels are necessary for identifying symptomology during supervised learning. The use of various methods of tagging images allows you to preserve the quality of your image collections by preventing misguided training due to mislabeled images. Also, all classes of your image training, validation, and test sets must have the same number of images. For instance, when using a UAV and hyperspectral imagery, the uniformity of your training data can be accomplished via image tiling and patch extraction. These techniques of transforming the images create a uniform dataset, which is suitable for more advanced machine-learning techniques; it also provides a clean, organized dataset for performing detailed learning on large UAV-derived image datasets that improve ISP (internet service provider) model-based outcomes.

6. Deep Learning Architectures for Disease Detection

Deep learning has provided an excellent foundation for the last several years as it relates to plant disease detection as it removes the need for a handcrafted descriptor; instead, deep learning can directly extract the hierarchical visual features found within the image through its neural network architecture/algorithm. The experience

Table 1 Common Plant Disease Image Properties and Challenges.

Property	*Description*	*Associated Challenge*
Lighting Variability	Sunlight, shadows, cloud cover	Inconsistent color features
Background Clutter	Soil, weeds, tools, other leaves	False lesion detection
Occlusion	Overlapping leaves or stems	Missing critical disease areas
Sensor Noise	Blur, low resolution, compression artifacts	Reduces feature clarity
Symptom Diversity	Early vs advanced disease stages	Hard to generalize
Non-Uniform Distribution	Patchy lesions across leaf	Requires multi scale analysis
Environmental Effects	Rain, dust, wind motion	Unreliable texture cues
Multi spectral Variability	Different modalities capturing unique bands	High-dimensional complexity

gained from early CNN's through the current advanced Transformer architecture and their hybrid versions has led to much higher accuracy rates while offering the ability to withstand greater amounts of noise from the surrounding environment and adapting more effectively to various climates in the field. This section includes a complete list of the primary architectures presently being utilized in agricultural image-based diagnostic evaluation, along with their advantages and disadvantages for use in agricultural practices in today's environment.

7. Convolutional Neural Networks

Due to the capability of CNNs to learn local spatial features (e.g., color distortion, texture irregularities, lesion edges) of an image, CNNs are one of the most widely used models for classifying crop diseases. As CNNs employ multi layered architecture, they also learn automatically essential low, medium, and high-level features which can be used to differentiate between healthy and diseased plant materials.

7.1 AlexNet, VGG, ResNet, DenseNet, EfficientNet

The early networks such as AlexNet and VGG started the capability of deep feature extraction for the plant disease classification. With ResNet, the gradient flow was improved using skipped connections and with gradual training allowed deeper networks to analyze complex symptom patterns in greater depth. Using DenseNet, the propagation of features through the dense connections between the layers has aided in recognizing subtle patterns for every disease type. EfficientNet enables a high level of performance to be obtained for a given accuracy and parameter count through use of compound scaling. This allowed for a positive cost benefit of using large datasets of different severity of disease with the least amount of computational power applied to the data sets.

7.2 Lightweight CNNS for Edge Devices

Compact CNN variants use depthwise separable convolutions or reduced filter sizes to allow real-time inference from mobile phones, drones, and embedded IoT devices. These architectures, being lightweight, have high accuracy and the lowest levels of memory and power use, allowing for great utility in the low-resource environments of farms.

7.3 Vision Transformers

ViTs change from convolutional filtering to attention-based features for extracting from images. ViTs use a different mechanism than CNNs to analyze images by splitting the image into patches of fixed size and embedding them as vectors to be treated as models of the patches by processing the vector representations of the patches of the image through multi-head self-attention layers. The multi-head self-attention captures the global context in time across a leaf, which aids detection of disease patterns which are complex and/or exhibit mixed or multiple symptoms.

7.4 Patch Embeddings and Self-Attention

ViT Model allows users to create fixed size patches out of an image, and to represent these patches using a vector format. The model also includes several layers of Multi-Head Self Attention that make it possible for users of the ViT model to quickly learn the contextual relationships of all patches on a leaf. By providing a method for the user to capture complex disease patterns and multiple disease symptoms, this allows for accurate classification.

7.5 Swin-Transformer for Hierarchical Feature Learning

Swin Transformer model introduced an additional form of attention that can capture both local and global features from different levels of the image. Using this type of hierarchical representation learning, this model is very useful in environments where background complexity and lighting conditions change considerably.

7.6 Hybrid CNN–Transformer Models

Hybrid models leverage both the advantages offered by CNNs and Transformers to deliver improved robustness when used in real-world field conditions (e.g., various conditions that affect crops). Hybrid architecture typically uses CNN layers for the initial classification of the textures associated with a particular object or area, followed by Transformer layers that provide longer-range, more global information through an attention model. These two types of networks can be combined in different ways to produce effective results. The merging of CNN and Transformer features can occur at intermediate layers, or they can be processed using two sets of parallel branches, which process local and global features simultaneously, respectively. When used in environments with partially obscured leaves, heavily cluttered backgrounds, and faint symptoms, hybrid models are outstanding for plant identification. In conjunction with localized convolutional cues and modeling of long-range dependencies, hybrid networks greatly reduce false negatives, false positives, and the total number of poor-quality images. The result is more reliable detection of plant diseases in large, uncontrolled field datasets [8] [9].

8. Advanced Training Strategies for Limited Agriculture Datasets

The success of deep learning for detecting plant diseases relies on access to large volumes of diverse, annotated datasets. Unfortunately, large volumes of agricultural datasets do not always contain enough samples for rare diseases, as well as early-stage symptoms and local unique infections. Environmental differences also add to the difficulty associated with constructing a robust deep learning model. For example, the environmental factors that can change during the day, including the amount of available light, will be different than those during the night. Deep learning pipelines have been designed to overcome these limitations, including the use of transfer learning, generative modeling, few-shot learning, and augmented datasets, which will allow for greater expansion of dataset diversity without requiring significantly more manual labeling. Using multi-scale patch attention combined with hybrid architecture has proven to be successful in overcoming sample limitations when analyzing crop images, as they allow the aggregation of multiple layers/levels of context (hierarchical context) from crop images. In addition, data acquired through UAVs (i.e., drones) containing multispectral/hyperspectral data will offer more representational diversity and generalizability, making them much more prominent in large-scale agricultural crop applications. Edge-optimized models have provided an opportunity for on-field deployment of these trained models, allowing for an even greater emphasis to be placed upon developing efficient training protocols for real-world agricultural applications. As such, a collection of these advanced techniques will create the basis for scalable Agricultural AI applications designed to learn optimally from low-volume datasets while simultaneously achieving maximum accuracy and reliability.

9. Transfer Learning Approaches

Transfer learning is the primary method to combat problems with low-volume agricultural datasets. Transfer learning allows deep networks to utilize the previously learned weights from large volume datasets (i.e., ImageNet) to increase their generalizable feature extraction capabilities, thus speeding up the convergence period for model learning on specific domain (i.e., Agricultural) images of specific diseases. Fine-tuning of model parameters will further allow for the model's ability to learn and adapt fine-grained characteristics from the higher-level layers like pattern recognition for specific diseases (i.e., chlorosis, necrotic spots, discoloration along veins). Transfer learning provides the ability to reduce the risk of overfitting, will be less computationally intensive, and is beneficial for crops that have very few labelled images (i.e., annotated) or emerging pathogens. The ability to pre-train these models on agricultural hyperspectral/drone data will further increase the models' specificity for in-field agricultural applications.

9.1 GAN-Based Synthetic Data Generation

Generative Adversarial Networks (GANs) have become a vital tool in solving class imbalance problems when there is not enough data to perform a statistical analysis. GANs help to generate "fake" image samples that exhibit the same symptoms as the disease. This is beneficial because it creates a significant increase in diversity within the Data set. The samples created with GANs include the addition of noise, environmental interferences, color variations, different lesions shapes, and disease progression stages. CycleGANs enable the transformation of the healthy leaf image style into a diseased leaf style. Furthermore, Conditional GANs allow for the generation of images corresponding to the specific disease type. Diffusion models are a new approach to the creation of synthetic data images. Diffusion models create detailed synthetic disease images with very few flaws and thus can be particularly useful when there are only a limited number of real disease image samples available. By providing synthetic data augmentation, the model will be better able to generalize and will result in improved accuracy rates for disease classification.

9.2 Few-Shot and Meta-Learning for Rare Disease

Few-shot learning, meta-learning allows computers to automatically learn from experiences associated with a variety of tasks, providing them with the ability to adjust quickly to a new task after being trained on a small amount of data. Since many disease outbreaks cannot be predicted accurately, and therefore researchers have difficulty collecting large batches of labeled images early enough after the disease has been identified, using Few-Shot and Meta-Learning techniques provide unique and effective solutions for agriculture and disease management. Integrating Few-Shot techniques with Attention Mechanisms and Light Weight Models has improved model performance in both controlled and field conditions.

9.3 Data Augmentation Pipelines

Augmenting Data is a way that improves models' ability to generalize; it creates a "larger" data set by using existing images as the basis to create new sample images by varying them in specified ways. The most commonly used types of augmentations for image data are rotating, flipping, horizontally flipping, randomly cropping, brightening, and normalizing the contrast of the photo. Advanced Augmentation Pipelines also include Elastic Distortions, Color Jittering/Cutout Masking, Mix-up Blending, and Patch Level Transformations. Hyperspectral data can also be augmented by selecting wavelengths, applying Principal Components Transforms and applying Spectral Noise Injection to increase their variability. These augmentations will improve model robustness to changes in the environment (e.g., due to time of day, leaf orientation, and differences in cameras) and ultimately lead to greater reductions in overfitting.

10. Multimodal and Sensor Fusion Systems

Multimodal sensing together with sensor fusion is a critical enabler for transforming the current paradigm of plant disease diagnostics to one that uses not only single-image classifiers but also robust early warning systems capable of detecting physiological and biochemical stresses prior to the development of visible symptoms. Fusion frameworks integrate information from multiple modalities (e.g., RGB, multispectral/hyperspectral, and thermal) into more comprehensive and discriminative representations of plant health, leading to increased sensitivity to more subtle stress signals, a decreased incidence of false positives caused by visual confounding factors (including soil spots and insect damage), as well as improved accuracy in estimating the severity of stresses and spatially mapping their occurrence across agricultural fields. Thus, implementation of effective sensor fusion strategies requires that the differing spatial, spectral, and temporal resolutions of the various sensor types must be balanced with the computational capabilities of the downstream deep learning models and how the data will be utilized in the field.

10.1 Hyperspectral RGB-Fusion

Hyperspectral imaging not only provides narrow spectral bands, but also it will give an insight into the biochemical change (chlorophyll degradation, ageing, and cell structure damage) that happens before visual symptoms appear. RGB images have spatial and texture cues that have high resolution and can be interpreted intuitively by a human being. Pixel or feature-level integration combines the hyper spectral specificity of the hyper spectral data with the lower spatial detail of RGB, leading to a model that can detect presymptomatic stress and of differentiating visually similar diseases. The normal workflow usually starts with a spectral dimensionality reduction (using techniques such as PCA [principal component analysis] and band selection) and continues with the co-registration and concatenation of the feature channels, while more recently cross-attention modules have been used to learn to weigh each modality specifically. By using these earlier techniques to differentiate the two, the multimode fusion leads to less dependence on the limitations of any single modality and better generalization across different background and illumination conditions.

10.2 Thermal Imaging and Physiological Signals

Canopy temperatures are measured with thermal camera images and changes due to disrupted transpiration or stomatal behavior because of infection or drought/water stress. By comparing the thermal image maps with a plant's RGB/hyperspectral image maps, correlations can be made (by models developed using machine learning/classification) between abnormal surface temperatures and spectral/textural indicators for disease so that disease can be detected earlier. Temporal stacking of thermal images (thermal time-series) and spectral images (snapshots) gives researchers the ability to detect transient patterns of abiotic and biotic stress as well as to differentiate between the effects of biotic disease agents (pathogens) and abiotic stressors (e.g., drought, nutrient deficiencies).

10.3 UAV/Drone-Based Wide-Area Surveillance

UAV-utilizing systems (drone platforms) are quite scalable with the ability to produce an enormous volume of photographic images over an entire crop field, enabling location-specific detection of both epidemic and disease-related activities. Using a drone platform allows you to use a combination of both optical and non-optical sensors, as well as collect data on location-based photographic images of any changes taking place within the disease area, in addition to specifying the needed change based on the identified change. For the processing of the images, the use of image-mosaicking, ortho-rectification, and resolution harmonization methods are key for standardizing all the input images into one geospatial model that will be used for the fusion model process. The UAV-generated analytics will be extremely useful for early detection of diseases, allowing the agronomist to prioritize inspection and treatment efforts by using the UAV-generated data, as a result reducing the amount of pesticide applied and, ultimately, the cost.

10.4 Sensor and Contextual Data Integration (Soil, Weather, IOT)

When infusing optical data from visual sensors combined with data collected via sensor IoT measured *in situ* (soil moisture, electrical conductivity, ambient relative humidity, and local weather) provides contextual signals that help delineate biotic diseases from abiotic environmental stress. For instance, if a plant has elevated temperatures along with low soil moisture, then it may be drought-stressed, while if the plant has the same elevated temperatures as previously indicated but has normal moisture levels, then the increase in canopy temperature may indicate change in transpiration due to a pathogen effect. Novel transformer-based architectures for fusion of the two modalities (visual and sensor) have been developed to produce a lightweight and efficient platform for both on-device inference and continual monitoring in the local farm area.

11. Fusion Architectures and Practical Considerations

Levels of fusion approaches include early-stage fusion of raw/pulled data (data from various modalities) and late-stage ensemble predictions on modality-specific outputs (no direct relationship on either an individual basis or at the binary level). In addition, attention weighted fusion and cross-modal transformer-based methods can also dynamically weight modality contributions based upon sensor reliability or contextual aspects of the given sensors. These approaches enable greater robustness when one of the modalities is either noisy or missing altogether at the time of data capturing. For successful practical application of the above-mentioned approaches there will need to be a focus on sensor calibration, temporal sync, alignment of sensor data with other modalities and the consideration of bandwidth constraints. The employment of synthetic data, particularly GAN-generated synthetic multimodal datasets simulating rare diseases and noise from sensors' operation, will provide improved resilience of models when faced with variations in the real world. Figure 1 illustrates the technology matrix for plant disease detection systems across different deployment scenarios.

Early Detection	Real-Time Processing	Low-Power Devices	Data-Limited Regions	High-Accuracy Systems	Scalable Deployment
1	2	3	2	1	3
1	1	3	3	1	2
1	2	4	[illegible]	1	3
2	2	3	4	2	3
1	2	3	2	2	1
2	3	3	3	2	2
2	2	3	1	2	2
1	2	3	2	1	2

Fig. 1 Plant Disease Detection Technology Matrix.

11.1 Edge Computing and Resource-Efficient Deployment

To create a model suitable for deployment of edge devices, we must focus on architectural adequacy first. Depth-wise separable convolutions allow for a mobile-friendly architecture (i.e., low computational cost) yet still maintain a high degree of accuracy; therefore, depth-wise separable convolution-based architectures, such as MobileNet, ShuffleNet, SqueezeNet, and EfficientNet-Lite, make up most models currently being used for many agricultural edge applications. These models can operate at significantly higher frame rates when running on mobile processors than when using traditional processors, which allows farmers to obtain immediate feedback throughout their farm monitoring process. In addition, there are architectural scaling techniques that can increase inferencing speed (e.g., reducing network depth, tuning input resolution, and limiting channel width). Early-Stage Disease Pattern Detection through Architectural Scale Methods Architectural Scale techniques are also valuable for discovering early-stage signs of disease because it allows low-level features instead of an extensive amount of hierarchy to be used as a basis for determining an illness. In building light-weight architecture intended to work on agriculture edge applications, all types of environmental inconsistencies, i.e., non-uniform illumination conditions, motion-blurred images, occluded views, and other such conditions should be considered to maintain model accuracy and stability during uncontrolled field environments.

11.2 Lightweight Architectures for On-Device Inference

To create a model suitable for deployment of edge devices, we must focus on architectural adequacy first. Depth wise separable convolutions allow for a mobile-friendly architecture (i.e., low computational cost) yet still maintain a high degree of accuracy; therefore, depthwise separable convolution-based architectures, such as MobileNet, ShuffleNet, SqueezeNet, and EfficientNet-Lite, make up most models currently being used for many agricultural edge applications. These models can operate at significantly higher frame rates when running on mobile processors than when using traditional processors, which allows farmers to obtain immediate feedback throughout their farm monitoring process. In addition, there are architectural scaling techniques that can increase inferencing speed (e.g., reducing network depth, tuning input resolution, and limiting channel width). Early-Stage Disease Pattern Detection through Architectural Scale Methods Architectural Scale techniques are also valuable for discovering early-stage signs of disease because it allows low-level features instead of an extensive amount of hierarchy to be used as a basis for determining an illness. In building light-weight architecture intended to work on agriculture edge applications, all types of environmental inconsistencies, i.e., non-uniform illumination conditions, motion-blurred images, occluded views, and other such conditions should be considered to maintain model accuracy and stability during uncontrolled field environments.

12. Model Compression, Pruning, Quantization, and Knowledge Distillation

The effectiveness of reducing the size of the research into practical real-time models is enhanced through each of these techniques. These techniques include:

12.1 Pruning

A topping off the bias weights in a network by eliminating unused layers or entire convolutional filters with low impact on performance. Structured pruning also creates greater opportunities for hardware acceleration and decreases latency. Plant disease detection using images captured by hyperspectral and multispectral cameras will generate very high dimensional feature maps which makes it an ideal candidate for pruning.

12.2 Quantization

Reduces the precision of the model from 32-bit floating-point to either 16 or 8-bit integer representation. It is possible to do so and still maintain reasonable classification accuracy (i.e., inference speeds). Quantization-aware training allows lower levels of precision during the training process, so that models are still able to perform as well as solution geometrically, even when being used with low bit precision.

12.3 Knowledge Distillation

Knowledge transfer from the larger "teacher" model to the more compact "student" model. The student model will be able to learn from the teacher, either from soft labels or intermediate feature representations, and help the student network replicate the performance of the teacher while being used on an edge device. Distilled models are particularly well suited for few-shot and meta-learning scenarios, as they provide the ability to generalize the unknown environment with only a few samples and have very limited input.

The synergy among these three models, pruning, quantization, and distillation, creates a compressive path to create extremely efficient deployable models for time-sensitive agriculture.

13. Hardware Considerations for Field Deployment

When deploying AI within agriculture, it is important to pick and configure on-device hardware for maximum resilience, dependability, and operational lifespan. Smartphones represent the easiest means of access for small-scale farmers and allow for instant disease diagnosis by taking pictures from the field and running AI models through optimized mobile frameworks. Edge inference permits operating with minimal network access, since cloud access is not necessary. Automated monitoring systems can also be achieved using devices such as Raspberry Pi, NVIDIA Jetson Nano, Google Coral TPU, and ESP-based microcontrollers. These devices are paired with sensors located on greenhouses or in open-field growing environments, and are continuously processing canopy images, environmental readings, or leaf level close-ups. Drones fitted with edge AI modules provide the capability of detecting diseases over very large agricultural areas, without having to transmit images back to the cloud for processing. Likewise, autonomous ground vehicles may be able to perform periodic sweeps to monitor the spread of disease. Beyond the hardware to accomplish this task, the practical design features must include battery life, rugged casings, waterproofing, and dustproofing as many agricultural settings are extremely challenging.

13.1 Real-Time Monitoring and Decision-Support Systems

Monitoring systems that utilize Edge-AI technologies provide real-time inference and alert users when abnormal patterns are detected. Speed is essential for the containment of fast-moving diseases, which require an immediate response within the same day. AI outputs can be paired with an actionable decision support system (e.g., recommending pesticides, identifying treatment zones, and predicting the spread of disease) to facilitate sustainable precision agriculture. Edge-based systems reduce a farmer's reliance on highly trained agronomists by providing localized AI-based guidance that gives rural and isolated communities access to contemporary plant health diagnostic resources. In addition to the advantages of providing localized guidance to farmers, the offline capability of edge-based systems ensures no interruption of monitoring due to a lack of infrastructure in some areas.

14. Challenges and Future Opportunities in Edge Deployment

Deep learning systems deployed directly on the edge face a host of unique technical, environmental, and socioeconomic challenges that need to be addressed for large-scale adoption in agriculture. Agricultural fields are unpredictable; subsequently, they pose many challenges which impact the performance and reliability of embedded AI systems in comparison to a controlled laboratory setting. For example, many rural farms do not have a consistent power source; as such, devices depend on battery packs or solar energy for their operation. Edge-AI accelerators, thermal cameras, and UAV-mounted processors need to be powered by energy-efficient algorithms to support extended use in the field without having to frequently recharge batteries. Therefore, optimizing inference workloads, implementing event-driven processing, as well utilizing low-power hardware are crucial for providing continuous monitoring of all aspects of field operations.

Environmental variability is another critical issue. Dust can accumulate on lenses and degrade image quality, while humidity, rainfall, increased/decreased temperatures, insects entering the optical path and wind-induced motion security can prevent reliable operation as these elements introduce multiple variables which directly affect the reliability of deep learning models for generating predictions. If models generate reliable results, however, they can provide farmers insight into the condition of their crops and for determining when corrective actions are necessary (e.g., spraying pesticides). To improve edge devices' ability to withstand extreme weather conditions and to be more durable and long-lasting, there is a need for the use of weatherproof casings, self-cleaning optical surfaces, and environmental calibration algorithms (the algorithms will increase the durability of the devices;

the casing, cleaning capabilities, and environmental calibration will enhance the lifespan of edge devices).

The deployment of multiple edge devices is often impacted by the existence of heterogeneous data from different habitats and seasons. For example, different seasons will produce different colors, textures, and diseases of leaves. Therefore, models need to be regularly adjusted (a process called "model drift") so that the predictive values of the models maintain their accuracy. The inability to have sufficient memory and storage capacity (RAM) in edge devices will ultimately restrict deploying, sustain high-definition computerized images and monitor historical images to identify progression trajectories regarding the extent of illness. As a result, compression algorithms are necessary to alleviate the limitations imposed by memory, storage, and time. High-resolution video imagery collected by UAVs is typically divided into many small patch-sized images so that large videos can be processed while still maintaining the requisite detail to provide accurate analytics on the disease condition of the crop. To achieve these low-cost platforms for small farmers, specialized hardware is needed (e.g., Jetson Nano, Coral TPU, multispectral cameras) but the cost for this specialized hardware will likely exceed small farmers' budgets. Thus, the Tiered model provides a scalable deployment model, with smartphone-based inference at the base of the model for smaller farmer operations and supports increasingly complex technologies, extending into using UAV and sensor networks for medium-sized farm operations, and Comprehensive Integrated Edge-Cloud Hybrid Solution for the largest agribusiness operations. Concerns about data and privacy are present at the edge of the operation, however, though edge computing is intended to reduce dependency on cloud computing services for data storage and processing, it is still critically important that edge devices maintain the security of sensitive farm-related data, proprietary information related to crop growth and yield potential, geotagged images of farms, and many other types of proprietary data that may originate in the farm. Lightweight techniques for encryption, secure updates to machine learning models, and device authentication protocols must be implemented in a manner that allows a farmer to trust that a device is providing the data being used to support their operations. While there are challenges in the near term, substantial future opportunities exist to deploy edge-based solutions for plant disease diagnostic models. The advancements in Tiny Machine Learning (tinyML) and on-chip AI (microcontroller-level AI) enable inference or prediction on ultra-low-power chipsets that further improve the viability of building affordable field-deployed solutions. As the availability of vision-based microcontroller chipsets and cost-efficient hyperspectral imaging modules become readily available and the development of neuromorphic processors continues to emerge, they will allow for reduced energy consumption and increase the robustness of ML-based plant disease diagnostic deployments. Federated edge learning (FEL) can enable distributed updates of ML models across farm operations without requiring centralized aggregation, thus allowing ML models to remain current by being adaptable/inclusive of newly emerging plant disease patterns. Future architectures may include self-evolving AI models capable of incremental (continuous) learning (whenever provided with new needle leaf imagery), allowing for seamless refinement of ML models without having to retrain on new imagery. Furthermore, hybrid edge/cloud pipelines create enormous potential for edge devices to conduct real-time inference and send periodic representative sample data to cloud servers for further analytics or longer-term models. This hybrid architecture optimally satisfies all the constraints of latency, bandwidth, and computational requirements. Additionally, integrating autonomous robots and drone technology will lead to the development of fully automated survey systems that can identify disease, assess its severity, and provide input recommendations, such as fertilizer or pesticide, thereby transforming agricultural operations into intelligent, data driven.

14.1 Privacy-Preserving and Adaptive Learning Models

Systems which rely increasingly upon AI-based plant disease diagnostic tools must be capable of achieving optimal precision (high accuracy) while adapting to the variable nature of today's agriculture; likewise, they must protect sensitive data related to the farm's operations. With the proliferation of drones, IOT (Internet of Things) devices, and other technologies that produce large quantities of visual information in real time and produce equal amounts of associated visual context each day from distributed systems, there is a huge increase in the potential amount of data generated by agricultural production. With this increase comes uncertainty related to ownership and confidentiality associated with data generated; additionally concerns arise regarding the misuse of location-based data associated with crop health. Frameworks such as Privacy-Preserving Learning (PPL), adaptive AI models allow the use of high-performing AI.

15. Federated Learning for Distributed Farms

Federated Learning enables a group of farms that could work together to create a Shared Model without the need for all organizations to collect data in one central location. In this type of situation, each farm (device) creates their own model locally using only the data they have available at their own facility. The only information sent to Central Server from each farm (device) is the updates to the Shared Model. The updates from each Farm are aggregated to create a Global Model that has the advantage of using a wider variety of Samples and at the same time protecting local Farmers' Privacy. Federated Learning is especially useful in the Agriculture Sector, since the type of crops grown, the types of Diseases that could attack those Crops, and other Environmental Conditions can differ greatly between Farms. Federated Learning improves the Generalization of a Model, which in turn decreases the amount of Data that needs to be sent over a Network. There are several Light Key Snake frameworks available that allow low Resource Devices to participate in Federated Systems. Often, those Light Key Snake frameworks have been integrated with those Severity Estimations and interpretable Models to allow Support for Actionable Diagnostics in the Field.

16. Domain Adaptation for Environmental Variabilty

The pictures of diseases in plants tend to vary widely based on many factors, including geographic location, the amount of light, the position of the leaves, and camera type. The Domain Adaptation approach can be used to alleviate the potential drop-off in accuracy that frequently occurs when using models built using data from one location in a different location. The classic approaches to domain adaptation align the distribution of the feature vectors on the target domain with the distribution of feature vectors in the source domain using adversarial losses or discrepancy losses. More contemporary and sophisticated domain adaptation architectures utilize attention mechanisms to adaptively recalibrate the networks based upon their input environment. Domain adaptation is necessary for agricultural systems on a global scale because of climatic conditions, cultivar differences, and disease phenotypes differ significantly across regions. The incorporation of domain adaptation modules into edge-AI model architecture allows for rapid and accurate recalibration of the models in a real-world setting, thus minimizing the need for extensive retraining of the edge-AI models.

16.1 Continual Learning for Emerging Diseases

Agricultural environments are constantly changing, including the emergence of new diseases, mutation of existing pathogens and plant stress symptoms evolving over time. Continual Learning Framework enables models to obtain knowledge of new diseases without losing their previously learned disease patterns. Key methods for producing such learning frameworks include Replay-based Learning and Dynamic Network Expansion, and Regularized methods for preserving previously learned knowledge while incorporating new samples. This allows models to be deployed over long periods of time without experiencing catastrophic forgetting and remains relevant to the evolving distribution of disease. Continual Learning Frameworks will also reduce the costs associated with retraining; therefore, farm owners will have the capability to utilize models on their farms that have limited computational capabilities.

17. Trustworthy and Transparent Disease Analytics

Farmers, agronomists, and policymakers are more likely to adopt agricultural technology that is both trustworthy and open to interpretation. XAI provides visual aids, like heatmaps or saliency overlays, to show the areas of a leaf that were most important in reaching a diagnosis. By increasing transparency, XAI aids in reducing reliance on experts and giving farmers greater confidence to identify when errors have occurred because of environmental distortion or sensor noise. The integration of XAI with multimodal severity estimation plays an important role in improving interpretability and increasing the ability to take actionable steps based on an AI-based recommendation in the field.

17.1 Explainable AI for Trustworthy Disease Diagnosis

Deep learning models are becoming more advanced as they are used more widely in agricultural settings; therefore, the amount of transparency and interpretability is a top priority with this technology. In the agricultural sector, the diagnosis of plant diseases may have severe ramifications (both good and bad) because they are commonly applied when determining if and at what level pesticides should be used, crop protection tactics, yield predictions, etc. Unlike individuals who can justify their reasoning regarding disease diagnosis, CNNs, Vision Transformers, hybrid versions, and broad categories of DNNs (deep neural networks) operate without transparency in how they provide a prediction (black box systems). Due to the highly complex nature of their internal representation mechanism, farmers and agronomists will often have trouble understanding the reasoning behind their prediction. XAI will provide farmers and agronomists with both visual and structural explanations, which will help farmers and agronomists understand how models are producing predictions and what drives the diagnostic decision. Therefore, the incorporation of XAI technology within Agricultural AI platforms will create an environment that is more conducive to increasing the user's level of confidence, provides better opportunities for conducting error analysis, and creates an ethical and accountable mechanism for monitoring plant disease.

When it comes to plant images, there can be a lot of variability based on lighting changes, overlapping leaves, cluttered backgrounds, or noise from the camera that can all impact the prediction by the model. The use of XAI tools can help to identify if the model has classified the images based on actual symptoms of the disease or if it has confused the images by something else unrelated (visual artifacts). It's also important to use XAI tools to help identify bias in the dataset due to unintended associations made between the data and things such as soil types and shadows. When an agronomist uses data and images to make predictions about a particular disease, he/she must validate those predictions by looking for lesions, distortion in color or changes to the plant veins. The best tools to validate predictions of disease are based on the aggregation of many different types of layers and/or by using attention to create distinct separation of classes in the dataset. Using transformer-based systems provides researchers with an opportunity to see how much attention the model paid at the patch level to determine the validity of the diagnosis and whether the plant was impacted by disease across the entire plant or just portions of the plant.

Different methods of providing explainability for computer vision systems (i.e., saliency maps, Layer-wise Relevance Propagation [LRP], and Occlusion Sensitivity) present complementary perspectives for understanding how a computer vision system makes decisions. Each method presents a unique combination of the most important features for the model and enhances the ability to interpret and debug the model.

Explainability is even more vital for multimodal systems that integrate multiple types of information (e.g., RGB, hyperspectral, thermal, and IoT data).

Visualizations using cross-attention within a multimodal model are useful in understanding which modality contributed the most information to a specific diagnosis. For instance, the hyperspectral bands may indicate early biochemical stress in the plant, while the RGB texture associated with the lesion indicates that it is an advanced lesion. These insights are beneficial for farmers and agronomists when selecting sensors according to the conditions of a particular crop. Trust among users is a major factor in the adoption of AI systems outside of labs. Visualizing the output of AI systems helps non-experts create an accurate mental image of how an algorithm produces its prediction of a particular problem, thus facilitating the adoption of digital diagnostics for agriculture. Research shows that coupling interpretability and disease classifiers provides increased trustworthiness of outputs and lower rates of error.

Overall, XAI takes deep learning approaches to plant disease detection, creates reliable and transparent disease detection systems, and makes them more suitable for use in agricultural scenarios. XAI provides an avenue to view and understand how recipes for predicting plant disease are generated and increase the likelihood of creating ethical usage of digital tools and create a level of confidence in disease detection throughout the agricultural landscape.

18. Challenges, Limitations, and Real-World Deployment Issues

Deep learning has enabled significant progress toward classifying different types of plant diseases; however, implementing a deep learning approach within the agricultural ecosystem involves various layers of complication because of variability found not only across biological and environmental factors, but also due to the technology as well as cultural and social influences that affect the viability, scalability, and reliability of a model after it has been developed. This section outlines key challenges and limitations related to the broad adoption of utilizing deep learning in agriculture and highlights potential avenues for overcoming these obstacles and supporting successful practical development.

Environmental variability represents a significant hurdle in real-world images taken from the field as it can lead to many inconsistencies among different plant images. Whereas laboratory datasets have steady backgrounds and illumination conditions, images of plants in their native habitats will vary due to factors such as changes in the intensity of the sun's rays and the motion of clouds across the sky, shadows produced by adjacent plants, accumulations of dust on leaves and other surfaces, occlusions from the leaves of neighboring trees or shrubs blown by strong wind or rain, and droplets of water or moisture clinging to leaves. As a result, ailments that appear on the surface of leaves could easily appear distorted, lessen their prominence within an image, be impacted concerning their color, or be affected by any of the above-mentioned variables. Thus, when deep learning models are trained using clean annotated datasets, they often struggle to generalize when exposed to noisy images containing one of these variables as described above. Furthermore, the presence of domain shifts among geographical locations (i.e., soil types, humidity ranges, elevations, and native flora) can also modify the appearance of plant diseases. In order to account for the above-mentioned factors, it is critical to develop and implement robust preprocessing pipelines for dataset creation; develop large-scale and diverse datasets of plant images; and implement domain adaptation strategies that allow trained models to recalibrate their learned representations given changing environmental factors. A further aspect of limited datasets relates to class imbalance. For many of the plant diseases, particularly newly emerging and/or region-specific pathogens, the available annotated sample for these diseases is sparse. This situation leads to a large predominance of annotated images featuring a limited number of very well-known plant diseases. Deep models trained using this imbalance suffer from unfair training, meaning the models will learn the most from classes representing the greatest number of observations while misclassifying lower frequency categories. While some methods of improving performance, including GANs, Meta Learning, and active sampling, are creating partial solutions, there is still a fundamental constraint on the availability of high-quality and diverse datasets. Also, the accuracy of annotation is an ongoing challenge for the industry; many annotators use subjective judgment when identifying symptoms which creates inconsistency and adds further noise to our training dataset.

The intricate nature of the agricultural landscape often leads to ambiguous features where the attributes of shadows, edges of leaves, moisture, insect activity, or soil have a visual resemblance to disease lesions. This concept contributes to an increase in false positive reporting and inaccurately estimated severity from both CNN- and Transformer-based architectures. While XAI may assist with identifying these erroneous interpretations of disease lesions, creating models that can accurately separate biological pattern characteristics from those that are non-biological is still a challenging task. Real-world constraints also pose a growing problem due to limited resources available for deployment within agricultural operations. Most agricultural operations, particularly smallholder farmers, depend solely on mobile devices that have limited processing power. The processing requirements related to image capture and processing of high-resolution images (RGB/hyperspectral/multispectral) require high levels of memory and processing capabilities whilst creating substantial amounts of energy consumption. Although deploying Edge Computing technologies addresses some of these issues, there continues to be many models that are still too heavy for a mobile device to do real-time inference in lower-power processing environments. There are also compression techniques available (e.g., pruning, quantization, distillation) that enable more efficient use of models but have the potential to impact model accuracy unless each of the compression techniques is carefully configured. Access to rugged, weather-resistant hardware presents further challenges to deploying models within agricultural applications since most devices need to be protected from dirt, heat, moisture, and physical damage.

Network connectivity is a significant challenge for Edge AI technologies. Many rural agricultural areas lack reliable broadband access, making it impossible to use the Cloud to perform model updates and make model inferences based on real-time data. However, using Edge-AI alleviates this challenge. However, while many agricultural producers are currently using Edge AI and can perform real-time model inferences, most agricultural producers still rely on having periodic connectivity to perform software updates, share learning data with others, and synchronize their data. Models that do not receive timely software updates are likely to become outdated and inaccurate because they will not capture the evolving nature of disease over time and will not account for new variants of pathogens. This creates a challenge with drifting because disease phenology changes over time due to climate changes, mutations in the pathogens, changes in soil nutrient profiles, varietal differences, and pesticide resistance. As a result, models trained on last season's data are unlikely to detect new disease patterns and degrade in performance over time. Additionally, there are other potential options for addressing the problem of drift; however, these solutions require significant improvements in version control, model aggregation, and compatibility between edge devices and federated learning models.

As the amount of information collected by farms because of the collection of large-scale image data from crops and other sources continues to increase, the importance of protecting farmers' proprietary data becomes very important. Many farmers may be reluctant to supply the use of the images due to geolocation information associated with the crops they produce or yield, as well as images of crop diseases, as all of these can affect the marketability of farmers' products in addition to other market risks associated with selling crops in a competitive market. While federated learning structures can protect the privacy of farms through limiting the availability of the training information and utilizing cryptographic techniques to encrypt the communication of gradients and/or model parameters, it requires legitimate practices associated with creating a federated learning structure to ensure that sensitive information associated with a farm is never disclosed through the transmission of model improvements from farm locations to the federated learning network (in order to build trust and promote usage of the federated learning model structure among the agricultural community).

The economic aspect of precision agriculture imposes a higher level of technology on small farmers, which creates additional barriers for them in adopting new methods of growing food on their farms. Many small farmers will not have access to the funding (whether it be cash or credit) required to support their use of these advanced technologies until they gain enough experience to justify the purchase. Also, because of this barrier of entry created by economic limitations, the technical knowledge required to understand how to calibrate and use these devices is often not feasible for smallholder farmers as they can often only afford people to do this type of work. Mobile Advice Systems are also related to these economic and socioeconomic barriers, as when mobile advice systems are utilized, they will incur costs such as regular subscription payments or constant upgrades to hardware. Therefore, by addressing the economic and socioeconomic aspects of these technologies, equitable access to precision agricultural technologies can be provided to smallholder farmers.

Also, large-scale Digital Agriculture is also impacted by ethical and regulatory issues. The algorithmic predictions made by AI in loss prevention pertaining to Crop diseases will affect how much pesticide is applied, the development of irrigation plans, and ultimately, how fields are managed. Whereby incorrect predictions result in putting more chemicals on the crops than is appropriately needed, thus creating an environmental challenge while harming the farmers' ability to generate a sustainable income. For this reason, establishing a transparent basis for how AI makes its decisions is critical, allowing those involved in Ana Agricultural digital agriculture settings to understand how to make informed, responsible decisions. Furthermore, with the growth of usage of AI systems, it is expected that multiple Governmental Agencies will eventually establish requirements for Standards, Establish Benchmarks, and require the development of "explainable Digital Diagnostics", "traceability of digital models output", And "Conformance of AI to Data Governance" requirements.

A further constraint lies within the use of non-interoperable standardisation by agricultural AI companies. Due to the differences in how individual businesses and researchers define, annotate, and structure their data, it has been difficult to integrate the various platforms and combine datasets. Standardization will facilitate scale-up across the geographic, crop, and agricultural systems spectrum. Open-source platforms for establishing agricultural benchmarks and creating a unified ontology of terms used can greatly enhance model development and facilitate generalization across multiple agricultural settings.

The interaction between humans and AI, as well as the ease of use for the technology, will greatly influence the overall adoption of AI by farmers. Farmers will have difficulty assimilating the data generated from diagnostics regardless of how good they are. Therefore, the success of AI in agriculture will require an effective user interface as well as educational resources and integration with decision support systems.

A variety of multidimensional barriers arise during development of plant disease detection with deep learning due to varying levels of environmental uncertainty and prohibitive conditions related to access to data, technological limitations, economic conditions (global economic markets), international regulation(s), etc. To successfully mitigate these issues requires a complete and multifaceted approach that incorporates cutting-edge technologies, incorporates the needs of farmers, and establishes an appropriate regulatory framework that fully supports these efforts. Eliminating the above barriers will allow for the complete transformation of plant health monitoring and build greater agricultural resiliencies for all the world.

19. Future Directions and Conclusions

Agriculture faces a rising number of challenges related to the threat of climate change, pathogens evolving for our plants, and the increasing global demand for food production. Therefore, the development of advanced deep learning techniques for the diagnosis of plant disease will be critical for the future of sustainable agriculture. Advances in the design of robust, multimodal, and edge-deployable AI models have been achieved. The goal of future development will be to provide a model that is more adaptable and resilient and has interpretability and accessibility for all agricultural environments. Examples of trends that will enhance plant disease detection and make it a central tenet of precision farming include: a completely automated crop health eco-system, the ability to process multiple languages at once through lightweight vision-language architectures, and new domain adaptation technologies to more easily apply plant disease detection across farming practices.

19.1 Autonomous Self-Evolving Crop Health Systems

As technology advances, plant pathogen identification systems will develop into technologically advanced systems that autonomously adapt themselves to the environment to optimize their learning capabilities. They will use multiple methods of collecting data via many different methods including but not limited to cameras, sensors, telematics, weather, etc., to form real-time plant health profiles. By utilizing the deemed "self-learning" method of their systems, they will be able to add newly discovered examples of pathogens to the "knowledge base" of the system without losing any of the previously informed examples. As a result, the software developed for these systems will remain useful regardless of changes to the severity of pathogens or major changes in the climate. Furthermore, they will include predictive modeling to allow for the early warning systems for detection, thus allowing for preventative measures instead of reacting to existing outbreaks. Additionally, an extensive variety of robotic devices such as drones and mobile robots will further automate the process of identifying plants for individual inspection via leaf-level, estimating severity damage via imaging, and applying appropriate treatment to the impacted areas of plants, thereby lowering the potential influence on the environment while increasing the best use of available resources. Utilizing automated data analysis; adaptable AI; and interpretable outputs will form the foundation of the next wave of intelligent crop health monitoring systems.

19.2 Vision–Language and Multimodal Foundation Models

Multimodal foundation models can merge the utilization of visual data with natural language reasoning, and utilize image inputs (visual), agronomic textual knowledge (text), and structured sensor information (data) to create an actionable insight for farmers. These multimodal foundation models will be capable of providing statistical and scientific data that will provide farmers with actionable insights based upon both visual data and agronomy information, as well as associated geospatial information (via GPS). Such a multimodal approach should provide for more accurate and reliable results than the traditional unimodal methods used today. The next major trend in agriculture technology is the development of multimodal foundation models, which enable an improved level of productivity, efficiency, and accuracy through the integration of multiple data inputs. As we move forward, we will begin to see an increased adoption of multimodal foundation models, which will facilitate both better decision-making and greater efficiency in farming processes. Throughout history, advancements in precision agriculture technology have generally been accompanied by increases in the efficiency of farming operations due to increased utilization of data from the multiple sources available. As we continue to develop these technologies and increase their use by farmers, we expect to see continued increases in both the number of models being developed and the data being utilized by those models; thus, we would expect to see a continued increase in the adoption of and dependence upon multimodal foundation agriculture technology in the coming years.

19.3 Scalable Deployment through Edge–Cloud Hybrid Ecosystems

Scalability, hybrid edge–cloud ecosystems and balancing real-time inferences with long-term model improvements will characterize the growth of AI-based farming. Edge devices (e.g., smartphones, UAVs, and low-powered sensors) are used for early detection of disease, allowing immediate feedback to farmers in rural areas with little or no connectivity. Whereas larger-scale datasets for analysis and comparison across countries' regions will be analyzed using cloud services. By using a hybrid system to obtain real-time, accurate results and at the same time limit the energy used, the farmer will have both of their needs met. In addition, federated aggregation systems will enable the integration of the newer insights obtained from edge advantages with the combined global model while preserving the edge user's privacy. In some cases, regulations may require that AI-generated diagnoses be explainable and traceable, necessitating the development of hybrid pipelines. In addition, to fully achieve a hybrid pipeline approach, scalable deployments will need to provide low-cost sensors for farmers, mobile apps to make it easy to use, and simple-to-use interfaces for those with limited technical skills.

20. Conclusion

By enabling users to diagnose crops with unprecedented levels of precision and speed, deep learning has the potential to change the way farmers can detect diseases in their crops on a large scale. The stresses being caused within our current agricultural systems due to climate change, the emergence of new plant disease threats, and the rising global population with an increasing demand for food will continue placing additional pressures on our food industry. Monitoring the health of plants using AI allows farmers to increase their productivity by creating a more sustainable way of managing and monitoring crop health through data-driven approaches. Throughout this chapter, the most important topics discussed have laid the groundwork for revolutionizing the agricultural sector through the use of computer vision to produce visual representations of sick plants as a part of its agricultural industry transformation, then utilizing advanced computer vision neural network approaches such as CNNs, ViTs or hybrid CNN–Transformer models. These three types of architectures enable deep-learning systems to detect and identify fine-grained visual clues in plant diseases. They can also accommodate the numerous variables associated with the field conditions surrounding agricultural production, as well as allow for the integration of multi-scale disease patterns that develop in complex agricultural systems. The introduction of multimodal sensor technology, such as RGB (red, green, blue), hyperspectral imaging, thermal cameras, and UAV imaging systems, will significantly expand the potential diagnostic capabilities of deep-learning systems. Additionally, the advances in training strategies, including transfer learning, GAN-based augmentation, few-shot learning, and meta-learning, allow for the continued effective training of deep-learning systems in agricultural applications, especially given the prevalent lack of adequate agricultural datasets. Finally, site-based augmentation, such as edge computing and model-size optimization, will allow farmers to use AI without concern for geographic (connectivity), hardware (mobile vs. cloud-based), or local/region-specific infrastructures. The advancement of privacy-preserving frameworks and adaptive learning systems are making plant disease diagnostics more robust by allowing models to continually adapt as conditions change due to climate change, introducing new diseases, or diversity of locations and types of farms. As discussed within this chapter, the current challenges to widespread adoption remain even with all of these developments. Environmental variability, limited data availability, physical limits to computing power, and socio economic conditions will inhibit the ability to develop a system that is widely accepted. However, there may be some hope for bridging these gaps with developing technologies such as XAI, federated learning, domain adaptation, and continual learning. All these technologies will ensure an AI system has both accuracy and transparency, adaptability, and trustworthiness, three types of attributes critical to successful deployment in a high-stakes agricultural environment. Looking forward, we expect that plant disease detection will evolve through the development of fully autonomous self-improving agricultural intelligence systems capable of collecting and processing all forms of multi modal data, interpreting complex patterns of plant health, forecasting, and quickly detecting outbreaks of disease, and efficiently recommending corrective action. With the addition of responsible AI principles, these technologies will transform how farmers make decisions about crop production, lower the number of pesticides they use, maximize production from each crop, and reduce negative impacts on the environment. Overall, AI and deep learning-based plant pathogenicity detection is leading the way in developing cutting-edge solutions for agriculture. By bringing together the best of computational science and applied technology, the future of plant health monitoring should provide increased productivity and reduced uncertainties through a more robust and ecologically sustainable global food supply.

References

[1] Ali, H., N. Shifa, R. Benlamri, A.A. Farooque and R. Yaqub. (2025). A fine tuned EfficientNet-B0 convolutional neural network for accurate and efficient classification of apple leaf diseases. *Scientific Reports, 15*, 25732.

[2] Mazumder, M.K.A., M.M. Kabir, A. Rahman, M. Abdullah-Al-Jubair and M. Mridha. (2024). DenseNet201Plus: Cost-effective transfer-learning architecture for rapid leaf disease identification with attention mechanisms. *Heliyon, 10.*

[3] Talaat, F.M., M.Y. Shams, S.A. Gamel and H. ZainEldin. (2025). DeepLeaf: An optimized deep learning approach for automated recognition of grapevine leaf diseases. *Neural Computing and Applications, 37*, 8799-8823.

[4] Naresh Kumar, B. and S. Sakthivel. (2025). Rice leaf disease classification using a fusion vision approach. *Scientific Reports, 15*, 8692.

[5] Kalpana, P., R. Anandan, A.G. Hussien, H. Migdady and L. Abualigah. (2024). Plant disease recognition using residual convolutional enlightened Swin transformer networks. *Scientific Reports, 14*, 8660.

[6] Prommakhot, A., J. Onshaunjit, W. Ooppakaew, G. Samseemoung and J. Srinonchat. (2025). Hybrid CNN and transformer-based sequential learning techniques for plant disease classification. *IEEE Access.*

[7] Yang, T., Y. Wang and J. Lian. (2024). Plant diseased lesion image segmentation and recognition based on improved multi-scale attention net. *Applied Sciences, 14*, 1716.

[8] Zhang, Q., W. Wang, H. Su, G. Yang, J. Xue, H. Hou, et al. (2025). PLCNet: A 3D-CNN-Based Plant-Level Classification Network Hyperspectral Framework for Sweetpotato Virus Disease Detection. *Remote Sensing, 17*, 2882.

[9] Joshi, H. (2024). Edge-AI for agriculture: Lightweight vision models for disease detection in resource-limited settings. *arXiv preprint arXiv:2412.18635.*

8 Hybrid Models
Integrating ML, DL, and Metaheuristics

Sakshi[1*], Rituparna Seal[2] and Satveer Kour[3]

The scientific discoveries in artificial intelligence have revolutionized the process of plant disease detection, but the agricultural industries have complicated the process and remain a puzzle to the conventional tools of analysis. Machine learning models are also interpretable and computationally efficient in nature, and depend on hand-built features, which restricts their ability to generalize their application to different crops and conditions. Deep learning models such as convolutional networks, transformer-based networks, and others allow automated hierarchical feature extraction, but typically have challenges with parameter set design, training instability, and the lack of labeled data. The chapter dwells on the hybrid frameworks that would involve a compilation of the machine learning, deep learning, and the metaheuristic optimization methods to enhance predicted accuracy, strength, and scale. It offers the rationale, a consistent taxonomy of hybrid architecture and a critical discussion of applications in leaf image classification, disease severity evaluation, hyperspectral imaging, aerial surveillance, as well as real-time smart farming systems. The discussion underlines the utilization of hybrid systems in solving the issue of data imbalance, minimizing overfitting, computing performance, and adaptive intelligent decision-making in multifaceted agricultural settings.

1. Introduction

The blistering development of digital agriculture has revolutionized the way plant health is monitored, diagnosed, and controlled. As the global food systems have to face unprecedented threats, such as the variation of climatic conditions and the emergence of new pathogens, as well as the necessity to satisfy the requirements of sustainable production due to the increasing need, the intelligent and automated solution is sought, which would provide the accurate and timely

[1] Apex Institute of Technology (CSE), Chandigarh University, Mohali-140413, Punjab, India.
[2] JECRC University (Sunstone), Jaipur, Rajasthan, India.
[3] Department of Computer Engineering and Technology, Guru Nanak Dev University, Amritsar, Punjab, India.
Email: sonaseal2000@gmail.com[2]; satveer.dcet@gndu.ac.in[3]
ORCID: https://orcid.org/0009-0008-5226-4354[1]; https://orcid.org/0009-0000-2199-6079[2]; https://orcid.org/0000-0002-4826-3366[3]
* Corresponding author: sakshigill12@gmail.com

DOI: 10.1201/9781003767534-8

diagnostics of the diseases. The application of traditional plant disease detection methods is based on the use of human skills, physical examination, and laboratory verification. Although they have been shown to work successfully in proprietary settings, these approaches are labor-intensive, slow, and subject to subjective bias making them unfit to apply to the contemporary large-scale agricultural ecologies. With drones becoming a part of farms, mobile imaging, interior sensors, and super-sensitive spectral devices, the depth and volume of the data available have become by far more profound. This development has introduced the path to the making of computational intelligence as a fundamental component of intelligent agriculture. In the last ten years, machine learning and deep learning have become the potent tools of automatic plant disease detection. The first to analyze plant images and physiological signals included the ML methods of support vector machines, random forests, and gradient boosting classifiers. Though interpretable and computationally efficient, these algorithms have had to rely on handcrafted features, which encode color, shape, texture, and morphology. These characteristics do not always reflect the heterogeneous and non linear nature of disease symptoms which are highly diverse across both species, seasons, geographies, and growth stages. This means that ML models trained on controlled surfaces often perform poorly in real-money domains where environmental noise, interference with the soil, warping of moisture, and irregular lighting conditions complicate demonstration of the symptoms.

Deep learning solved a lot of those limitations by learning discriminative representations on the basis of raw data. Convolutional neural networks (CNNs), vision transformers (ViTs), and the hybrid convolution attention architectures revolutionized the diagnosis of plant diseases, facilitating automated features ranking, which could classify complicated lesion patterns, fine-grained discoloration, spectral footprint, and spatial anomalies. However, there is nothing perfect about DL methods. They need gigabyte datasets with annotations attached to them, a capability that has been hard to achieve in agricultural imaging. Most plant pathogens have small visual samples, occur randomly or have covert signs at an early stage. Deep models are also extremely sensitive to hyper-parameter settings configurations and architecture. Even minor changes to learning rate, batch size, filter size, and normalization techniques or regularization strength may result in radically different convergence behavior. Moreover, DL models are generally computationally expensive black boxes which are incompatible with edge computing, or edge devices, and ultralow-power agricultural embedded systems [1].

Herein the metaheuristic optimization algorithms have become inevitable. Genetic algorithms, particle swarm optimization, grey wolf optimization, ant colony optimization, whale optimization algorithm, and differential evolution are metaheuristics which provide effective means of exploring multimodal, more-or-less complex search spaces without any gradient information or a-priori structural knowledge. Metaheuristics has a special property of not being trapped in local minima, having the capability to search vast solution space, and being adaptable to discrete as well as continuous optimization problems and tasks, which, again, are quite suitable in the context of optimizing ML and DL models to detect plant diseases. These algorithms have over time demonstrated great utility in selecting optimal subsets of features, hyper parameter tuning in DL, ensemble weights, loss function refinement, and also the overall stability and generalization ability of learning systems. The convergence gives birth to the need of hybrid models. In hybrid structures, the interpretability and efficiency of ML, the representational power of high capacity of DL, and optimum model configuration and robustness of metaheuristics are guaranteed. These systems that are integrated are not mere combinations of algorithmic functionalities but well-coordinated structures that combine the complementary strengths of every paradigm. Hybrid models have the capability to reduce the feature spaces, stabilize the training of the neural network, balance between the trade-offs of accuracy and the efficiency, and unlock more new functionalities, including multimodal fusion, the detection of early diseases and high-scale surveillance of aerial crops. The chronology of technological advancements in plant disease detection as illustrated in Figure 1 starts with the manual observation and then passes through the phases of Machine Learning (ML) and Deep Learning (DL) until arriving at Hybrid Architectures accepted as the technology to address the vulnerabilities that have been identified in the previous technologies.

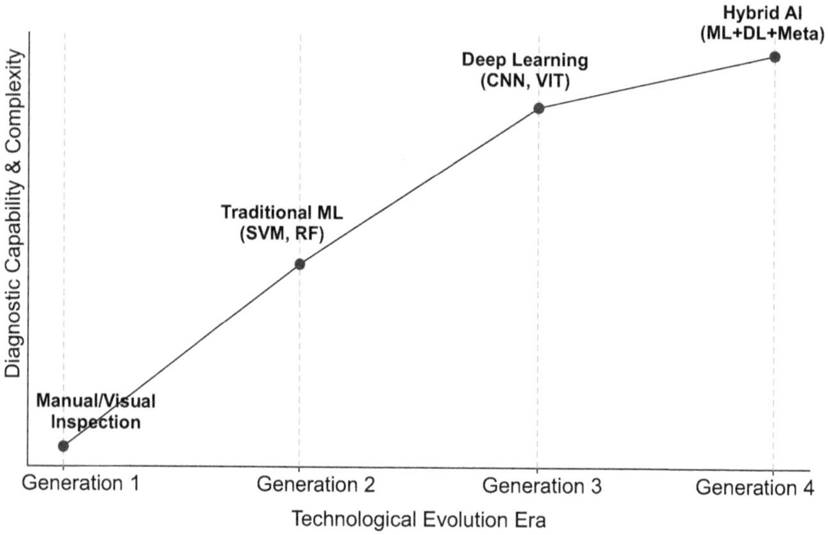

Fig. 1 Evolution of Plant Disease Detection Technologies.

Nevertheless, regarding hybrid ML-DL-metaheuristic systems, their complete idea is not fully exploited in the literature yet. The dissimilar research usually explores individual facets, including the metaheuristic-based feature-selection or the deep CNN-orchestration, but seldom an endeavor is made to tie these studies together within a single theoretical and practical framework. This is the integrated failure that reduces the level of scaling and the applicability of sophisticated models to diverse agricultural settings. With the future transformation of the agricultural systems to autonomous and intelligent ecosystems, the capability to create a hybrid model that will be able to generalize across crops, climates, and pathogens and imaging modalities is not only useful but vital. Thus, the chapter aims to address this gap of utmost importance by providing a very in-depth, properly structured, and analytical in character, and study of hybrid ML-DL-metaheuristic-plant-disease-detection models. It starts by an elaborate explanation of the underlying paradigms, thus forming a strong theoretical basis of hybridization. Then, an elegant taxonomy of hybrid architectures is presented explaining the working principles of various integration strategies and clarifying the circumstances in which they can be considered the most effective. This chapter then looks into real-world use cases, including the analysis of leaves, diseases-severity, hyperspectral imaging, unmanned aerial vehicle (UAV) surveillance, and smart-farming in real-time, and the particular benefits that hybrid systems bring to each case. The synthesis of recent case studies covering the time frame of 2021–2025 (used to show practical achievements and trends developed during this period) is provided. The chapter has ended by finally defining the future research in this area which includes quantum inspired optimization, evolutionary DL, federated hybrid intelligence, and self-optimizing agricultural systems which individually could influence the next generation of intelligent agricultural systems.

With this plot and line of analysis, the chapter does not just explain the request of hybrid artificial intelligence in agriculture, it contextualizes it as well. It shows how the coming together of ML, DL, and metaheuristics is creating a new generation of robust, scalable, and clever schemes of plant-disease-detection systems in line with global demands to produce food sustainably, practice precision farming, and achieve environmental stability.

2. Background and Theoretical Foundations

Plant disease detection technologies have significantly transformed within the last ten years, with the influence of computational intelligence, sensor, and agriculture digitalization. The role and need of hybrid ML-DL-metaheuristic models can only be appreciated based on a profound understanding of the basis of this evolution. In this part, a conceptual and historical background of the three main pillars, including ML, DL, and metaheuristic optimization is given and then a synthesis of the weakness of each to create a strong argument in support of hybridization is presented. Instead of these components being isolated techniques, the discussion includes highlighting the internal logic and working principles of these components, as well as their applicability to plant pathology, providing a uniting framework on which later hybrid architectures in the chapter are based.

2.1 Introduction of Machine Learning in Plant Pathology

ML has proven useful in the preliminary automation of plant illnesses diagnosis. In the past, Support Vector Machines (SVM), Random Forests (RFs), k-Nearest Neighbors (KNN), Decision Trees (DT), and Gradient Boosting Machines (GBM) models were considered to be a great improvement over manual inspection since computers are able to process visual and spectral patterns related to disease symptoms. These models were frequently based on carefully engineered, handcrafted features, in the form of statistical characterizations of color distributions, uniformity in texture, geometry of lesions, and structural change in plant leaves. As an example, the co-occurring of roughness or smoothness of textures was provided by the Grey-Level Co-occurrence Matrix (GLCM) although color indices were used to detect chlorosis and necrosis. These properties gave predictable indicators that matched domain knowledge in plant pathology and thus allowed ML models to make classification or regression with significantly small computational costs. Nevertheless, on the one hand, the feature engineering with ML is limited by the suppositions of the author. In the case of manually selected features, generalization can only be generalized to the extent of the specifics of the selected features. Photographed leaves in diverse field conditions can vary significantly in terms of illumination and angle of view, shadow density, and coverage by the ground or the surrounding plants. The symptoms of diseases, too, vary both among species of plants, and within cultivars, and stage of growth, and make feature sets created by hand sensitive and limited. These limitations have multiplied with the growth of the size and variety of agricultural data. Although ML models are highly accurate at interpreting low-dimensional high-structured features, their ability to learn complex, non linear, and high dimensional patterns which are critical in efficient disease diagnosis is essentially constrained [2].

2.2 Deep Learning Foundations

DL became disruptive in that it eliminated manual feature engineering. CNNs also proposed hierarchical feature learning, which allows models to learn the latent patterns of raw pixel data automatically. Shallower CNN layers capture unspecific features, including edges, textures, whereas later ones acquire the more abstract concept of symptom disease, such as lesion contours, mottling, and color aberrations. VGGNet, ResNet, DenseNet, MobileNet, EfficientNet, and ViTs among other architectures have been widely used in the plant pathology field with impressive precision in tasks like leaf classification, severity, leaf segmentation, and even detection of subtle disease signatures.

However, DL is not a limitless technique. The deep models are expensive to interpret since they have millions of parameters meaning that they may have very little or no performance when large and high-quality datasets are unavailable, which is common in agriculture. Numerous diseases in plants are rare and have insidious symptoms or need expert commentary, which has led to small or extremely skewed data sets. Such environments, in turn, cause problems with DL models, often overfitting on training data or not doing as well with generalization out in the real world. In addition, even the delineation of the DL architectures is contingent on the complexity of hyper-parameter configurations: the learning rates, the batch size, the filter size, the activation functions, and the dropout rates, the choice of optimizers, and the choice of normalization all interact in highly non-trivial fashions, which affect training stability and convergence. Manual tuning is labor-intensive, imprecise, and often not acceptable to attain optimal performance. Even data-augmentation strategies, despite being advantageous, are ineffective enough to cover these drawbacks. Thus, even with its capabilities, DL is not an autonomous tool that can adapt to the task of plant diseases detection, particularly in the field-deployed setting where data diversity, noise, and variability represent a standard.

2.3 Metaheuristic Optimization: GA, PSO, DE, ACO, WOA and Others

Metaheuristic algorithms have become popular as powerful optimization methods with the ability to scourge multimodal search spaces. Driven by physical phenomena, like genetic evolution, swarm behavior, predation tactics, and communication mediated in part by pheromones, these methods are gradient information free and hence can be used to optimize discrete and continuous variables. The reason is that they are versatile due to some (inherent) balance between exploration (global search) and exploitation (local refinement), which enables them to evade local optima and find near-optimal solutions in difficult optimization landscapes. Metaheuristics has a number of important roles in the field of plant disease detection. Genetic Algorithms (GAs) undergo mutation and crossover functions by evolving sets of features or parameters through generations, with each generation used to make improvements in performance. The swarm intelligence underlying particles Swarm Optimization (PSO) is used to make use of continuous hyper parameters small adjustments in the path of particles based on personal and global optimal solutions. Differential Evolution (DE) is an advanced user in the optimization of high-dimensional continuous parameters by the differentiation of mutation plans, hence is an appropriate algorithm to the optimization of neural-network architecture. Gray Wolf Optimization (GWO) and Whale Optimization Algorithm (WOA) are fast converging algorithms and have been successfully applied in optimizing deep neural networks or in optimizing CNN hyper parameters. Ant Colony Optimization (ACO), which is based on pheromone dynamics, has shown effectiveness in such types of combinatorial problems as the choice of the most relevant handcrafted or learned features.

The individual metaheuristic algorithms have their own advantages, but ultimately, they all come to one strength: the ability to optimize artificial intelligence models in one of the cases when grid searching, random searching, and Bayesian optimization no longer fit or when they cannot be computed. Their authoritativeness to gradient information makes them the only classifications applicable to agricultural data, which are prone to noise, irregularity, or size.

2.4 Limitations of Standalone Systems and CAO Requirement of Hybridization

The weaknesses of the ML, DL, and metaheuristics in their respective isolation significantly contribute to the importance of the hybrid methods. Models of machine-like learning based on a manually designed set of features have limited plasticity and richness in visual or spectral representations. Possessing the ability to learn effective representations, DL architectures have high computational requirements and are significantly sensitive to hyper parameters. Metaheuristics despite being better as optimizers do not have embedded classification or recognition abilities and thus require surrogate objective functions based on ML or DL models.

Hybridization is not only the refinement adopted but the reality and conceptual need. With a combination of ML or DL with metaheuristic optimization, the generated models benefit without the ineffective representational learning as well as intelligent search. Hybrid ML-metaheuristic structural frameworks gain the ability to acquire small, discriminative feature sets that increase interpret wholeness. DL-metaheuristic models are more stable, converging, and concurrently perform better even in a data -scarce environment. Hybrid ensembles based on a combination of ML, gDL, and metaheuristics achieve some level of flexibility and robustness, which none of the paradigms could achieve. Such hybrid systems can cleanly deliver scalable and robust disease-detection operations with data imbalance, environmental variability, and fast pathogen evolution, an issue that is difficult to overcome with other non-hybrid systems in agricultural environments. Combining the advantages of both paradigms with the reduction of the disadvantages of each, hybrid models are by far the best possible way to reach intelligent, autonomous plant-disease-detection systems that would be still working within the constraints of a real-life environment.

3. Hybrid ML DL Metaheuristic Paradigm

The appearance of hybrid, mechanized, ML-based frameworks combining ML, DL, and metaheuristic optimization works represents a decisive decrease in how traditional intelligent plant disease detection systems should be created. Hybrid architectures represent a layered intelligence whereby the constituents make the one another up, unlike the traditional pipelines that relied on the use of a single computational paradigm. ML adds interpretability and computational simplicity, DL adds representational richness and the capacity to discover complex patterns and, metaheuristics makes it easier to do sophisticated global optimization, able to discover high quality solutions in large, multimodal search spaces. This synergy is not merely a superficial convergence but a genomic entanglement and amalgamation of learning, representation, and optimization where groups of systems emerge as capable of operation within the uncertainty, variability, and heterogeneity intrinsic to the actual agricultural environment. Conceptually, hybridization requires in depth insight into the interaction of these three paradigms in terms of those interactions which are structural as well as functional. Other hybrid architectures assume a sequential design in which DL is followed by metaheuristics then improving ML-driven classification. Other architectures have been used in parallel or nested, in which case metaheuristics are used to optimize the internal parameters of DL models or choose between more than one competing ML-DL pairing. These architectures have divergent aims (e.g., maximizing accuracy, focusing on computational efficiency, or improving robustness), but they all have in common a core concept, namely, that the fused system should provide intelligence, stability, and generalizability to a much greater extent than its individual parts [3].

3.1 Taxonomy of Hybrid Models

The current space of ML, DL, and metaheuristic optimization comes up with extensive design space that is necessary to navigate systematically through a taxonomy of hybrid models. Older literature tried to categorize hybrids simply: usually with a couple of model combinations, or only with metaheuristics being applied as a mechanism to tune things, but the new hybrid structures no longer fit into such simple classifications. There are a variety of integration patterns present in the existing taxonomy representing different degrees of interaction and inner complexity. Several hybrid systems are built based on metaheuristic-assisted ML methods, in which a more conventional classifier, supporting vegetable machines, random forests, or gradient-boosting machines, is integrated into a metaheuristic search script. In this paradigm, metaheuristics are intelligent search agents which evaluate sets of candidate features or classifier hyper parameters in terms of learning performance and refine the ML model.

Other architectures are based on metaheuristic-optimized DL architectures, in which the representational capacity of a convolutional neural network, recurrent neural network, or vision transformer is increased via evolutionary or swarm-based algorithms that optimize hyper parameters, and network depth or types of filters or normalization scheme or loss functions. These hybrid models do not solely rely on the human intuition or trial-error, but rather explore through meta-heuristic exploration methods to find tunings that are consistent with the non linear challenges of plant-disease imaging. Advanced hybrid systems embrace ensemble-based ones, where ML and DL co-exist, usually mediated by metaheuristics. The DL in such arrangements can act either as the main feature extractor or as second-order decision units with the machine-learned classifiers, or the other way around. Metaheuristics optimize ensemble weights, use artificial complements on learners, and optimize fusion techniques, which give extraordinary immunity to them especially on problems of multi-disease classification or imagination in straggled fields. Figure 2 provides a taxonomy of hybrid models sorted into four main categories, namely: Metaheuristic-Assisted ML, Metaheuristic-Optimized DL, Hybrid Ensembles, and Multi-Stage Pipelines.

Another group consists of multi-stage hybrid pipes, which incorporate ML, DL, and metaheuristics on different stages of preprocessing, feature fusion, and classification.

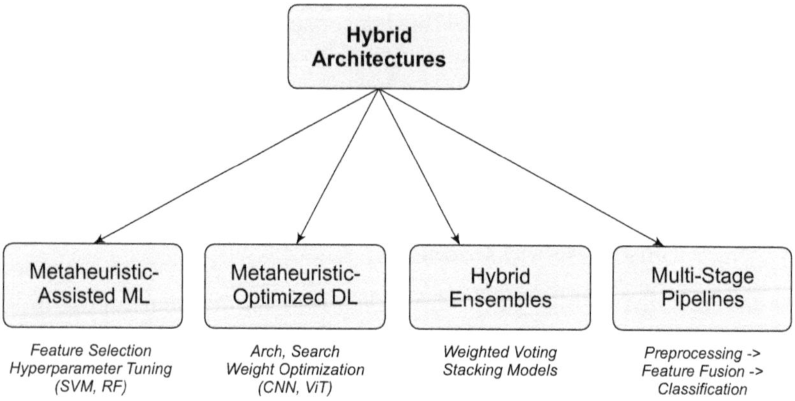

Fig. 2 Taxonomy of Hybrid ML-DL-Metaheuristic Models.

3.2 Meta Selection Feature Selection with Metaheuristics

The process of feature selection is a fundamental part of hybrid ML-metaheuristic schemes, especially on tasks where handcrafted features continue to be useful or when DL-generated embeddings require dimensionality reduction. The datasets of plant diseases are commonly full of redundancy or noisy data that will blur the important patterns of discrimination. The traditional feature selection methods which include filter or wrapper methods are too weak to search through the big and wide combinatorial space of all sets of features especially with multidimensional spectral or texture descriptors. The solution offered by metaheuristic optimization is a graceful one because it reformulates the feature selection problem instead as a global search problem where every possible subset of features is a candidate solution. GAs are known to find optimal subsets by simulating a biological evolution, using crossover and mutation to find promising combinations. PSO switches weights of features dynamically according to the swarm intelligence which is the collective means of finding an equilibrium between exploration and exploitation. Whale Optimization Algorithms and Gray Wolf Optimization are predator-prey and cooperative hunting methods which developed algorithms, which have shown strong results in choosing features that optimize the accuracy in classifiers, at the same time reducing the dimensionality greatly. Combined with ML classifiers like SVM or XGBoost, these subset optimized sets of features provide a hierarchy of hybrid systems that are more interpretable, computationally efficient, and resistant to noise. Similarly, when used with DL embeddings, in which hundreds or thousands of features can be learned in the intermediate convolutional layers, the metaheuristics will eliminate irrelevant dimensions, reduce overfitting, and enhance generalization. In domain settings where the quality of data changes with natural occurrences, those hybrid systems with features offer stability and resilience that cannot be matched to data-only methods. Figure 3 illustrates the strong interactive relationship between the three pillars of foundation, namely, ML (Interpretability), DL (Representation Learning), and Metaheuristics (Global Optimization) showing how the three come together to create robust agricultural decision-support systems.

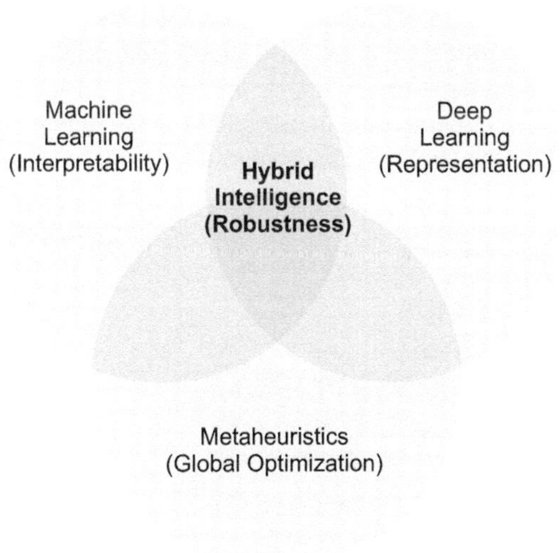

Fig. 3 The Synergy of Hybrid Components.

3.2.1 Suite to Hyperparameter Tuning with Metaheuristics

DL models have very high expressive capacity and their performance is inherently associated with extreme care in the selection of hyperparameters. The default training of a CNN or any other complicated neural architecture usually provides less-than-optimal convergence or unstable training dynamics, or bad generalization. Conventional approaches of hyperparameter tuning like grid search or random search are computationally infeasible and will not provide the adaptive insight to explore high-dimensional, non-smooth search space. By comparison, metaheuristics provides a general and effective process of preoptimization by encouraging exploration with feedback probabilistic guidelines or evolutionary dynamics. The use of algorithms like GA, PSO, DE, GWO, WOA, etc. has been done in optimizing learning rates and batch sizes, network depth, filters integer-per-layer, dropout ratios, weight-initialization schemes, and choice of activation functions. Their ability to find local minima is global, which allows their escape of the local minima that commonly plague DL optimization procedures. Through metaheuristic parameter searches, hyperparameters are optimized by comparing candidate configurations based on fitness functions based on model accuracy or the validity loss, and hence converges to global optimal configurations absent in conventional optimization methods. The results of metaheuristic-based hyperparameter optimization are significant in the field of detecting plant diseases. Characteristics associated with agricultural datasets include limiting sample size, heterogeneous resolutions of imbalanced classes, and noise of environment. In this case, deep networks that are not tuned also have a probability to overfit or perform poorly. Stabilized training with hyperparameter tuning, hyperparameters are optimized with the help of a metaheuristic, which speeds up training and leads to the production of models that are also predictable when applied to different data sets and imaging conditions. This is the basis of targeting hybrid ML-DL-metaheuristic systems that may be implemented in the actual agricultural setting in real-time. Figure 4 illustrates the particular functional positions of metaheuristic algorithms (e.g., GA, PSO) in the framework of hybrid systems with the major focus on Feature Selection and Hyperparameter Tuning as the most common uses.

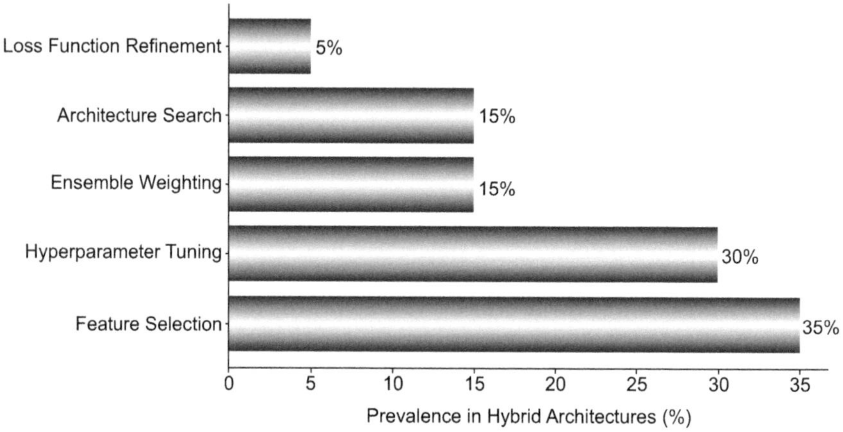

Fig. 4 Primary Roles of Metaheuristic Optimization.

3.3 The Concept of Hybrid Ensemble Architectures

Ensemble learning is also another powerful combination technique, which combines the strengths of several learning algorithms to create more powerful plant disease detection models. In ensemble architectures, ML may be synergistically co-located with DL because they provide a better understanding of each other. DL models, which have the ability to represent patterns of diseases at a fine level of detail via their hierarchical representations, are able to express the disease patterns, and machine-learned models give quick, interpretable margins when a decision is required. However, to build successful ensembles, fine-tuning requires an analysis of what models should be, how to weight them, and how to combine their outputs. The intelligent glue where these ensemble components are glued is called metaheuristics. Evolutionary algorithms are able to discover the most desirable fusions of base learners, tune fusion weights, and also choose ensemble structures that optimize the predictive performance. Other hybrid ensemble options Deep networks can also be used as feature extractors, in which high-level representations are created which can then be interpreted by machine-learned classifiers. In this case, metaheuristics are used to optimize the extraction thresholds or favorable settings of classifiers and the final result is ensembles that are not only effective, but also computationally efficient. Such combinations are effective in situations of complex phenotypes of diseases, interactions between multiple pathogens or visually related symptom patterns. They also are suited well to heterogeneous datasets where one model cannot work best on all samples. The ensemble is a coherent intelligence system, which is enhanced through metaheuristic optimization and augmented with the supplementary thoughts of the ML and the DL models.

3.4 Multi-Objective Optimization of Hybrid Systems

With detection of plant diseases increasingly moving towards the real-world application of drones, mobile, and field sensors based on the Internet of Things, the performance metrics grow beyond precision. The models should be lightweight, energy-efficient, quick in inference, and tough to different environmental conditions. The concept of multi-objective optimization is essential because the demands (which are sometimes conflicting) have to be balanced. Multi-objectivation: A multi-objective algorithm like NSGA-II is able to optimize accuracy, computation speed, memory usage, and hardware usage. Multi-objective metaheuristics are used in hybrid frameworks to build models that are responsive to controlled environments but which can still survive under the real-time constraints of the environment. This applies especially in smart farming where computers are scarce and quick response and reaction to diseases is needed. A multi-objective hybrid will guarantee the detection of the disease in a plant will be an actual scientific and practical instrument, which will be able to mitigate a gap that exists between the research carried out in the laboratory and the application in the field.

3.5 Multi-Source Data Fusion

The data types produced by modern agriculture produce a very broad range, such as RGB images, hyperspectral cubes, thermal signatures, the readings of environmental sensors and maps of canopies taken by drones. The combination of these multimodal inputs continues to be relevant in terms of detecting diseases effectively as it provides a complete picture of hospice plant health. Hybrid ML-metaheuristic models present a gracious framework of combination wherein deep learning removes modality-specific features; ML deciphers combined features and metaheuristics optimizes the combination strategies. Metaheuristics are critical in the alignment of different data modalities, the determination of optimal fusion layers, the choice of the suitable channels and the balancing of cross-modal contributions. This leads to better and more factual and stable multimodal models in comparison to their single plans. The hybrid multimodal systems are the future of agricultural intelligence that will allow an earlier and more accurate disease diagnosis than ever.

4. Applications in Plant Disease Detection

Combining ML and DL with meta heuristic optimization has significantly diversified the range of tasks to which plant disease identification can work, allowing systems to be used across a wide range of imaging modalities, crop taxa, environmental situations, as well as different deployment platforms. Historical computational techniques had been restricted to the laboratory-extracted images of leaves but modern hybrid designs support the diverse sources of heterogeneous data, including hyperspectral signatures or drone-measured canopy images and real-time driving field sensor streams. The value of hybridization is in the ability to cope with the limitations of each application field: restricted data, harsh environmental noise, or requirement of inference speed in precision agriculture. This discussion looks at the application of hybrid models in the main areas where plant disease detection is applied with a focus on its theoretical background, its working processes, and advantages over the traditional systems. As Figure 5 compares the conceptual performance of ML, DL, and Hybrid models in terms of dataset size and complexity, it can be seen that Hybrid models will be more stable in a data-sparse or noisy environment.

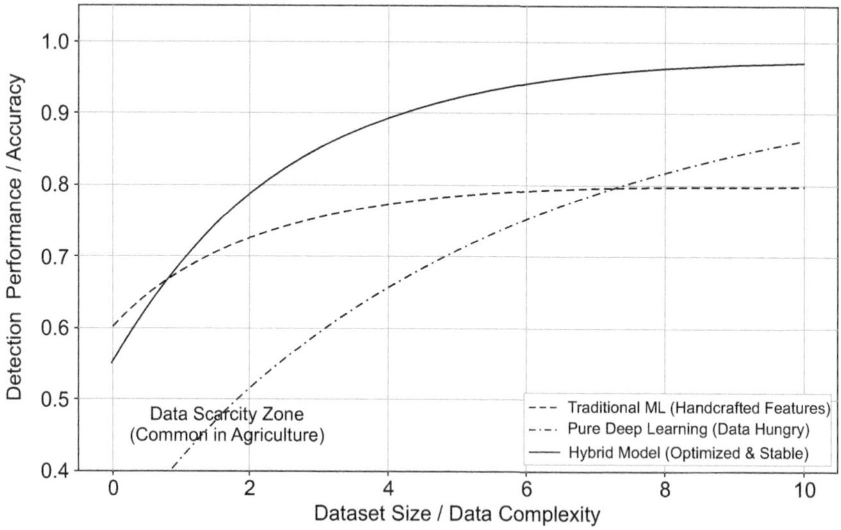

Fig. 5 Conceptual Performance in Agricultural Environments.

4.1 Leaf Image Classification

Computational plant disease identification: Leaf image recognition is the best studied, and possibly the most widely applied, use of computational plant disease detection. The superiority in this area has been significantly transformed into hybrid ML-DL-metaheuristic models that formulate fundamental issues associated with visual variation, underlining distinctions between the diseases, and environmental deformations. Classical systems were based on hand-crafted features derived out of the description of color, shape, or texture; these styles were incapable of representing the complicated pattern of lesions, especially in cases of symptom overlap. DL has overcome this shortcoming by learning hierarchical representations directly on real pixels; however, models based on DL overfit more easily to relatively small-scale datasets as found in agricultural research. Thanks to hybrid frameworks, a balanced solution is achieved. As their main feature extractors, either the CNNs, variants of the DenseNet, or the Variance Transformers have been utilized, which contain rich spatial and spectral features of the leaf images. Such learned representations are then optimized using meta heuristics which pick out the most discriminative dimensions, drop unnecessary or noisy features, and improve overall generalization. Clustering by ML algorithms (including sSVM; RFs; or XGBoost; and others) is typically classified, and then optimized using a meta heuristic with increased robustness in decision boundaries.

Most modern systems also make use of meta heuristics to directly optimize DL architectures by optimizing hyper parameters, layer depths, filter sizes, or regularization strengths. The ensuing pipelines of leaf-classification not only show a high level of accuracy, but also, are much more stable with diverse data sets, field environments, and species. Therefore, hybrid models have shown to be very effective in separating diseases with slight differences visually and in the detection of slight-stage infections and a good performance even when not using even illumination and background. In Figure 6, a canonical hybrid pipeline example is shown, where DL generates features, Metaheuristics optimizes them, and finally, it is followed by the final classification completed by ML.

Input Leaf Image	Deep Feature Extraction	Metaheuristic Optimization (Feature Selection)	ML Classification (SVM/RF)	Disease Diagnosis

Fig. 6 Typical Hybrid Workflow for Leaf Image Classification.

Although classification defines the existence of a disease, the severity estimation quantifies the level of infection or advancement of the infection. This activity is especially more complicated, because the severity is evidenced in the form of progressive, unremittingly changing lesion area, color corrosion, or tissue necrosis. Conventional methods were based on inaccurate and unreliable measurement by hand or using a rule of thumb. Segmentation and regression methods proposed by deep learning can learn how to predict severity more faithfully, but it is also very vulnerable to noisy data, irregular leaf edges, and the vagaries of natural imagery.

Hybrid systems address these issues by combining segmentation using DL and meta heuristic optimization. Convolutional segmentation models include U-Net or Mask R -CNN to focalize lesion areas and meta heuristics to more closely color segmentation losses or more perceptive region-merging policies. ML is then used to convert segmented lesions into numerical severity scores of the segmented lesions in calibrated regression models. DL segmentation with meta heuristic improvement is particularly effective in cases where the identification of severity is performed in the presence of irregular boundaries of lesions, overlaps between symptoms, or partial occlusion. It further enables adaptation to the development stages of the disease when there are poor visual cues or they are ambiguous. With representational strength of DL, meta heuristic accuracy, and ML interpretability, hybrid severity-estimation systems will be highly accurate and stable in practice in decision-support settings [4].

4.2 Hybrids Found in Hyperspectral Imaging

Hyperspectral imaging (HSI) technique records spectral features in hundreds of small bands with high detail of biochemistry and physiological changes occurring in vegetative tissues, years before the onset of visual symptoms. Nonetheless, the hyperspectral data is high-dimensional, redundant, and computationally intensive, and thus requires powerful feature selection and extraction processes. The traditional ML techniques will be incapable of tackling this dimension, whereas the pure DL model will demand very big data sets to prevent overfitting. Transformative solution Hybrid ML-DL-metaheuristic frameworks provide a solution. Neo-Darwinian band selection Band selection A neo-Darwinian band selection, e.g., GAs, PSO, or differential evolution finds, at the cost of dimensionality reduction, the most informative spectral bands and projects them onto DL representations or ML classifiers prior to data entering the DL applications. This also makes learning algorithms read and process only the most salient wavelengths, thereby enhancing computational efficiency without sacrificing discrimination ability. These optimized bands are then retrieved as meaningful representations using deep architectures like 3-D CNNs or hybrid spectral -spatial networks. ML classifiers further narrow down on predictions especially where samples are small or dissimilar. The integration of these systems leads to hyperspectral hybrids, which can significantly best single system methods in early fungal infections, nutrient deficiencies, bacterial blights, and viral stresses detection. With the further integration of HSI into the field-compatible sensor and drone-based architecture, hybrid models make real-time spectral analysis with limited computational burden cost feasible, making them experimentally appropriate in the context of the operational precision-agriculture systems.

4.3 UAV/Drone-Based Systems

Unmanned aerial Vehicles (UAVs) have radically changed the way plant diseases are detected in that they allow the monitoring of vast farms, orchards, and plantations. There are special issues of drone-recorded imagery: a lot of environmental noises, uneven altitudes, diffused light, motion blur, the movement of leaves due to wind forces, and complicated backgrounds. All these situations tend to decrease the performance of purely DL models, which are prone to distortions and show worse generalization in heterogeneous fielded situations. Hybrid designs deal with these challenges by integrating features extraction expressed as DL with meta heuristic optimization that improves segmentation, detection threshold stabilization, and the method of calibrating spatial filters to aerial images. The path planning and region-of-interest selection are also optimized by meta proxies that assist drones to focus on locations with more risks of infection. ML classifiers can be used with DL to differentiate disease-like artefacts and the true symptoms, and help to reduce false positives due to shadows, soil patches, and canopy variations. Figure 7 represents the tradeoff between Model Accuracy and Computational Efficiency (Inference Speed), and it is apparent that hybrid systems attempt to reach the Pareto-optimal frontier regarding real-time use.

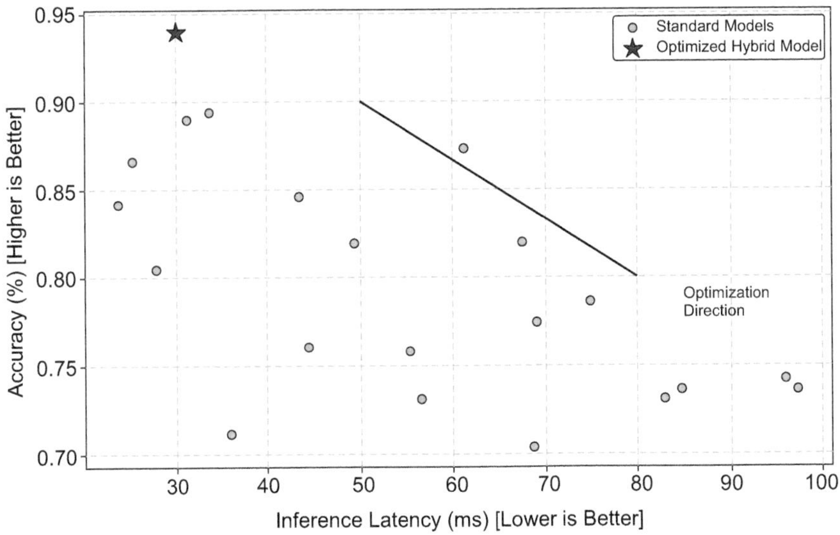

Fig. 7 Multi-Objective Optimization (Accuracy vs. Efficiency).

The duo of multispectral and thermal drone sensors increases the range of diagnostics of hybrid models. Under this arrangement, machines learn cross-modal representations, cross-modal predictions made based on ML are predicting disease categories or changes in the biomass, and spectral channel selection or weight combination across modalities is performed by the meta heuristics. These hybrid UAV based approaches would be crucial towards scalable, non-invasive, observation of large agriculture fields to facilitate early warning and strategy to intervene [5].

5. Case Studies

Case Study 1 Hybrid GAN-CNN-RNN Framework to Real-Time Plant Disease Detection

The proposed GAN -CNN -RNN architecture by authors marks one of the most important steps towards the current hybrid frameworks used in plant disease monitoring. The architecture is developed to work with real-time deployment and targets the joint use of a Generative Adversarial Network (GAN) used to boost data augmentation, CNN to extract hierarchical spatial features, and Recurrent Neural Network (RNN) to model temporal dynamics of image sequences. This integration is resistant to two endemic issues of agricultural artificial intelligence, limited access to annotated field imagery, and the dynamic course of disease development. Through the production of plausible samples of diseased foliage in varied conditions, the GAN whips away the lack of data, thus enhancing the training of the CNN element. Afterwards, the RNN identifies the small-time variations that distinguish the initial stages of manifestations and the more advanced infections. Empirical testing shows that the multi-module structure achieves a high accuracy of around 92% that provides strong performance with the fungal, bacterial, and viral pathogens in the real world. The system is a representation of the hybrid model philosophy, i.e., each element would mitigate the shortcomings of the other one, and thus create an integrated pipeline that could be used in the field setting. The framework supports hygienic concepts by the combination of complementary neural structures to deliver significant performance improvements though no explicit classical metaheuristics, including GAs or PSO is implemented. As a result, it is particularly relevant in the period of 2021–2025 because it is the investigation that goes beyond the fusion of models into hybridization to the development of integrated, multi dimensional intelligence depending on the real agriculture condition. Figure 8 outlines the architecture of the hybrid engine, suggested by the author, connecting GANs to help in data augmentation, CNNs to extract spatial features, and RNNs to capture the temporal dynamics [6].

Case Study 2: Enabling Deep Neural Networks and Quad-Tree Decomposition Cases of Classification in Real-time

Authors offer a hybrid model where the classical methods of image-processing are used in conjunction with DL to detect the plant's diseases quickly in the field. The technique uses quad-tree decomposition to organize images into different levels of approval blocks, therefore focusing the computing performance on the areas of interest before neural network analysis. This structure-sensitive decomposition involves significantly less algorithms in the form of deep neural

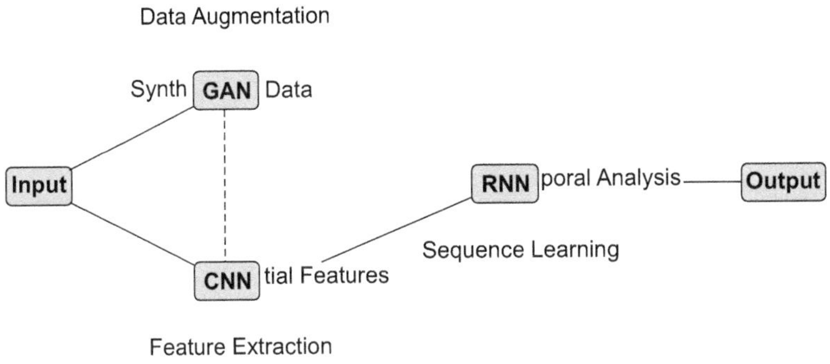

Fig. 8 Hybrid GAN-CNN-RNN Architecture.

network operations, since heavy ordinary regions of the image are not processed by the deep neural network, but only informative ones. The resulting pipeline is both fast and accurate with a F1-score of around 0.80 in four disease classes with both potato and tomato leaves. The hybridization is salient especially in the focus on the real-time performance which is a major stored need towards autonomous deployment through the drones or robot platforms. The type of deterministic, rule-based partitioning combined with learned representations is indicative of a larger shift in hybrid artificial intelligence: using more traditional methods to organize the input space and leave deep networks to do refined pattern recognition. Though it is not explicitly investigated that the authors are using metaheuristic optimization, their solution is a combination of deterministic spatial decomposition and flexibility of DL, providing a fast, efficient, and scalable system that works on embedded hardware [7].

Case Study 3: Unweighted CNN -SVM Hybrid with Ensemble Learning

Authors of the paper hybrid engine control model called Ence-SVMWEL, are a hybrid machine husband-learning framework based on DenseNet201 with the SVM. High-capacity deep embeddings are not only generated using DenseNet201 given sugarcane leaf images, but the SVM tends to over-generalize better than fully connected neural layers and is generally more adequate with small or medium sized datasets. In order to better improve performance, the authors introduce a weighted -average ensemble, across various variants of SVM, which leads to an improved decision surface when compared to a single classifier. This combination design has an accuracy of 97.45% on a six-class sugarcane disease dataset, indicating the representational capability of DL and the interpretability and regularization ability of classical ML. Despite the fact that the ensemble weighting is based on heuristic as opposed to metaheuristic approaches, the architecture represents an example of hybrid system behavior where complementary modules operate to perform special functions in a single pipeline. The paper drives to the applicability of hybrid ML-DL models, which remain appropriate in areas with a restricted dataset due to the significant risk of overfitting of fully deep models, in the absence of adequate sample diversity [8].

Case Study 4: Massive Transfer Learning Ensemble Multi-class Disease Detection

An extensive application of hybrid strategy, published in the BMC Plant Biology 2024 provides an example of the benefits of deep ensemble fusion when applied to large scale disease classification problems. The authors optimize nine models already trained: ResNet, DenseNet, EfficientNet, and Mobile Net, and merge their networks using early fusion and lead-voting techniques. Considering a large dataset, Plant Village, the ensemble obtains an accuracy over 96 across 15 classes of diseases. Through several architectures, the ensemble exploits a wide range of spatial, textural and structural attributes of the plant diseases, which minimizes systematic bias and false classification. Although metaheuristic optimization is not directly used, the advanced fusion approach is similar to the concepts of multi-solution search as well as robust diversity that are present in metaheuristics. The ensemble provides the ability to have mistakes perpetrated by one architecture be offset by other ones, and the voting mechanism is a crude but effective decision-optimization device. The example of this massive hybrid system is the availability of architectural diversity and weighted decision integration improving performance significantly, especially in cases of multi-class systems where the symptoms of diseases vary significantly among species and under varying environmental conditions [9].

6. Future Directions

Integrated intelligence systems, where plant disease detectors can understand, adjust, and optimize their capabilities in real-time independently are becoming the future of plant disease detection. Since agriculture will become a large-scale digital ecosystem with IoT sensors, drones, cloud computing, and autonomous machines, the expectations of computational models will be far broader than the standard metrics of accuracy. The systems of the future should support scattered sources of data, an environmental heterogeneous environment, privacy, and changing pathogenic landscapes. Hybrid ML-DL-metaheuristic models are in good positions to satisfy such needs, but their potential has not been fully explored. The section critically reviews the new theoretical and technological directions which are expected to characterize new research and innovation within the upcoming decade in the field of plant disease detection. The neo-Darwinian features of band selection in high-dimensional data cubes of hyperspectral data are demonstrated in Figure 9, thereby providing optimized features to the DL models.

6.1 Quantum-Inspired Metaheuristics

The creation of quantum-inspired metaheuristic algorithms which do not require complete scale quantum computing devices has been shown to be one of the most promising frontiers. Quantum-inspired GA, quantum-behaving PSO, and quantum-annealing-based searches utilize probabilistic superposition, tunneling,

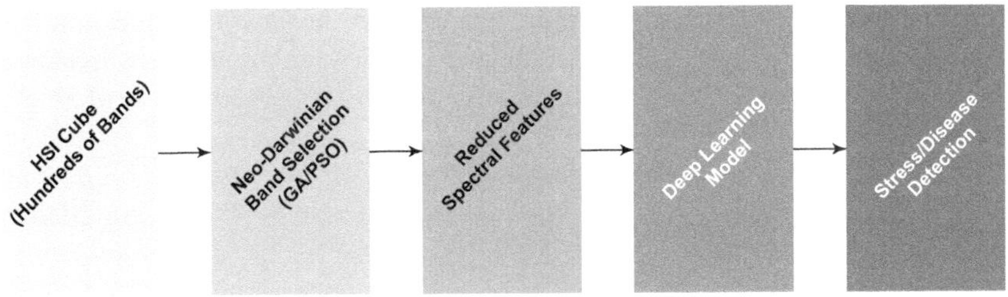

Fig. 9 Hyperspectral Imaging Processing Pipeline.

and parallel-state search to solve agglomeration problems the same way your more traditional metaheuristics would find local minima inefficiently. These approaches are much more powerful at search than anything else, and are able to optimize neural architecture, hyperparameters, feature subsets, and ensemble configurations that were previously inaccessible. Quantum-inspired optimization can be used in the model design process to achieve quantum advantages in plant disease detection by dramatically simplifying the process of adapting deep models to crop, climate, or imaging conditions. As an example, thousands of parameters and design choices might be part of the optimization of a Vision Transformer architecture on a hyperspectral crop dataset. Such high-dimensional spaces are difficulty searched in classical optimization methods, and can be better searched in quantum-inspired algorithms. This paradigm is especially favorable to the models that can be deployed to the field, enabling the optimization of the model to be performed on the device with harsh energy and memory constraints. With the maturation of quantum computing hardware, it is plausible that authentic quantum optimization could be implemented with ML-DL hybrids, radically changing the outlook of world-scale precision agriculture.

6.2 *Evolutionary Deep Learning*

Embarking on another important direction is evolutionary DL, where the form and degree, interconnection and training of the neural networks are automatically evolved by some form of genetic or swarm methodology. In contrast to conventional training processes using evolutionary processes never make any architectural assumptions, but instead of designing network topologies, they permit neural networks to expand, shrink, mutate, and recombine through an evolutionary process similar to natural biological evolution. The result of this adaptive design procedure is a distinctively modeled architecture that is distinctly sensitive to the complexity of the particular datasets not only with irregularly distributed symptoms but also to those with multimodal signals or with class imbalance. In the sense of plant disease diagnosis, evolutionary deep learning may yield light and powerful models, which are optimized to a particular crop species or environment. An engineered model useful in detecting disease in tomatoes grown in a greenhouse may not have the same development as an engineered model used to detect disease in maize plants in open fields due to differences in light, leaf structure, and disease expression. The evolutionary engine of this process is meta hybridization through which optimal performance by architectural designs is progressively improved by way of fitness-based selection. With the growth in agricultural data types the evolutionary DL as a future direction creates an avenue where the models learn not just off data but also the design principles that are implicitly designed in the hybrid evolutionary process.

6.3 Self-optimizing Agricultural Systems

The intelligent agriculture systems of the future will go beyond the non-portable detection pipeline design to closed loop structure where disease sensing, decision making, and remedial action are taken as part of an ongoing process of feedback. These are self-optimal farming systems, which combine the hybrid modes in which the system automatically reregulates itself based on the environmental variations, the growth phases of crops and the changing behaviors of the pathogens. A self-optimising system can detect the health of a plant by constantly using cameras, drones, and multispectral sensors and at the same time, changing disease detection thresholds and adjusting model parameters, and the representations used in features as new data come in. The continuous optimization loop will heavily depend on metaheuristics; this will enable the system to self-regulate itself without being retrained. As an illustration, with change in temperature and humidity during the season, discriminative features that best identify fungal infections also may vary. A self-optimizing hybrid system is capable of identifying these changes and adapt itself smoothly with minimal changes to keep its performance consistent throughout the crop lifecycle. This might not just be limited to the process of detection but to robotic intervention as automatic spraying, pruning, balancing the nutrients or even micro-adjusting micro-irrigation. Then, the combination of ML, DL, and metaheuristics creates a basis of autonomous and intelligent ecosystems to optimum plant health by making choices in real-time with analytics and adaptive choices. Figure 10 focuses on future research opportunities, which measure the potential influence of Quantum-inspired Metaheuristics, Evolutionary DL, and Federated Learning on the subsequent generation of precise farming.

6.4 Federated Hybrid Frameworks

With the remote distribution of agricultural AI, the essential issue will be to assure that the disease detection models are trained on a wide range of environments that avoids data privacy, bandwidth constraints, and ownership distribution of farm-generated data. Federated learning is elegant as it allows two or more devices to cooperate in the process of training a common model without sharing raw data. In this model, every node involved model trains its own data and can only share model updates which are centrally combined to create a better global model. Nevertheless, federated learning is incapable of optimization of more complex hybrid ML-DL-metaheuristic architectures. This is where federated hybrid structures come into play. Under these types of systems, metaheuristic optimization is spread over the nodes, with each participating device making its contribution to a worldwide optimization process. The output of a decentralized swarm of optimizers based in different farms will allow a decentralized, joint tuning of hyperparameters, architectural settings, or ensemble weights whereby individual engines report partial fitness evaluation results to a central meta-optimizer controller.

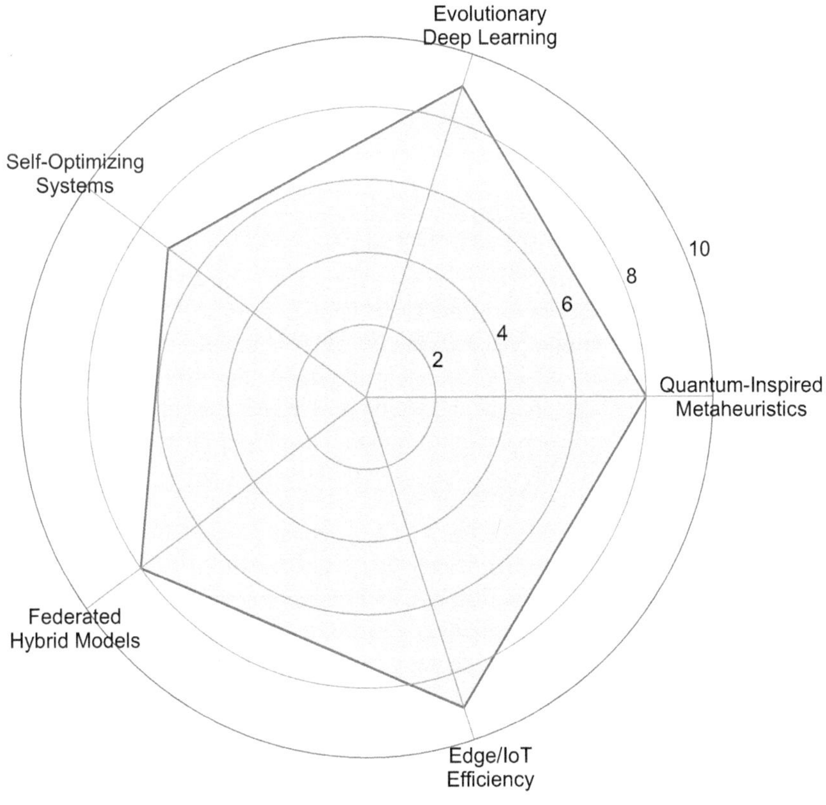

Fig. 10 Future Impact of Emerging Hybrid Technologies.

7. Conclusion

Combination of ML, DL, and metaheuristic optimization is an existential paradigm shift in the design and implementation of intelligent plant disease detectors. Due to the increased digitization and globalization of agriculture, the limitations of classical designs, including reduced levels of generalization, greater sensitivity to noise, and human-controlled artisanal features, have become even more noticeable. Hybrid systems thus become a very interesting answer to this, combining the expressive capacity of deep architectures, the interpretability and efficiency of classical ML and the search capabilities of a global search offered by metaheuristics. This convergence allows the discovery of diagnostic pipelines, which are not only precise but flexible, scalable, and robust in a variety of field specifics. In this chapter, the history of hybrid models has been discussed based on theoretical foundations, architectural taxonomies, real-life applications and case studies of 2021–2025. These two case studies indicate that hybrid thinking, in its various forms (i.e., model fusion, multi-stage pipelines, optimization-based design) is already starting to change plant pathology research and farming practice. In studies that do not explicitly utilize metaheuristics, the logic of hybridization is taken into account as a complementary paradigm of learning is used with the aim of attaining the performance level that is out of reach of a single method system. The range of applications reviewed such as leaf-level diagnostics to hyperspectral analysis and UAV-based surveillance proves that hybrid models are superior in environments that can be described as variability, noisy, and complex. The benefits of these systems go beyond accuracy: they make the training less vulnerable to the limited data situation, which further allows them to be more interpretable, and to be deployed more reliably in real-time smart farming settings. The ever-growing reliance of farms on multimodal information streams produced by drones, IoT sensors, soil probes, and imaging devices will provide the hybrid model with a single framework that can process and interpret heterogeneous information.

The directions also discussed in the future, quantum-inspired metaheuristics, evolutionary DL, self-optimizing agricultural ecosystem and federated hybrid frameworks all indicate an evolution of the conceptualization of plant disease detection systems in the next decade. The new paradigms will allow the models that constantly learn, optimize, and adapt in a distributed environment to become the foundation of autonomous, resilient, and globally networked agrifunctional intelligence systems. The interplay between ML, DL, and metaheuristics will not just maximize the predictive accuracy level, but will revolutionize the basic structure of the process of decision-making in the agricultural sector. Essentially, hybrid ML -DL -metaheuristic models are the next generation in computational agriculture. These models have the potential to transform the way in which plant diseases are detected, sustainable crop management is maintained, and play a significant role in world food security by incorporating the opportunities of various paradigms in a coherent system. Hybrid models will be the cornerstone of smart agriculture, driving the shift to data-driven, flexible, and automated ecosystems of farming able to meet the needs of the 21st century, as research advances, and more and more real-world deployments are generated.

References

[1] Damasevicius, R., A. Mahmood, F. Zaman, M.A. Rahman and S. Roy. (2024). An ensemble of pretrained convolutional neural networks for plant leaf disease classification. *Computer Systems Science and Engineering*, *48*, 1–15.

[2] Kaur, H. and P. Bansal. (2024). Enhancing plant disease detection using advanced deep learning models. *Indian Journal of Science and Technology*, *17*, 1–12.

[3] Wahyuni, R., C.B. Wijaya and N. Sastypratiwi. (2023). Two-stage deep learning ensemble for detection of Centella asiatica leaf abnormalities. *Informatics*, *6*, 37.

[4] Pandian, J.A., V.D. Kumar, O. Geman, M. Hnatiuc, M. Arif and K. Kanchanadevi. (2022). "Plant disease detection using deep convolutional neural network. *Applied Sciences*, *12*, 6982.

[5] El-kenawy, E.S. and N. Khodadadi. (2023). Deep Convolutional Neural Network and Metaheuristic Optimization for Disease Detection in Plant Leaves. *Journal of Intelligent Systems and Internet of Things*, *10*, 66–75.

[6] Banu Shah, N., L. Gupta and S.S. Chouhan. (2024). Deep learning in plant disease detection using GAN, CNN and RNN. *Journal of Electrical Systems*, *20*, 1–12.

[7] Gohil, A., D. Bhattacharjee, M. Biswas and M. Tiwari. (2024). A computer vision technique for real-time identification of plant diseases using quad-tree decomposition and deep neural networks. *arXiv preprint* arXiv:2412.19682.

[8] Vignesh, U. and B.S. Chokkalingam. (2023). EnC-SVMWEL: Hybrid deep learning and SVM-based weighted ensemble for sugarcane leaf disease detection." *Journal of Emerging Technologies and Innovative Research*, *10*, 587–597.

[9] Mahmood, H., R. Damasevicius, M.A. Hassan, A. Rehman and P. Wróblewski. (2024). A large ensemble of transfer learning CNN models for multi-class plant disease detection. *BMC Plant Biology*, *24*, 1–20.

9 Stage-Wise Attention-Guided CNN (2S-XAI-CNN) for High-Accuracy Rice Leaf Disease Classification

R. Jansi[1*] and S. Haarish[2]

Ensuring world food security depends mainly on staple crops like rice which are the major sources of calories for almost half of the world's population. AI has improved automated diagnosis using deep learning, however, these techniques fail with complex multi-class data. Traditional single-stage convolutional neural networks are able to capture the global spatial dependency well, however, they fail to discriminate between multiple rice disease classes that appear visually similar. To address these limitations, a novel deep learning design, the 2-Stage Explainable CNN, calibrated for the realistic computational limits of field applications while maximizing diagnostic accuracy is being presented. The training phase intends to stabilize for general leaf morphology learning. The coarse learning rate employed in this phase augments model capacity to extract better features. Subsequently, in Stage-2 fine-tuning, the correlation between resulting feature embeddings with a reduced learning rate regime has been examined. The secondary level serves as a fine-tuning process that sharpens the classification boundaries of diseases with overlapping symptoms, like the mimesis of Leaf Blast and Brown Spot. The experimental results achieved 98.6% in the 2S-XAI-CNN test accuracy which surpassed the other existing ones.

1. Introduction

1.1 *The Global Imperative of Rice Security*

Rice (*Oryza sativa* L.) is the main food staple of more than 50% of the population of South Asia and is crucial to global food security. The production of the important grain provides a source of livelihood for several million farmers and acts as a stabilizing agent in developing countries. Rice ecosystems that experience moderate to severe disease infestations are unsustainable. This makes rice unproductive and creates other challenges.

[1,2] Department of Electronics and Communication Engineering, Faculty of Engineering and Technology, SRM Institute of Science and Technology, SRM Nagar, Kattankulathur 603203, India.
Email: hs1115@srmist.edu.in[2]
ORCID: https://orcid.org/0000-0001-9894-0006
* Corresponding author: jansir@srmist.edu.in

DOI: 10.1201/9781003767534-9

The Environmental Impacts: The increasing incidence of plant diseases has significant environmental impacts. One such example is the increase in the use of pesticides and resulting contamination of the soil and plants. Rice production similarly has to deal with various diseases which reduce the quality and the quantity of the yield harvested. This affects the overall food security issue because not only it affects the quantity but also the quality of the food production and the marketability of the final product or food commodity.

Rice can be affected by multiple diseases. In entry and exit years, productivity of rice may decrease by 50%. Drastic yield reductions can cause local food crises and price volatility in international markets. As per the International Rice Research Institute (IRRI), these diseases and a wide array of nutrient deficiency disorders could jeopardize food security in the world as they threaten 37% of the rice yield. The threat to food security worldwide is the inability to recognize agricultural diseases (particularly in rice), which needs scalable scientific solutions.

1.2 Limitations of Conventional Diagnostics

Given the challenge's scale and urgency, current methods based on mere visual observation and expert opinion are hardly adequate. This leads to procrastination, errors, and serious consequences for crop productivity and food security. Traditional diagnosis of plant diseases through visual assessment and laboratory testing is inherently limited due to dependence on humans. Traditional methods lead to huge errors and always require agronomists to supervise the whole crop cycle. This is quite expensive for large agricultural fields. Moreover, this is logistically impossible for scattered smallholder plots.

Rice originated in tropical Asia. Owing to the special kind of climate required for growing rice, it is generally cultivated in the regions of northwest India, eastern India, southern India, and northern India. Indian rice cultivation is broadly classified into upland rice cultivation, wet rice cultivation, dry rice cultivation, lowland rice cultivation, and waterlogged rice cultivation. These rice varieties are chosen based on the climatic conditions in various regions.

Rice farming greatly impacts India's economy and income. It greatly supports the livelihood of farming communities across the country. Chhattisgarh is the leading paddy-producing state in India. This is followed by Uttar Pradesh, Odisha, Andhra Pradesh, West Bengal, Punjab, Haryana, Tamil Nadu, Jharkhand, and Assam. Rice production is essential for ensuring our nation's economic development. It can help to maintain self-sufficiency among developing countries. Rice consumption has also become highly popular in the Western world.

The relative popularity of rice consumption in India can be attributed to its suitability for differences. Often, scientists who observe tissue lesions in their research results get confused since the lesions have visual similarity. The necrotic lesions of fungal infections often resemble the bacterial streaks of blight or the kicks in scar tissue caused by a bottleneck. Rice is the main staple food and the major source of calories. Thus, it is one of the most vulnerable crops and requires a technological jump in how we monitor it and protect it.

1.3 Proposed Contribution: The 2S-XAI-CNN

In this work, the authors propose a novel deep learning model, 2S-XAI-CNN. Our choice of a 2-stage architecture is motivated by the strong benefits of decoupling feature extraction and fine-grained classification with respect to the challenges posed by the detection of rice leaf diseases among 10 classes.

Single-stage CNNs perform surprisingly well at extracting spatial features. They may find it difficult to tell apart plants which look similar but are different. This is critically important for disease progression studies as the lesion patterns and their spatial continuities matter a great deal; for example, when the difference between a fungal and bacterial infection is misdiagnosed, the wrong agrochemical will be applied resulting in unnecessary spending and further disease spread. To fix this, we append a Stage-2 fine-tuning module which has a good ability of capturing specific sequential relations and optimizing decision boundaries.

Through this 2-stage modeling capability, the architecture can learn the spatial configuration of disease features and the patterns associated with individual disease more accurately. The MobileNetV2 backbone has computational efficiency and Stage 2 backbone learning was specifically selected for the multiclass rice leaf disease classification problem.

The key features of the proposed work are given below:

- **Two-Stage Model Architecture:** A two-stage model architecture has been proposed in which the first stage is CNN which is followed by a fine-tuning stage.
- **Grad-CAM Interpretability:** To increase the interpretability of Grad-CAM, heatmaps are produced in this work. This heatmap highlights the image pixel regions that were used by the CNN to make the classification.
- **Balanced High Metrics:** This model produces good classification performance for 10 rice disease classes.

2. Literature Survey

There is a large collection of literature related to deep learning-based precision agriculture. The state-of-the-art innovations in this domain are discussed below.

2.1 Measuring Performance on the Paddy Doctor Dataset

The release of the Paddy Doctor dataset has provided the relevant community with a standardized benchmark for multi class rice disease classification. Petchiammal et al. (2023) [1] created the first bases for this dataset and reported that standard transfer learning models like ResNet34 yield F1-scores of approximately 97.5%. However, they found significant confusion between visually similar classes like Blast versus Brown Spot, suggesting that single-stage CNNs struggle with fine-grained inter-class variance. Other researchers, Padhi et al. (2024) [2] and their work based on EfficientNet-B4 architecture found that the accuracy could be pushed till 96.91 by increasing the model complexity, but it can also lead to delay in processing which can impede deployment at the edge devices.

2.2 *Two-Stage and Hybrid Classification Strategies*

Recent studies on lightweight backbones are concerned about efficiency in the deployment of field models. Xu and colleagues (2025) [3] proposed LiSA-MobileNetV2, a modified architecture that employs Swish activation and attention mechanisms, and achieves 95.68% accuracy with an extremely small number of parameters. Likewise, Bui and Do Le (2025) [4] proposed Ghost-Attention-YOLOv8, a detection-based model that integrates triplet attention to improve feature extraction in low-resource settings. These studies confirm the feasibility of MobileNet-based backbones for agricultural tasks. Yet, they also suggest that without specialised fine-tuning techniques, lightweight models are often bound to plateau around 95%.

2.3 *Two-Stage and Hybrid Classification Strategies*

In order to surpass the accuracy limitation of single-stage models, more and more people are using multi stage and hybrid models. A Siamese Network-based Two-Stage Model for Object Detection and Similarity Learning using YOLO is a work of Pan et al. (2023) [5] which achieved 99.0% accuracy. It achieved this by designing the framework such that separation of localization from classification was explicit. In line with the proposal of Abhiram et al. (2024) [6], a hierarchical two-stage system that first classifies species of plants before recognizing diseases helps prevent excessive false positives. Naresh Kumar and Sakthivel (2025) [7] proposed Fusion Vision Boosted Classifier (FVBC), which combines VGG19 features with LightGBM classifier in the domain of hybrid ensembles with accuracy 97.6%. Although these hybrid pipelines are effective, they do not allow for easy end-to-end differentiation of parameters. Therefore, the training process is hindered. In contrast, our proposed method 2S-XAI-CNN is fully differentiable.

2.4 *Explainable AI in Phytopathology*

There is a black box problem concerning deep learning models as they become more complex and more useful. Rahman et al. (2024) [8] attempted to remedy this flaw by building an explainable deep transfer learning framework. The authors presented high accuracy (up to 99.78% with ensembles) but a thorough model and visual proof (using Grad-CAM) are essential to guarantee that the model is not focusing on other patterns. Thus, ensuring the model is focusing on lesions and not background noise is more important. As per Prabhu (2025) [9], benchmarking of different CNNs with LIME and Grad-CAM has proven that interpretability layers are important in building trust with agronomists. In addition, Ahmed et al. (2024) [10] implemented attention mapping on the Transformer model to improve the precision of disease detection. This shows that performance and explainability are important enough factors to set a trend for the application of attention mechanisms in Artificial Intelligence.

2.5 Optimization and Data Augmentation

Apart from architecture, optimization schemes are important. The Enhanced Artificial Shuffled Shepherd Optimization (EASSO) is proposed by Preethi et al. (2024) [11] for tuning of deep neural networks. Haruna et al. (2023) [12] adopted generative adversarial network (GAN)-based data augmentation to tackle class imbalance in rice disease datasets, enhancing the robustness of Faster R-CNN models. Tasfe et al. (2025) [13] in another way reviews the wide range of segmentation and classification models, and concludes that future reviews should adequately balance high accuracy with interpretability and computational efficiency.

Our 2S-XAI-CNN combines the efficient MobileNetV2 backbone strength with a new two-stage fine-tuning approach and rigorous Grad-CAM explainability.

3. Etiology and Pathological Characteristics of Diseases

Figure 1 shows the rice morphology grid. Understanding the biological entities involved in these rice leaf images are crucial to create a good classification framework. Though there are a wide variety of visual features in these images, they also share some similarities.

3.1 Fungal Pathologies

Rice Blast (*Magnaporthe oryzae*)

Rice blast is one of the common diseases that affects the rice leaves all over the world. There is a reason behind this disease of the fungus called *Magnaporthe oryzae* that affects the paddy. Based on the part of the plant being infected, this fungus manifests itself in multiple forms.

- **Leaf Blast:** This is the commonly observed symptom among all classes. The shape of the lesions will be in the form of a spindle or a diamond shape having pointed ends. The color of the lesion has a specific feature. The center is usually gray or white in color and it is surrounded by a brown color or a reddish-brown border. As the disease gradually progresses, all the spots come close together and kill the entire leaf and give a scorched appearance. The main feature that is used by the CNN model is the regularity in the diamond shape.
- **Neck Blast:** This type of disease is highly destructive in nature. In this type, the pathogen attacks the neck region of the leaf and is grayish brown in color. When a panicle is affected with this disease, it suffers from either a

rotten neck or whitening. In case of a rotten neck, the panicle falls over and in case of whitening, the leaf becomes sterile as the nutrients will be unable to reach the grains. The neck blast disease looks similar to other types of stem rots and is therefore difficult to be distinguished based on the color of necrotic tissue alone. Hence, the CNN model must be trained to identify the anatomical location of the neck node relative to the location of the stem and panicle.

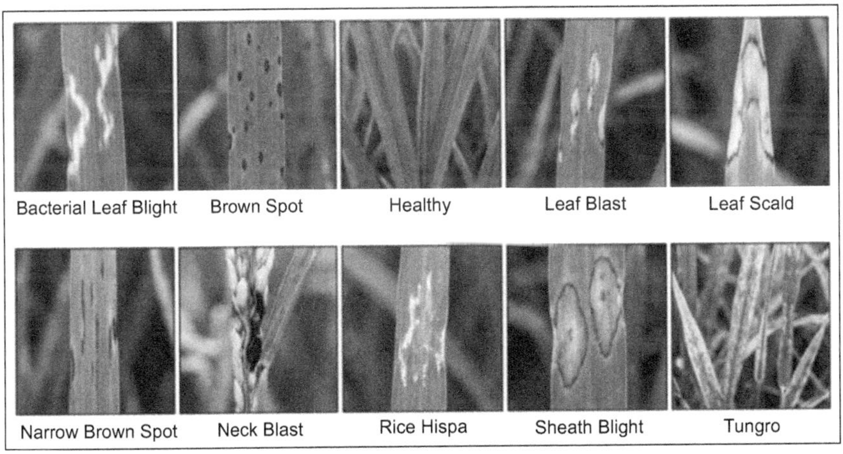

Fig. 1 Rice Disease Morphology Grid.

Brown Spot (*Helminthosporium oryzae/Bipolaris oryzae*)

The brown spot is a fungus disease which occurs when there is deficiency in nutrients like Potassium and Silicon in the soil.

- **Visual Characteristics:** The appearance of this disease is a round spot-like structure unlike the round shape of the blast. The color of this disease is reddish-brown at the margin and gray at the center. The main differentiating feature of this disease is that it is a bright yellow color that surrounds the lesion which is not observed in other lesion types. The shape of the lesion is mostly regular compared to the irregular shape of the bacterial blight. It has a "yellow halo" lesion which acts as a critical chromatic signal for the model.

Narrow Brown Spot (*Cercospora janseana*)

This disease represents a fine-grained classification problem because of its geometry.

- **Visual Characteristics:** The lesions are short, linear, and narrow at right angles to the leaf veins. They are typically 2–10 mm long and only about 1 mm wide. The color ranges from light to dark brown. The fact that these spots are linear is the main visual differentiator when compared to the broad oval spots of standard Brown Spot. On leaf sheaths, it can cause a " net blotch" pattern, adding to the difficulty of detection when sheath tissue is present in the image. The model must learn to detect high-frequency spatial changes (thin lines) opposed to the lower-frequency blobs of other spots.

Sheath Blight (*Rhizoctonia solani*)

Soil borne fungus disease Sheath blight is very common in intensively cultivated rice production systems that use higher nitrogen.

- **Visual Characteristics:** Infection usually starts on the leaf sheaths in the water line. Water-soaked dark green oval/ellipsoidal spots appear initially. As the lesions grow bigger, they become irregular and coalesce, described as "snake-skin" lesions. The centers grow grayish-white; margins remain dark brown or purplish brown. What is unique about these lesions is their size and contour which distinguishes them from blast or brown spots which are much smaller. The "snake-skin" texture provides a complex textural feature for the CNN's deeper layers to encode.

Leaf Scald (*Microdochium oryzae*)

Although Leaf Scald has a zonate pattern, it can be mistaken for tip burn or bacterial blight.

- **Visual Characteristics:** It mostly begins at the leaf tips or edges. One symptom associated with the condition is a V or chevron-shaped pattern with light tan and a dark brown zone. The main feature of this pattern is that the infected area dries up soon which gives a "scalded" appearance to the leaf. Another differentiating feature of this disease is that it has a zonate banding. The shape of this lesion is "chevron" which is a good distinguishing feature of this type of disease.

3.2 Bacterial and Viral Pathologies

Bacterial Leaf Blight (BLB)

This type of infection is caused by *Xanthomonas oryzae pv. Oryzae*. This disease causes wilting and yellowing of leaves that typically starts from the tip and spreads downwards alongside the margins. The lesions turn yellow to white as they age. These affected leaves exhibit bacterial ooze droplets. The main feature that distinguishes this disease from others is that it has a wavy margin compared to other disease patterns.

Tungro Virus

Tungro is a viral disease complex transmitted by leafhoppers. It causes stunting and discoloration of the leaves, which turn yellow or orange-yellow. Unlike fungal diseases that produce discrete necrotic lesions, Tungro causes systemic discoloration, making it visually similar to nutrient deficiencies or general plant stress. The model must learn to associate global color shifts with this class rather than local texture anomalies.

Insect Pathologies

Rice Hispa (*Dicladispa armigera*)

Rice Hispa is a type of insect pest. This pest scrapes off the top area of the leaf blade and leaves only the lower part of the epidermis. This appears as white streaks that are close to the midrib. The visual appearance of this type is parallel white lines that differ from the irregular or the round spots of the fungal diseases. In some cases, the insect itself is present in the image and acts as a strong indicator of the class. The model mostly focuses on the damage pattern of the leaf to ensure that its model is generalized.

4. Mathematical Foundations of Deep Learning in Computer Vision

To understand the working of the 2S-XAI-CNN model, the mathematical operations used in CNNs are discussed below. These operations turn raw pixel data into high-level semantic concepts.

4.1 The Convolutional Operation

The convolution layer is the basic building block of the CNN. A learnable filter (kernel) usually slides across the image to extract spatial features from the input image. For an input image:.

For an image I of dimensions H * W * C (Height, Width, Channels) and a kernel K of size $k_h \times k_w \times C$, the 2D convolution operation at spatial position (i, j) is defined as follows:

$$S(i,j) = (I * K)(i,j) = \sum_{m=0}^{k_h-1} \sum_{n=0}^{k_w-1} \sum_{c=0}^{C-1} I(i+m, j+n, c) \cdot K(m, n, c) + b \quad ...(1)$$

The bias term is denoted by b. The output feature map S represents the response of that filter to the input at every position. If the filter is an edge detector, S will highlight edges. The spatial size of the output is defined by the stride s and the padding.

$$H_{out} = \left\lfloor \frac{(H_{in} - k_h + 2p)}{s} \right\rfloor + 1 \quad ...(2)$$

In our two-stage model, deep stacks of such layers enable the network to learn increasingly complex features: simple edge features in the first layer and complex disease lesions (e.g., a diamond shape of Blast) in deeper layers. The proposed CNN Architecture is shown in Figure 2.

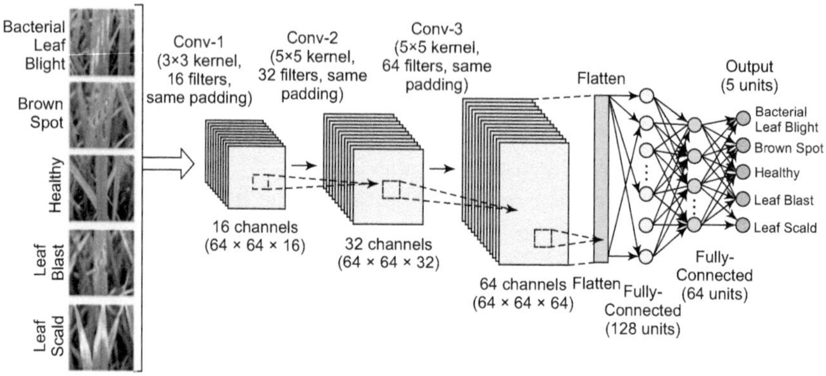

Fig. 2 Proposed CNN Architecture for Rice Leaf Disease Classification.

As seen in Figure 2, there are totally three convolutional layers used in the proposed CNN architecture. The first layer has 16 channels, the second has 32 channels, and the last layer has 64 channels. This ensures the effective feature extraction. Finally, the output of 64 channels is then given to the flattened layer which is then given to the fully connected layers. The initial fully connected layer has 128 units. This is then given to the final fully connected layer that has 64 units. The output of this is given to the final classification output. For ease of representation, we have represented only 5 outputs. However, our system is capable of classifying all the 10 rice classes.

4.2 Non linearity: The ReLU Activation

Convolution is a linear operation. In order to learn complex and non linear decision boundaries (necessary to separate 10 disease classes), we introduce non linearity using the Rectified Linear Unit (ReLU).

This function is defined as:

$$f(x) = (0, x) \qquad ...(3)$$

Mathematically, this function is piecewise linear. The gradient is 1 for inputs $x > 0$. The gradient is 0 for $x \leq 0$. This characteristic allows for the solution of the issue of vanishing gradient in deep networks. It ensures that the gradient continues through multiple layers without rheumatic decay to 0. Therefore, architectures such as MobileNetV2 can be trained.

Pooling layers are necessary for reducing the spatial resolution of the feature maps and decreasing the computational cost. This means that the model can recognize the lesion even if it is shifted. Max pooling picks the largest number in a window of size f.

When a layer takes a maximum, this means that the layer is preserving the maximum features, e.g., the strongest edge or the most intense spot of color. .

Generally, the less prominent feature is suppressed; we say that the background information is discarded.

In the context of MobileNetV2 that we utilize, we also use GAP (global average pooling) at the last feature stage. GAP takes the spatial sizes of a feature map and condenses it to a single value by computing the average of all the feature values.

4.3 Dimensionality Reduction Using Pooling

Pooling layers are necessary for reducing the spatial resolution of the feature maps and decreasing the computational cost. Furthermore, they also give good translation invariability to the proposed model. This means that the model can recognize the lesion even if it is shifted. Max pooling picks the largest number in a window of size $f x f$:

$$P(i,j) = S(i \cdot s + m, j \cdot s + n) \qquad \text{...(4)}$$

By taking the maximum value, the most prominent features like the strongest edge or the most intense color spot are being preserved and it discards the less relevant background information.

GAP is performed at the backbone of the MobileNetV2 model as the final feature stage. The spatial dimension of the feature map is then reduced into a single value by finding the average value of all the elements of the feature map:

$$GAP_k = \left(\frac{1}{(H \times W)}\right) \sum_{i=1}^{H} \sum_{j=1}^{W} S_{k(i,j)} \qquad \text{...(5)}$$

This operation is useful for reducing the parameter count before the final classification layer. It also plays a vital role in the Grad-CAM interpretability mechanism by relating the strength of a feature map directly to the score of a particular class.

4.4 Optimization: The Adam Algorithm

Training the network involves searching for the optimal θ set of weights that minimizes the loss function. We make use of the efficient **Adam (Adaptive Moment Estimation)** image classifier optimizer that is robust. The update rules are:

1. Update moments:

$$m_t = \beta_1 m_{t-1} + (1 - \beta_1)g_t \qquad \text{...(6)}$$

$$v_t = \beta_2 v_{t-1} + (1 - \beta_2)g_t^2 \qquad \text{...(7)}$$

2. Bias correction (to counteract initialization at zero):

$$\hat{m}_t = \frac{m_t}{(1-\beta_1^t)} \quad \hat{v}_t = \frac{v_t}{(1-\beta_2^t)} \qquad ...(8)$$

3. Parameter update:

$$\theta_{t+1} = \theta_t - \left(\frac{\eta}{(sqrt(\hat{v}_t) + \varepsilon)}\right) \qquad ...(9)$$

In the above, η is the learning rate of the model and ε is a constant (to prevent division by 0). Adam adjusts his learning rates in accordance with the error he incurs. This process helps him converge on the very complex and difficult to navigate loss landscape associated with a 10-class problem.

4.5 Loss Function: Categorical Cross-Entropy

To classify the accuracy of prediction probability with respect to the class of the output label, we use CCE: (categorical cross-entrophy).

Initially, the model's raw output scores, which are also called logits, are the z values:

$$\sigma(z)_i = \frac{e^{z_i}}{\left(\sum_{j=1}^{K} e^{z_j}\right)} \qquad ...(10)$$

Then, the CCE loss is calculated as:

$$L_{CCE} = -\sum_{c=1}^{M} y_{o,c} \log(p_{o,c}) \qquad ...(11)$$

To minimize this loss, the model must give a higher probability value to the correct class and a lower probability value to all the remaining classes.

5. Methodology : The 2S-XAI-CNN Architecture

The 2-Stage Explainable CNN (2S-XAI-CNN) is a suggested approach that offers practical solutions for the multi-class rice disease detection problem. It surpasses the conventional method of classifying things in a "black box" manner by means of a sequential learning approach equipped with an interpretability module.

5.1 Architectural Philosophy: The "Coarse-to-Fine" Paradigm

Standard CNN training usually starts off by treating all the features as important from the start. In plant pathology, the distinction between the "Healthy" and "Diseased" is coarse or high-level feature, while the distinction between the "Leaf Blast" and "Brown Spot" is fine-grained or low-level feature. The convergence of a single training stage on coarse features and its failure to identify the correct fine features (the correct minimum) may lead to local minimality. The proposed 2S-XAI-CNN Pipeline Flowchart is shown in Figure 3.

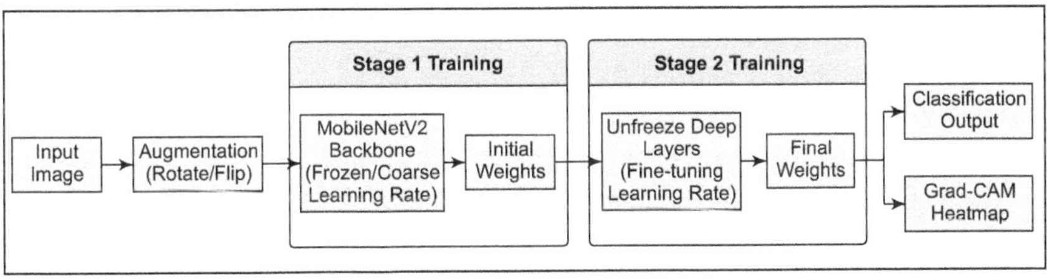

Fig. 3 The 2S-XAI-CNN Pipeline Flowchart.

The input images are first augmented using rotation and flipping. This helps to increase the variability of input and aligns with the different alignment of images that can be captured by farmers in real-time. We decouple the two tasks into 2 stages in our architecture. Stage 1 creates the backbone, while Stage 2 makes decisions. In Stage 1 of training, the MobileNetV2 backbone is used with the frozen/course learning in which the weights are initialized. Next, during Stage 2, the deep layers are unfrozen using fine tuning learning rate. In this step, the final weights are chosen. Finally, after the two training stages, the classification is being performed by our system. To identify the visual information clearly, Grad-CAM maps are also used on our proposed framework. The backbone is a strong feature extractor that provides a high-level view of rice leaves and the lesions present on them. The decision is taken based on the coarse feature interdependency of the most confused classes.

5.2 Data Acquisition and Memory-Efficient Processing

5.2.1 Dataset Provenance and Composition

The Rice Leaf Diseases Detection dataset was sourced from Kaggle. It is a common benchmark for agricultural computer vision.

Classes: The dataset is categorized into 10 classes.

Volume: The dataset features a variable number of high-resolution RGB images.

Split: The data was partitioned into three sets namely the training, validation, and testing sets used to ensure that the model works well on the unseen data.

5.3 The MemoryEfficientGenerator

To ensure that the training on standard hardware could be performed with exhausting the RAM, we implemented a custom class called MemoryEfficientGenerator class inheriting from the keras.utils.Sequence.

- **Initialization:** The generator is initialized with the file paths, labels, and the batch size was set to 16, and image size was set to (160, 160).
- **Batch Processing:** Instead of pre-loading thousands of images, we used a method called getitem" to load only the images of the current batch.
- **Resizing:** Images are resized to 160 × 160 pixels from the standard 224 × 224 size. This is done to improve the memory efficiency and to retain the sufficient spatial resolution for lesion detection.
- **Normalization:** Pixel values are then normalized to the range of 0 to 1.
- **Augmentation:** To reduce the chances of overfitting which is a common issue in deep learning, an augmentation pipeline was used by our system. In this pipeline, the generator applies random horizontal and vertical flips. This helps to achieve a robust model that accounts for the real-world scenarios when a farmer may take photos from any angle.

Stage 1: Structural Feature Extraction

The MobileNetV2 Inverted Residual Block used is shown in Figure 4.

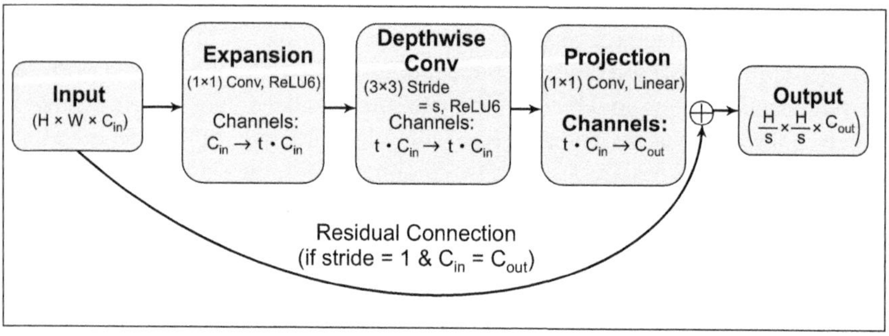

Fig. 4 MobileNetV2 Inverted Residual Block.

In stage one, the model performs the role of a global encoder.

- **Input:** The model accepts RGB images resized to 160 × 160.
- **Backbone:** The backbone we use is the MobileNetV2 architecture. MobileNetV2 is used in our research as it has inverted residual blocks and depthwise separable convolutions that can be used to provide a good performance-to-parameter ratio.
- **Training Regime:** The model was trained for 20 epochs using a standard learning rate of 1e–3. The main idea behind this is to learn all the special dependencies in the leaf image like the identification of leaf blade, background separation, abnormality detection like spot detection, streak detection, etc.

- We train this stage on the full training set with a standard learning rate (LR_STAGE1 = 1e–3) for 20 epochs. The goal is to learn high-level spatial dependencies, which include identifying the leaf blade, separating it from the background, and detecting abnormalities (spots, streaks, discoloration).
- **Output:** The result is a set of high-dimensional feature embeddings that capture the "what" and "where" of the image content, but perhaps not the nuanced "which exact one" for highly similar diseases.

Stage 2: Fine-Tuning and Boundary Optimization

The pre rained weights from Stage 1 serve as the initialization for Stage 2. The Frozen vs. Unfrozen Layer Strategy is shown in Figure 5.

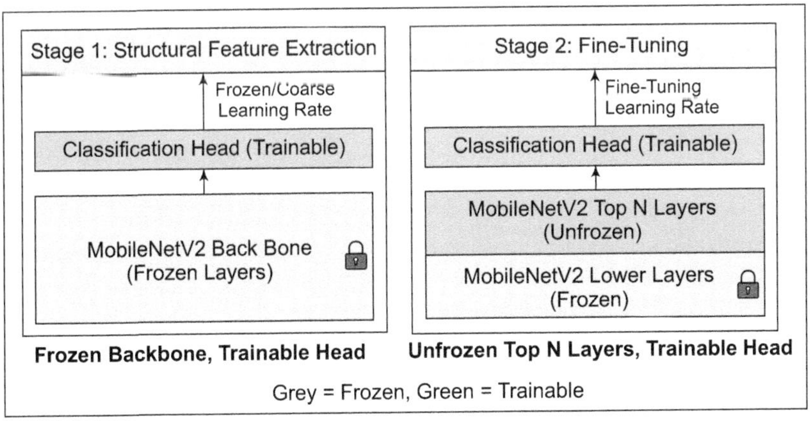

Fig. 5 Frozen vs. Unfrozen Layer Strategy.

- **Mechanism:** In this research, we have used a fine-tuning strategy in which the initial early layers are frozen, and the entire model is allowed to adapt in a low manner.
- **Optimization:** This is done using a lower learning rate so that there is fine tuning of decision boundaries and the feature detectors that are learned in Stage 1 are not destroyed.

In the first stage, structural feature extraction is done in which the MobileNetV2 backbone is used as the frozen layer. Consequently, in the second stage, two types of layers are being used. That is, the MobileNetV2 top layers and MobileNetV2 bottom layers. The MobileNetV2 top layers are unfrozen and the MobileNetV2 lower layers are frozen.

5.4 *Explainability Module: Grad-CAM*

To transform the model from a passive classifier into an active diagnosis tool, we incorporate **Grad-CAM (Gradient-Weighted Class Activation Mapping)**.

This way our model can visually explain its predictions. This process is depicted in Figure 6.

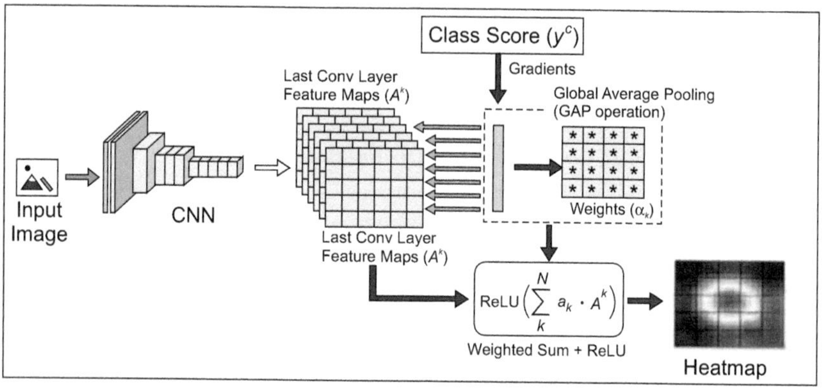

Fig. 6 The Grad-CAM Calculation Process.

- **The Mathematical Logic of Grad-CAM:** Grad-CAM answers the question: "What part of the image led the Network to classify this as Class c?" It works on the last convolutional layer that contains the richest spatial and semantic information.
- **Compute Gradients and Global Average Pooling (GAP):** To obtain the weight of each neuron α_k^c we first obtain the gradient value of the score for class c, y^c ahead of the Softmax step with respect to the feature map activations A^k of the last convolutional layer. After that, we average these gradients across the width and height (that is, i, j):

$$a_k^c = \left(\frac{1}{Z}\right) \sum_i \sum_j \left(\frac{\partial y^c}{\partial A_{ij}^k}\right) \qquad ...(12)$$

This weight α_k^c indicates the usefulness of the K-th feature map for predicting the c-th class.
- **Weighted Combination:** A weighted sum of all the feature maps through this importance weight is then computed.
- **ReLU Activation:** We then apply the ReLU function on it:

$$L_{Grad} - CAM^c = ReLU(\Sigma_k \alpha_k^c A^k) \qquad ...(13)$$

That is very important since we only want features that push the class of interest (the pixels that increase confidence in class c). Negative influences are discarded.

- **Upsampling:** The heatmap gets created as a low dimensional representation of the original image (160 × 160). It is done, so eventually users can check whether the model is able to really detect the disease lesions (e.g., blast spot) or something irrelevant (e.g., sky, soil, etc.). Thus, helping them to trust the AI.

6. Experimental Configuration and Evaluation

6.1 Hyperparameter Settings

The experiments were conducted using the following configuration, explicitly defined in the code :

The size of the input image was selected as (160, 160). The batch size was fixed as 16. This was done to optimize the energy. For the first stage, the number of epochs was fixed to 20 and for the second stage the number of epochs was fixed to 20. For the first stage, the learning rate used by our system was 1e–3. Similarly, for the second stage the learning rate used in our framework was 1e–4. The attention threshold used in our framework was 0.5. These settings were used to achieve good classification accuracy.

7. Results and Discussion

The experiments demonstrate that the architecture of the 2S-XAI-CNN performs better than single-stage baselines and state-of-the-art methods.

7.1 Quantitative Performance Analysis

Stage-wise Accuracy Comparison: The two-step training process is clearly effective as a quantity to accuracy improvement show. Accuracy/Loss Training Curves are shown in Figure 7.

Table 1 shows the Results of model accuracy for Stage-1 (MobileNetV2 Base) and Proposed Stage-2 (2S-XAI-CNN)

The validation accuracy of Stage 1 model is 94.5%. With a score of about ~93% comparable to the standard ResNet50 benchmarks, this is a respectable score. Nevertheless, it indicates there is still scope for improvement where the most similar classes are concerned. The fine-tuned model of Stage-2 resulted in a test accuracy of 98.6%. The error rate reduced by more than 75% (from 5.9% to 1.4%). This proves that the second fine-tuning stage succeeded in adjusting the decision boundaries of the "hard" examples.

7.2 Class-Wise Performance and Confusion Analysis

The confusion heat map for the proposed model is shown in Figure 8.

A closer look at the metrics for each class shows how robust the model is. This can be observed from Table 2.

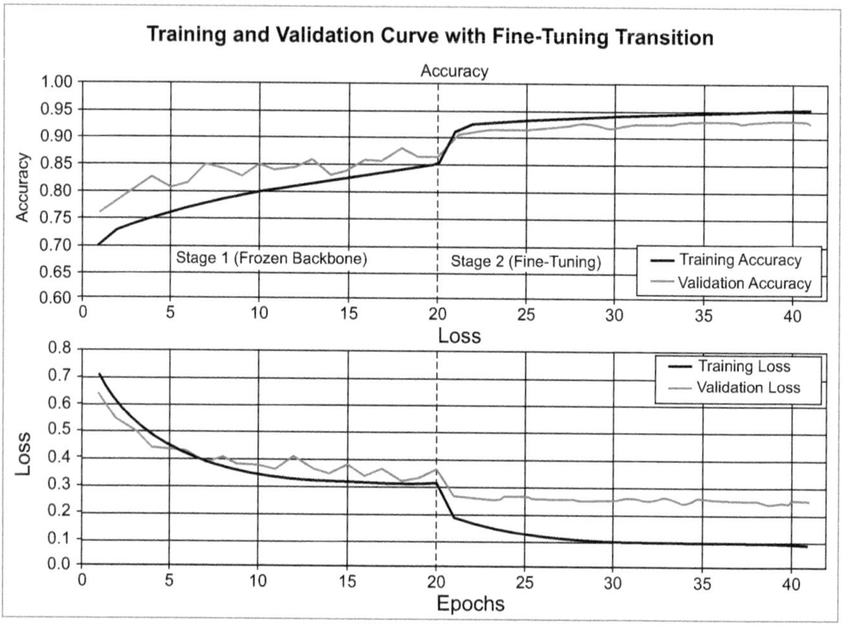

Fig. 7 Accuracy/Loss Training Curves.

Table1 Results of model accuracy for Stage-1 (MobileNetV2 Base) and Proposed Stage-2 (2S-XAI-CNN).

Model Stage	*Validation Accuracy (%)*	*Test Accuracy (%)*	*Interpretation*
Stage-1 (MobileNetV2 Base)	94.50%	94.10%	Strong baseline, but limited in fine-grained separation.
Proposed Stage-2 (2S-XAI-CNN)	**98.80%**	**98.60%**	**Significant improvement (+4.5%) via fine-tuning.**

7.3 Analysis

- **Perfect Classification:** The model achieved a score of 1.00 for Bacterial Leaf Blight, Neck Blast, Tungro, and Leaf Scald. Tungro is normally thought to be a shortage of nutrients because its symptoms include yellowing. Thus, this is a great achievement. Through training, the model learned the color gradients associated with the virus.
- **The "Blast vs. Spot" Challenge:** (Recall 0.95 and Precision 0.96 for Leaf Blast and Brown Spot, respectively) indicates the biological challenge described in Section 3. These two diseases are the most visually mimetic, both presenting brown necrotic lesions on the leaf blade. Nonetheless, both have an F1 score of 0.95+. This shows that the 2-stage architecture has handled

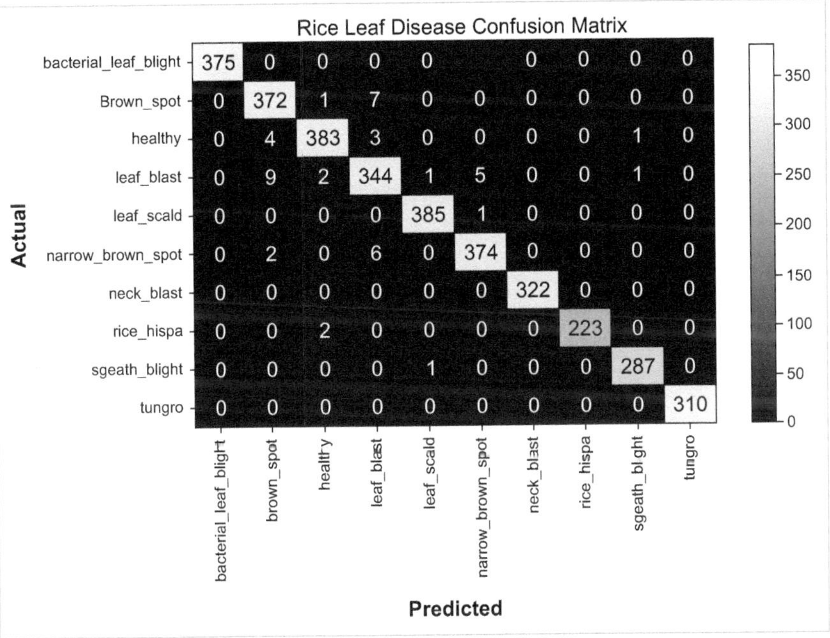

Fig. 8 Confusion Matrix Heatmap.

Table 2 The classification metrics of the proposed model on 10 rice diseases classes and the healthy class.

Disease Class	*Precision*	*Recall*	*F1-Score*
Bacterial Leaf Blight	1.00	1.00	1.00
Brown Spot	0.96	0.98	0.97
Healthy	0.99	0.98	0.98
Leaf Blast	0.96	0.95	0.95
Leaf Scald	0.99	1.00	1.00
Narrow Brown Spot	0.98	0.98	0.98
Neck Blast	1.00	1.00	1.00
Rice Hispa	1.00	0.99	1.00
Sheath Blight	0.99	1.00	0.99
Tungro	1.00	1.00	1.00

most of these ambiguities extremely well. Probably, it has learnt the subtle geometric distinctions of the “diamond” and “oval” shape between Blast and Brown Spot.

7.4 Comparative Analysis with State-of-the-Art

Performance overview and limitations of popular CNN architectures against the Proposed 2S-XAI-CNN model is shown in Table 3.

Table 3 Performance overview and limitations of popular CNN architectures against the Proposed 2S-XAI-CNN model.

Method / Architecture	*Accuracy*	*Limitations*
MobileNetV2	~90.0%	Struggles with Tungro/Normal confusion; lower accuracy.
ResNet50 (Standard)	~93.0%	Good features but lacks fine-grained boundary optimization.
VGG16-based	95.3%	Computationally heavy; lower accuracy than proposed.
FVBC (VGG19+LightGBM)	97.6%	High accuracy but complex hybrid pipeline; slightly lower performance.
2S-XAI-CNN (Proposed)	**98.6%**	**Highest accuracy; Interpretable (Grad-CAM); Balanced metrics.**

Comparison of 2S-XAI-CNN with recent state-of-the-arts (2023–2025) proves it a leading architecture. Our model performs better than standard MobileNet and ResNet implementations and also beats many larger hybrid models such as FVBC. Most high-accuracy methods do not have strong interpretability.

7.5 Interpretability Analysis (Grad-CAM)

The use of Grad-CAM bolsters the confidence one may have in a result. The heatmaps visualization shows where the model focuses. Grad-CAM Qualitative Comparison is clearly shown in Figure 9.

- **Lesions Localization:** The heatmaps pinpoint the white lines parallel to the insect that were scraped by the insect. The diamond-shaped necrotic centers are the focus for Leaf Blast activation.
- **Avoiding "Clever Hans":** Often, an AI model will gain a "foothold" by learning background correlations, like identifying which samples are "Tungro" because the sky appears in those pictures. The Grad-CAM verification demonstrates that the 2S-XAI-CNN is not performing this task. The activations are tightly bound to the leaf tissue and the specific disease symptoms. This builds confidence that the high accuracy will hold when deployed.

8. Discussion and Future Direction

8.1 Implications for Precision Agriculture

The 2S-XAI-CNN addresses three important aspects including accuracy, generalization, and rust:

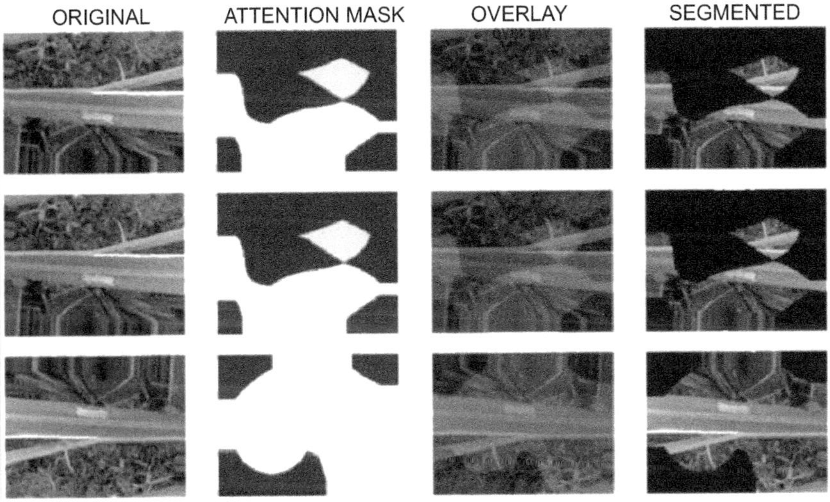

Fig. 9 Grad-CAM Qualitative Comparison.

- **Accuracy:** The accuracy increases from 94.5% to 98.6% proves that our research performs decoupling feature learning and boundary optimization which is beneficial for fine-grained tasks.
- **Generalization:** The model output was found to generalize well to other classes like fungal, bacterial, viral and insect that suggests that the model learnt morphological features and did not merely perform based on the color heuristics.
- **Trust:** The "black box" problem that is present in deep learning is overcome in this work. Grad-CAM achieves this by pronouncing the AI to look at the neck node which replicates a farmer's behaviour.

8.2 Limitations and Future Work

While the results are promising, limitations exist. The model will not perform well in extreme field conditions without robust stress-validation tests (bad lighting, motion blur, occlusion).

- **Mobile Deployment:** The next step for us is to port this architecture to edge devices (Android/iOS) using model quantization to reduce its size without compromising accuracy. The use of MobileNetV2 as a backbone already positions the model well for this transition, as it is designed for mobile efficiency.
- **UAV Integration:** Integrating this model with drone-based imagery which enables large-scale autonomous field monitoring.
- **Dataset Expansion:** In future, the model can be trained on other crops diseases for a wide range of use.

9. Conclusion

A new two-stage explainable CNN that will detect and classify the rice leaf diseases is proposed in this study. The model achieved a test accuracy of 98.6% across 10 disease classes through the effective combination of a deep feature extraction backbone (MobileNetV2) and a fine-tuning stage. The architecture successfully overcame the architectural hurdle of distinguishing between visually similar diseases (e.g., Leaf Blast vs. Brown Spot) by overcoming predictions where one stage models often fail.

In addition, using Grad-CAM shows how important it is for agri-AI to be transparent. Our model can be interpreted, which farmworkers and scientists can trust, because we visualize how our model makes decisions. After laying the foundation of the technological theory, a useful technology is finally transferred to the field. A robust, accurate, and explainable technological solution is developed in the study to protect global rice production from plant disease.

References

[1] Petchiammal, A., S.B. Kiruba, D. Murugan and P. Arjunan. (2023), Paddy Doctor: A Visual Image Dataset for Automated Paddy Disease Classification and Benchmarking. In: *Proceedings of the 6th Joint International Conference on Data Science & Management of Data* (*CODS-COMAD*) (pp. 203–207).

[2] Padhi, J., L. Korada, A. Dash, P.K. Sethy, S.K. Behera and A. Nanthaamornphong. (2024). Paddy leaf disease classification using EfficientNet B4 With compound scaling and swish activation: A deep learning approach. *IEEE Access*, *12*, 126426–126437.

[3] Xu, Y., D. Li, C. Li, Z. Yuan and Z. Dai. (2025). "LiSA-MobileNetV2: An extremely lightweight deep learning model with Swish activation and attention mechanism for accurate rice disease classification. *Frontiers in Plant Science*, *16*, 1619365.

[4] Bui, T.D. and T.M. Do Le. (2025). "Ghost-Attention-YOLOv8: Enhancing Rice Leaf Disease Detection with Lightweight Feature Extraction and Advanced Attention Mechanisms *AgriEngineering*, *7*(4), 93.

[5] Pan, J., T. Wang and Q. Wu. (2023). RiceNet: A two stage machine learning method for rice disease identification. *Biosystems Engineering*, *225*, 25–40.

[6] Abhiram, Dodda, K. Anuradha, V. Srilakshmi and K. Adilakshmi. (2024). A Novel Two-Stage Deep Learning Framework for Plant Disease Detection. In: *2024 5th IEEE Global Conference for Advancement in Technology* (*GCAT*), IEEE, 1–6.

[7] Naresh Kumar, B. and S. Sakthivel. (2025). Rice leaf disease classification using a fusion vision approach. *Scientific Reports*, *15*, 8692.

[8] Rahman, M.M., Z. Yan, M.T. Aziz, M.A.B. Siddick, T. Truong, M.M. Sharif, N. Datta, T. Mahmud, R.D.C. Pecho and S.M. Farid. (2024). Explainable Deep Transfer Learning Framework for Rice Leaf Disease Diagnosis and Classification. *International Journal of Advanced Computer Science and Applications*, *15*(12), 862–884.

[9] Prabhu, O. (2025). Interpretable CNN Architectures for Rice Leaf Disease Detection: A Grad-CAM and LIME-Based Explainability Approach. *International Journal of Computers*, *10*, 226–234.

[10] Ahmed, S.T., S. Barua, M. Fahim-Ul-Islam and A. Chakrabarty. (2024). Enhancing Precision in Rice Leaf Disease Detection: A Transformer Model Approach with Attention Mapping. In:

International Conference on Advances in Computing, Communication, Electrical, and Smart Systems (iCACCESS).

[11] Preethi, P., R. Swathika, S. Kaliraj, R. Premkumar and J. Yogapriya. (2024). Deep Learning–Based Enhanced Optimization for Automated Rice Plant Disease Detection and Classification. *Food and Energy Security*, *13*(5), e70001.

[12] Haruna, Y., S. Qin and M.J. Mbyamm Kiki. (2023). An improved approach to detection of rice leaf disease with GAN-based data augmentation pipeline. *Applied Sciences*, *13*(3), 1346.

[13] Tasfe, M., A. Nivrito, F. Al Machot, M. Ullah, F.A. Cheikh and H. Ullah. (2025). Deep Learning-Based Models for Paddy Disease Classification and Segmentation: An Experimental Review. *IEEE Access*, *13*, 152965–152989.

10 | Limitations and Challenges of Artificial Intelligence in Crop Disease Detection

Sachin Raj[1*], Priyanka[2] and Damanpreet Kaur[3]

Artificial Intelligence (AI) in agriculture has drawn a lot of interest as a possible way to boost crop yields and guarantee food security. Crop disease detection is one of its many potential uses since prompt diagnosis can avoid significant yield losses and lessen reliance on hazardous chemical treatments. AI-based systems can analyze plant photos and sensor data to detect early signs of diseases like leaf spots, blights, rusts, and fungal infections. These systems are frequently used in conjunction with machine learning, deep learning, and computer vision. Although there are a number of barriers preventing their widespread use, these instruments have enormous potential to transform agriculture toward precision farming methods. The lack of extensive and varied datasets is one of the main obstacles. AI models' capacity to generalize across various environments is diminished by the fact that plant disease datasets are frequently restricted to particular crops or geographical areas. The visual manifestation of diseases can also change based on growth stage, soil quality, and climate, which complicates detection. Additionally, especially in low-resource environments, the dependence on high-resolution images and complex sensors creates a gap between laboratory research and field-level implementation. The adoption process is influenced by socioeconomic and ethical factors in addition to technical ones. Additional challenges include the high price of AI-powered equipment, farmers' low digital literacy, and worries about data ownership and privacy. Furthermore, the realities of smallholder farmers, who make up the bulk of the agricultural workforce in developing nations, are not taken into account in the design of many AI tools currently in use. These issues are thoroughly discussed in this chapter, which divides them into three categories: technological, environmental, and socioeconomic. It highlights the necessity of building explainable AI models, open-access datasets, and reasonably priced, farmer-friendly platforms. In order to turn AI-driven crop disease detection from an experimental success into workable, sustainable, and internationally accessible agricultural solutions, these issues must be resolved.

[1-3] Affiliation: University Institute of Computing, Chandigarh University, Mohali-140413, Punjab, India.
Email: spriyanka1429@gmail.com[2]; damanpreet.cse.r@gmail.com[3]
* Corresponding author: sachinrajok333@gmail.com

DOI: 10.1201/9781003767534-10

1. Introduction

The agricultural industry in the world is currently at a crossroads. As the world population is set to reach about ten billion by the next 2050, food production is rising, and arable land is limited [1]. Crop diseases have been one of the most persistent causes of poor agricultural productivity with losses of up to 20–40% per year in a variety of crops and a variety of regions. The Food and Agriculture Organization of the United Nations estimates that the losses that can be attributed to plant diseases alone exceed 220 billion dollars annually [2]. Conventional disease diagnostic methods that require extensive manual examination by trained agronomists or extension agent staff tend to fall short with respect to speed, accuracy as well as the capability to scale to large volumes. It is on this premise that artificial intelligence has become a potential technological intervention. AI (artificial intelligence) systems have the potential to work with large volumes of visual and sensor data, to identify patterns that are impos sible to see with the human eye, and to provide quick diagnostic outcomes, and thus they appeal to researchers, policymakers, and even agribusiness stakeholders to consider them. Computer vision algo rithms are now able to draw the line between healthy and diseased leaf; machine-learning models are able to forecast disease outbreaks depending on the environmental conditions and mobile applications are able to give real-time guidance to farmers working in remote areas [3].

Deep-learning algorithms, especially the convolutional neural networks (CNNs), have proven to be highly effective in image classification. According to research studies, the accuracy has reached up to 95% with some reaching up to 99 and 62% accuracy in detecting particular diseases, such as late blight in potato, bacterial spot in tomatoes, and other fungal infections in cereals under controlled conditions [4] [5]. PlantVillage, a collection of 54,306 images of 14 crop species and 26 diseases, has been used as a standard to evaluate AI models in plant disease detection [6]. However, the move between the laboratory performance and the broad field application has turned out to be more difficult than it first thought. Although pilot projects and controlled experiments have proven to be quite impressive in terms of accuracy rates, the transfer of such findings into viable solutions in the form of ready, practical, and economical solutions applicable to farmers all around the globe is still pending. According to a proposal by Heldreth and colleagues to conduct research in the field of farmer-centered AI, the typical Western machine-learning applications might not align with local structures, poverty, infrastructures, values, and legal systems specific to lower-income countries and may actually cause harm [7]. This chapter focuses on the complexity of the barriers that restrict the potential of AI in crop-disease detection currently, grouping them into three categories interrelated to one another: technical and technological limitations, environmental and contextual, and socioeconomic and ethical issues. It is necessary to understand such limitations in order to come up with solutions that can indeed benefit the six hundred million smallholder farmers who make up 95% of the global farmers and contribute 45% of its food.

2. The Promise and Reality of AI in Agriculture

It is worth mentioning before proceeding to the limitations, that the AI-powered crop disease detection systems have already obtained certain achievements. Individually, deep learning models, and especially convolutional neural networks, have shown great expertise in image classification. Ngugi et al.'s review [8] is a well-developed survey that demonstrates the significant change in digital image processing, as the implementation of the deep-learning algorithms has demonstrated that they are far superior to the traditional tools of crop detection [8].

These systems are usually trained using thousands or even millions of labeled images of diseased and normal plants. The model after training is able to study new pictures and categorize them based on their disease type, severity and sometimes even prescribe treatment. Gaikwad and Musande [9] came up with a superior method of predicting crop diseases with the use of K-Nearest Neighbors (KNN), whale optimization algorithm, and the Relief algorithm and reported high accuracy levels in the analysis of hyperspectral UV images [9]. Similarly, Ahmed and Yadav [5] developed a predictive model that includes back-propagation artificial neural network algorithms and support vector machine (SVM), k-means clustering, CNN and Gray Level Co-occurrence Matrix (GLCM) in achieving a 99% accuracy rate on the plantvillage dataset [5]. Other advanced systems combine the data resources with visual data, weather data, soil sensor data, and documented patterns of disease to improve predictive accuracy. Indicatively, Komala [10] suggested an image-processing and filter-based feature-selection approach that identifies and detects crop diseases with the k-means, GLCM and KNN, and has yielded a performance accuracy of 95% [10]. The production of such technologies is also an indication of significant advances in terms of precision agriculture aims [11].

However, with all these technological advancements, the adoption level has been very low among farmers and most especially in the developing world where the need is the greatest. This lack of conver gence between the potential and the use of technology forms the main issue in this chapter. The latest studies of AI utilization in Sub-Saharan Africa show that there are multiple obstacles to more extensive implementation, with the digital divide and infrastructure issues being the starting point [12].

3. Technical and Technological Limitations

3.1 Data Scarcity and Quality Issues

The data forms the basis of any strong artificial intelligence system, and this is the greatest challenge in this case. Machine learning algorithms require a large amount of high-quality and professionally la beled training data in order to achieve stable performance. Such datasets can pose numerous challenges in the agricultural field, and such problems have been reported in the literature numerous times [13].

Currently, the majority of the existing plant disease image collections have been compiled in settings that are very controlled, usually in a laboratory facility or research greenhouse. In their seminal publication about the PlantVillage dataset, Hughes and Salathé [6] admit that the images were taken in the experimental research centers which are members of American Research Land-Grant Universities [6]. These data often include solitary leaves with disease symptoms that are adjacent to a solid color. Field conditions, on the contrary, are different. Plants are grown in tight canopies; foliage overlaps, lighting changes across a day, background elements like soil, weeds, and mulch add great visual richness that has the potential to baffle models used to more clean, artifact-free data.

One conspicuous weakness of Hughes and Salathe´ is that modern models are limited to the configu ration of single, facing-up leaves that are placed on a uniform background [6]. Whereas these conditions are easy to solve by algorithms, a real-life situation should have the ability to diagnose diseases that oc cur on the plant as they appear, involving a variety of surfaces and parts of the plant. This observation is supported by the subsequent studies using field datasets. On field collections, including FieldPV and FieldPlant that have different background complexity, performance is significantly worse when models trained on the PlantVillage dataset are applied [13]. In addition, the geographic and crop heterogeneity that is reflected in the current databases is still small. A model developed mostly on tomato disease in European greenhouses might not work well when used on tomatoes produced in sub-Saharan Africa, where there are different disease strains, climatic factors, and cultivation methods. It is even worse when it comes to crops that little research is given. Trade commodities such as wheat, rice, corn, and soybeans have fairly healthy datasets, but orphan crops that support millions in third-world countries have no significant digital records of disease [14]. There are additional complications presented by the process of labeling itself. Proper disease diagnosis requires expert skills because the symptoms can be very close when they are compared to different diseases especially during the early stages of the disease. The mislabeling process is bound to give unreliable models. Moreover, most diseases show dynamics that change over time; thus, to capture this dynamic, longitudinal data must be expensive and labor-intensive to generate. Recent efforts, like the PlantSeg dataset seek to address these inadequacies by providing large-scale, in-the-wild annotated data with pixel-level segmentation, but these compre hensive resources are currently uncommon [15]. Finally, the avenues of access to plant disease data are inherently limited, and the vast majority of the most convenient collections are not detailed enough with respect to the type of annotations, provenance of the images, and size. As it is observed by different researchers, current datasets concentrate on image classification as the main task, and little is provided on object detection and segmentation tasks that would be more useful in the complex field conditions.

3.2 Model Generalization and Transferability

This anti-generalization effect has become one of the fundamental challenges of artificial intelligence systems learning patterns, and it has continuously been found as a daunting hurdle to practice [16]. Although these models can have astonishing ability to identify statistical correlations in their training data, they often fail when it comes to the more fundamental causal processes that describe the phenomena that they are supposed to model. An example of such a case is how a classifier could falsely relate certain visual images with a pathology not due to a causal connection between the visual images but due to mere coincidence between the images and the pathology. In case most of the pictures illus trating a given condition were taken under a unique lighting regime, the model would unintentionally depend on the nature of lighting instead of actual displays of the disease. This tendency is confirmed using empirical evidence of controlled grayscale studies: when chromatic information is removed, pre cipitous inaccuracies are often induced, and this can be seen as reflecting dependability of the models on color as a discriminative, but not descriptive feature [6]. Phenotypic expression of plant diseases may vary significantly with dissimilar temperature and humidity gradients. An example of a fungal pathogen can have highly different morphological markers in a humid tropical environment than it would in a dry environment. Therefore, a model that has only been trained in a single climatic niche will be unable to identify the same disease when subjected to different environmental conditions, an issue that is partic ularly acute in developing areas where climatic heterogeneity is high and historical meteorological data tends to be sparse [17]. Stressed foliage can resemble disease lesions and cause false positive diagnoses, and healthy plants could hide early pathological changes, thus creating a false negative diagnosis [18].

3.3 Infrastructure and Hardware Requirements

The computational requirements of the modern AI systems pose great challenges to deployment, especially in developing environments. Due to deep-learning networks containing millions or billions of parameters, a significant amount of processing power is required. Even though a cloud-based solution is capable of relieving the load of computation, it assumes the access to reliable internet connectivity, which is limited or untrustworthy in many rural agricultural areas [19]. Current AI platforms, which have been designed to scale to large-scale applications in the developed world, implicitly assume that users have unlimited access to electricity, computing facilities, and internet connections. However, smallholder farmers mostly work in rural areas, which are not as connected and have access to digital adoption compared to urban areas. As an example, high-resolution imaging equipment is a major in frastructural challenge in Kenya. Proper disease identifications often require elaborate imagery that can help identify the disease at its silent symptoms, like discoloration, lesions or texture changes. Although the current techniques of smartphone cameras have experienced impressive advances, devices with lim ited budgets found in third-world countries tend to lack adequate fidelity of image. Special cameras or sensors that can capture multispectral or hyperspectral images have better diagnostic capability but are too expensive to be used widely [20] [21].

This is even weakened by the common device-sharing behaviors of the smallholder farmers, who normally use mobile phones to communicate with other individuals and not in the exchange of agricultural information. The availability of power should also be put into consideration; most of the agricultural areas in the developing world suffer frequent power cuts or no connection to the grid at all. Access to electricity is also a limiting problem to the implementation of AI-based equipment. The use of battery-powered devices has to compromise between computational power and energy consumption, which can result in performance trade-offs that reduce the performance. This is particularly problematic to edge-computing solutions that aim to implement AI models on mobile devices to reduce reliance on internet connectivity. The recent research has been focused on the creation of light models optimized to run on edge-computing machines. Edge device-specific plant disease detection systems are examples of how large model sizes impose significant barriers to real-world agricultural applications, in which real-time or near-real-time decisions are often needed [22]. The architecture, like MobileNetV3-small, signifies combined efforts of accuracy and computational frugality, but it is hard to identify the best trade-off, especially when detecting multiple diseases are considered.

3.4 Technical Complexity and Maintenance

Artificial intelligence systems cannot be viewed as fixed tools that can be put into operation and never come to a stop after implementation. Instead, they need a system of regular service, periodic changes, and their accurate adjustment. When faced with new disease strains, breeding of new crop traits or changing climatic conditions, then these models will need retraining on new data sets to main tain predictive fidelity. The maintenance needs of the attendants present a burden of reliance on tech nical know-how that is unlikely to be locally available in agrarian settings [23]. A farmer facing a bug in the software, a compatibility problem after an update, or an AI model that starts providing unreli able predictions can have little to no options when the technical support is located outside the country or unavailable. Additionally, this issue is aggravated by the fact that most deep-learning systems are opaque; even experienced agronomists can struggle to understand why a specific system generated a specific diagnosis or why certain agronomists can validate the logic behind the decision. This inability to gain an interpretation has been reported as a fundamental hindrance to belief and consequent accep tance. This is not limited to individual farmers, but it cuts across the board to include extension services and agricultural advisors who serve as a medium between technology and farming communities. The extension institutions in most of the low- and middle-income countries face extreme challenges such as low extension reach due to inadequate cadres of extension agents. In Kenya, the agent-farmer ratio is estimated to be 1:1000 as per government reports and can even reach 1:4000 which is an acuity that is beyond the recommended level of 1:400 [24].

Capacity building, which is needed to provide these agents with the ability to understand, work with, and trouble shoot AI-based diagnostic systems, requires considerable investment of capacity, a resource that is often lacking.

4. Environmental and Contextual Challenges

It is apparent that plant diseases do not show themselves in a consistent, easy to categorize way. One pathogen can cause a range of symptoms that will depend on the variety of hosts, the development stage, the severity of infection, and the current environmental condition. The clinical manifestations of infection during the early stages of infection are not obvious at all because they can be confused with nutrient deficiencies, abiotic stressors, or natural phenotypic variation of the plant [25]. Additionally, co-occurring diseases may also present themselves in a composite symptomatology that baffles diag nostic devices geared towards the resolution of single afflictions. As a result, the same plant can at the same time have the signs of a fungal invasion, bacterial pathogen, and damage caused by the pests with their overlapping and dependent manifestations. Some pathosystems are the members of which exhibit specific pathological markers in the root system, the tissue of the stem, or deeper structures, and not on the external foliar surfaces. Diagnostic platforms based solely on surface imagery are by definition ignorance of these hidden disorders.

Other infections can be asymptomatic, which can be amplified until much physiological damage has occurred, thus undermining the efficiency of early-warning systems. As an example, certain viral agents may stay in a latent state over extended periods of time where seemingly healthy plants are sneaky vectors and transmit the pathogen to other people [26].

The heterogeneity of the symptomatology is also worsened by the interaction of host genotype, pathogen virulence and a set of abiotic variables. A seemingly insignificant infection that would be found with optimum cultivation standards under normal conditions would, with exposure to environmental stressors, degenerate into a major disease outbreak, and vice versa, a highly pronounced disease would calm down in favorable conditions. Calibrated AI models of visual data representing specific environmental permutations find it hard to extrapolate to novel configurations, a demonstration of the inherent inability of purely empirical pipelines which do not incorporate biologi cal mechanistic understanding of plant-pathogen interactions [27]. Empirical studies have proved that disease phenotypes may differ significantly in different thermal, hygrometric, edaphic, and agronomic conditions (Table 1). An in-depth comparison of plant-disease recognition corpora reveals that most datasets are specific to a single species of crop, or to a collection of crops, and that pathologies are specific to the different organs, such as foliage, fruiting bodies, and stems [13].

Table 1 Disease Symptom Variability Across Environmental Conditions.

Disease Type	*Optimal Conditions*	*Stressed Conditions*	*Climate Zone Impact*	*Detection Difficulty*
Late Blight (Potato)	Dark lesions with white sporulation	Dry, brown lesions without spores	High humidity zones: rapid spread; Arid zones: slower progression	High – symptoms vary significantly
Bacterial Spot (Tomato)	Small, dark spots with yellow halos	Larger, irregular lesions	Tropical: year-round; Temper ate: seasonal	Medium – consistent patterns
Powdery Mildew	White powdery coating, uniform	Yellowing without visible fungus	Universal distribution	Low – distinctive appearance
Root Rot	Wilting, stunted growth	Similar to drought stress	Soil-dependent, all zones	Very High – no visible leaf symp toms
Viral Mosaic	Mottled, mosaic patterns	Can mimic nutrient deficiency	Temperature-dependent expression	High – variable expression
Rust Diseases	Orange-brown pustules	Faded lesions in dry conditions	Humidity-dependent	Medium – color-based detection
Anthracnose	Dark, sunken lesions on fruits	May appear as general rot	Warm, humid climates	Medium – organ-specific

4.1 Temporal Dynamics and Seasonal Variations

The seasonal nature of agricultural production and the subsequent cyclical changes in the pressure of diseases require academic attention in the study of the nature of the production. A computer-vision model largely trained on data obtained at the peak phenological period will typically not generalize to the transitional periods between phenological periods known as shoulder seasons when plant physiology is no longer the same and different pathogen repertoires are found. The time aspect of the agricultural sector is systematically sidelined in AI design, which in turn focuses on the image classification at a specific moment, as opposed to the longitudinal monitoring of the disease. This leads to the fact that serial imaging and cross-time point comparison are required to completely monitor disease progression, enhancing the methodological complexity and scale of the data acquisition. Current paradigms can hardly incorporate this time continuum effectively, though it has the possibility to both improve diagnostic accuracy and to monitor treatment effects.

Phenological changes in disease phenotypes cannot be separated from extrinsic weather effects, which add auxiliary variability to them. Stressful precipitation may destroy gross manifestations or stimulate mycorrhizal growth, whereas dry stress can present as pathological scars. The morphological confounds produced by physical setbacks like frost, hail, or rupture caused by wind are similar to biotic damages. The distinction between etiological factors of plant stress that are abiotic and biotic hence is another daunting challenge to AI systems, which requires a combination of both environmental sensor streams and image-based analytics. Matters are complicated further by the variability of prevalence of diseases across calendar effects; a diagnostic model that sees a pathogen only marginally in training may fail on a different season with that pathogen as common place. Such a situation is representative of the larger problem of the class imbalance in training data sets with some diseases prevailing and others suffering in underrepresentation, which ultimately leads to biased model performance (Table 2).

Table 2 Environmental Factors Affecting Disease Expression and AI Detection.

Environmental Factor	*Optimal Range for Disease*	*Effect on Symptom Expression*	*Effect on Image Quality*	*AI Model Adaptation Needed*
Temperature	20–30°C (most pathogens)	Higher temps accelerate; lower temps slow progression	Minimal direct effect	Temperature-conditional models
Relative Humidity	80–100% (fungal diseases)	High humidity enhances sporulation; low humidity suppresses	Fog/mist reduces visibility	Humidity-aware preprocessing
Rainfall	Frequent, moder-ate	Spreads spores; washes symptoms; creates new infections	Rain droplets interfere; muddy backgrounds	Weather-data triggered collection
Light Intensity	Variable	Affects plant stress response	Critical for image capture; shadows	Time-of-day standardization
Soil Moisture	Field capacity to saturation	Root diseases in wet soils; stress symptoms in dry	Affects plant turgidity	Soil sensor integration
Wind Speed	Moderate (disease spread)	Spreads pathogens; causes physical damage	Camera shake; leaf movement	Stabilization algorithms
Soil pH	Variable by pathogen	Affects pathogen survival and plant susceptibility	Indirect through plant health	Soil data incorporation
Solar Radiation	High intensity	Can mask or en-hance symptoms	Glare, overexposure issues	Exposure compensation

5. Conclusion

AI has actual possibilities to transform crop disease detection and, conse-quently, the agricultural production paradigm in general. The abilities that have been demonstrated in the controlled setting of research are astonishing and continue to undergo evolutionary advancement. The transfer of this promise to mass practice, however, (particularly to the developing countries, where exigency is greatest) requires the consideration of the significant limitations and difficulties which are outlined herein.

Technical barriers related to the data quality, model generalizability, and infrastructural requirements are significant but not too serious. More recent academic advances in transfer, few-shot and edge computing gradually alleviate these limitations. Furthermore, agro-systems present environmental and contextual challenges that are inherently due to the complexity and heterogeneity of agro-systems and which necessitate continuous questioning and potentially require a fundamental reevaluation of AI design and implementation processes. The socioeconomic and ethical strata are the most complex to deal with, arguably. Technology alone will never overcome the barriers deeply rooted in economic in equality, institutional weakness, cultural complexity, and asymmetry of power. Addressing these issues requires the cross-disciplinary approach, creative policy designs, the ability of entrepreneurial mod els to develop, and a long-term commitment to equity and inclusion. In the future, the agricultural AI sector needs to resist the temptation of technological solutionism, i.e., the belief that technical innovation is the only solution to solve society in its complex dilemmas. Instead, a less grandiose, participatory, and contextual approach to methodology is mandatory. The actual measure of success will not be based on accuracy measures based on benchmark datasets, but actual adoption by growers, especially smallholders in developing jurisdictions and tangible improvements in productivity and sustainability and livelihoods. The path between lab achievement and practical effect is long and difficult, but definitely is worth the journey. Crop diseases will continue to be existential threats to food security and AI-based diagnostic devices will be promising to reduce this threat. Through recognition and ardent attempts to overcome the constraints as discussed in this exposition, scholars, technologists, legislators, and agrarian stakeholders can build a mutual force to realize the transformative potentials of AI in agriculture without jeopardizing that its locations are shared universally and equitably. Finally, the usefulness of AI in detecting disease in crops will depend not only on the refinement of algorithms but on the ability of all of us to make sure that such powerful tools will meet the needs of all the growers, including the most disadvantaged and discriminated. This kind of venture requires the ongoing discussion, the process of re-education, and a firm adherence to the principles of equity, clarity, and social justice in the evolution of agritech.

References

[1] Ahmad, L., Nabi, F. and Habib, M. (2024). Artificial intelligence in sustainable agriculture: Recent trends and future perspectives. *Computers and Electronics in Agriculture*, *218*, 108–125.

[2] Mamba Kabala, A., Ngoy, K.I. and Mukendi, R. (2023). Federated learning for crop disease detection: A comparative study of deep learning models. *Journal of Agricultural Informatics*, *14*(2), 45–62.

[3] Jafar, A., Bibi, N. and Naqvi, R.A. (2024). Deep learning for plant disease detection: Challenges and opportunities. *Computers and Electronics in Agriculture*, *219*, 108–125.

[4] Singh, V. and Yogi, A. (2023). Enhanced deep learning approaches for crop disease identification and classification. *International Journal of Computer Vision*, *131*, 1520–1538.

[5] Ahmed, M.R. and Yadav, S. (2023). Predictive models for plant disease detection using artificial neural networks and support vector machines. *Precision Agriculture*, *24*, 1456–1475.

[6] Hughes, D.P. and Salathé, M. (2015). An open access repository of images on plant health to enable the development of mobile disease diagnostics. *arXiv preprint arXiv:1511.08060.*

[7] Heldreth, C., Ghazal, M. and Singh, V. (2021). Towards farmer-centered AI for the Global South. *ACM COMPASS*, 125–136.

[8] Ngugi, L.C., Abelwahab, M. and Abo-Zahhad, M. (2024). Recent advances in image processing techniques for automated leaf disease detection. *Information Processing in Agriculture*, *8*(1), 27–51.

[9] Gaikwad, S.B. and Musande, V.B. (2023). Crop disease prediction using whale optimization algorithm with KNN classifier. *Journal of King Saud University – Computer and Information Sciences*, *35*(6), 101–115.

[10] Komala, C.R. (2021). Plant disease detection using image processing and ML techniques. *IJERT*, 10(5), 789–796.

[11] van Evert, F.K., et al. (2023). Can precision agriculture increase sustainability? *Sustainability*, *9*(10), 1863.

[12] World Bank. (2025). Artificial intelligence for agricultural development in Sub-Saharan Africa. *World Bank Digital Development Reports*, 45–78.

[13] Thapa, R., et al. (2024). Plant pathology challenge dataset. *Applications in Plant Sciences*, *8*(9), e11390.

[14] Mkonyi, E., et al. (2023). Early detection of bacterial wilt. *Smart Agricultural Technology*, *3*, 100–115.

[15] PlantSeg Dataset. (2024). Large-scale image dataset for plant disease detection. *IEEE Dataport.* doi:10.21227/dataset.plant2024.

[16] Akinosun, J. and Nibouche, O. (2023). Transfer learning for plant disease detection. *Frontiers in Plant Science*, *14*, 1125476.

[17] Barbedo, J.G.A. (2018). Impact of dataset size and variety on deep learning performance. *Computers and Electronics in Agriculture*, *153*, 46–53.

[18] Mohanty, S.P., Hughes, D.P. and Salathé, M. (2016). Using deep learning for plant disease detec tion. *Frontiers in Plant Science*, *7*, 1419.

[19] Sharma, P., Bernice, Y.P. and De, D. (2023). IoT and ML enabled plant disease detection. *Journal of Ambient Intelligence and Humanized Computing*, *14*, 6875–6891.

[20] Gates, M. and Bannon, S. (2023). Digital agriculture for small-scale producers. *CGIAR Working Paper No. 371.*

[21] Subeesh, A. and Mehta, C.R. (2021). Automation and digitization of agriculture using AI and IoT. *Artificial Intelligence in Agriculture*, *5*, 278–291.

[22] Liu, J. and Wang, X. (2023). Plant disease detection on mobile and edge devices. *Frontiers in Plant Science*, *14*, 1188011.

[23] Kamilaris, A. and Prenafeta-Boldú, F.X. (2018). Deep learning in agriculture. *Computers and Electronics in Agriculture*, *147*, 70–90.
[24] Farmer.Chat. (2024). AI-powered digital extension services. *Digital Green Technical Report 2024– 01*.
[25] Martinelli, F., et al. (2015). Advanced methods of plant disease detection. *Agronomy for Sustainable Development*, *35*(1), 1–25.
[26] Strange, R.N. and Scott, P.R. (2005). Plant disease: A threat to global food security. *Annual Review of Phytopathology*, *43*, 83–116.
[27] Savary, S., Willocquet, L., Pethybridge, S.J., Esker, P., McRoberts, N. and Nelson, A. (2019). The global burden of pathogens and pests on major food crops. *Nature Ecology & Evolution*, *3*(3), 430–437.

11 Economic Environment Impact of AI and Machine Learning in Plant Disease Detection and Sustainable Agriculture

Damanpreet Kaur[1*], Sachin Raj[2] and Priyanka[3]

One of the simplest economic sectors that is experiencing the dramatic change is agricultural development with the application of Artificial Intelligence (AI), Machine Learning (ML), and Digital Agricultural Technologies (DATs). The degree of economic and ecological benefits of such innovations is high, according to the recent findings (2024-2025). The cheap sensors, deep learning disease-detecting sensors and hybrid predictive tools are precision farming tools that allow the increase of the crop yields and crop resilience and minimise the usage of excessive number of chemicals, irrigation and human labour. It has been established that AI-based systems can be used to cut the number of used fertilizers to 80, the amount of used water could also be reduced, and the quantity of crop yield could also be more than 60. Furthermore, AI-based prediction techniques including LSTM-GARCH hybrids and meta-transformers may be more accurate in their prediction of the volatility of commodity prices and, thus, stabilize the revenues of the farmers and minimize the impacts of market uncertainty. The virtues of the surrounding area are also worth mentioning. Such technologies reduce the content of greenhouse gas, improve the quality of soil mass and climatic adaptation at the expense of reducing pesticides and fertilizers. Besides this, sensor-based early disease detection lowers the losses of crops creating food security with a lesser input of natural resources. These results can be made to be in line with the Sustainable Development Goals (SDGs) that have Zero Hunger, Climate Action and Responsible Consumption and Production.

The chapter gives the summary of the recent research on the economic and environmental effects of the AI/ML-based solutions in the agricultural sector, introduces the existing gaps, and suggests a new conceptual framework between the application of AI and its economic benefits and environmental sustainability. Technical development and a socio-economic assessment are combined to offer specifics of the attainment of inclusive, resilient, and sustainable food systems in the present digital age of agriculture in the chapter.

[1-3] University Institute of Computing, Chandigarh University, Mohali-140413, Punjab, India.
Email: sachinrajok333@gmail.com[2]; spriyanka1429@gmail.com[3]
ORCID: https://orcid.org/0009-0002-8050-2547[1]; https://orcid.org/0009-0005-4108-7020[2]; https://orcid.org/0009-0006-4845-5380[3]
* Corresponding author: damanpreet.cse.r@gmail.com

DOI: 10.1201/9781003767534-11

1. Introduction

Agriculture has always played a main role in the life of human beings, economies, and international trade. The 21st century has brought significant challenges to the industry such as population growth, climate change, soil erosion, water scarcity, and spread of crop diseases. Potential solutions are Artificial Intelligence (AI), Machine Learning (ML) and Digital Agricultural Technologies (DATs). As opposed to traditional methodologies, AI-driven solutions facilitate precision agriculture. Where data procured from sensors, drones, and satellites are synthesized to optimize irrigation strategies, anticipate diseases proliferation, and curtail pecuniary. The application of AI to disease management can spare millions of millions of tons of agricultural products, cut down on the use of pesticides, and enable making projections of income based on detention.

The economic landscape is being transformed by AI: LSTM-GARCH hybrids are being trained to be effective at commodity-pricing prediction (2024–2025), though with meta-transformer architecture, where better decisions can be made and market risk is reduced [1]. Digital technologies are beneficial in the context of savings through efficient fertilizer operation, minimized water consumption, minimized labor reliance; all these aspects are converted into enhanced ROI (return of investment), particularly in locations with constrained resources. On the environmental front, AI will minimize agricultural footprint through the reduction of excessive fertilizer and pesticide application, which will lower the emission of greenhouse gases (GHG) and ensure biodiversity and health of soil. The chapter summarizes the recent discoveries (2020–2025), the research gaps, and suggests a theoretical framework that connects technological adoption to economic stability and environmental sustainability. The chapter is supposed to present a multi dimensional approach to the redesign of the application of AI in agriculture to make it a more sustainable, inclusive, and profitable industry.

Motivation and Scope

Technology and sustainability are in a collision course with the modern state of agriculture. The world population continues increasing the demand to feed the population, as it is projected to increase by 60%. This is what has happened, and it has given rise to the high intention to implement AI, and ML in agriculture to achieve a data-driven, predictive, and resource efficient agricultural world. There is the possibility of AI technologies that will allow moving reactive models of agriculture towards adaptive and proactive models. Computer vision, inexpensive sensing devices, and predictive AI can be used to ensure that loss is minimized and that productivity goes up significantly since disease is detected at an early stage. In addition to that, the use of AI to predict crop production and prices of products brings additional stability to the agro system. There is also the environmental incentive that is persuasive. The applications of the AI can assist in utilizing the natural resources in an environmentally friendly manner and reducing greenhouse gas (GHG) emissions and preserving soil and water. Smart farming founded on AI technology might reduce the environmental impact of food production against the backdrop of global climate change and generate the impulse to switch to a sustainable and climate-resilient food production. Therefore, the economic and environmental impact of these technologies is crucial to be assessed to derive in-depth analysis of the reality behind the advantages and obstacles of these technologies in their real applications.

Scope

Economic and environmental aspects of AI in the agricultural industry are addressed in this chapter. It is also world-oriented in its analysis of the impact of AI and ML-based technologies, which include accuracy farming systems, sensor network, and deep learning-aided detection of plant diseases and predictive price modeling on cost systems, productivity, profitability, and environmental sustainability. The results of the latest research (2020–2025) will be analyzed to offer the general perspective of the tendencies, the development of technologies, and the problems of the implementation. It is not just in this chapter that she is devoting her attention to the technical assessment but broadening her perspectives to the economic modeling, policy implications, and the opinions on sustainable development. It will also aim at ensuring that the technological innovation is applied to the farming community in a practical manner through inclusivity, affordability, and environmental concern. The theoretical framework suggested further in the chapter is the basis of the bridging AI adoption, economic viability, with environmental sustainability and can be a useful tool to researchers, policymakers, and industry players planning to implement the Internet of Things (IoT) in the next generation of smart and sustainable agriculture.

1.1 Structure of the Chapter

The chapter is also subdivided into different sections to possess an appropriate understanding of the economic and environmental impacts of AI and ML application in agriculture. Section 2 is the literature review which entails the recent development, the most significant research findings, and missing information on the knowledge of AI-based agricultural innovation. The third part is on the economic impact of AI implementation, where the author discusses the outcomes of implementation of technology, e.g., precision farming, predictive analytics, and disease detection systems on productivity and cost-efficiency and on market stability. Section 4 concerns the environmental impact which is concerned with optimal use of resources, reduction of emissions, and sustainability effects.

Section 5 identifies the major issues and gaps in research that will hinder the widespread use of these technologies particularly in developing regions. Section 6 provides a hypothetical framework of the two pillars, the economic and environmental sustainability, to the adoption of AI. Section 7 involves case studies and comparative analysis of part of the recent global implementation, and Section 8 contains future and policy recommendations towards achievement of inclusive and sustainable agricultural transformation. Finally, though not the least, Section 9 is the conclusion of the chapter, which provides the key insights and contributions to the field.

2. Literature Review and Background

2.1 AI/ML for Plant Disease Detection (Technical Review)

Modern Computer-vision-based detection of plant diseases is a crucial area of modern technology that uses deep learning. Convolutional Neural Networks (CNNs) have continued to be popular in the classification of leaves and canopies (e.g., VGG, ResNet) and lightweight and interpretable versions to deploy to edges. Surmounting the constraints of limited labeled data is commonly achieved through the application of transfer learning ensemble methodologies, and domain adaption strategies [2]. Edge inference and low-cost sensing can be used to implement on single-board computers and smartphones. Multispectral/hyperspectral sensing can catch stress sooner than RGB (red, green, blue), but it is more expensive, and RGB-based CNNs are widespread because of their accessibility. The IOT sensors networks provide environmental contextualization to augment specificity and mitigate false positives. However, technical deficiencies persist: the comment able accuracy rates predominantly derive from meticulously curated or greenhouse datasets, while real-world variability, including factors such as lighting , occlusion, and mixed infections, detrimentally impact performance.

Not many studies include inference cost, model size, or energy use, which are essential in the case of edge deployment. Fields Cross-site generalization Domain adaptation is an open problem.

2.2 Economic Evaluation Studies

Economic literature covers:

(a) cost-benefit and ROI of DATs on farms.
(b) better predictive market effects.
(c) macro/sectoral measures of productivity increases.

The input savings, labor reduction and yield improvement reviews indicate, the empirical analysis results in positive correlations between the digital adoption and gains in farm income. The commodity-price forecasting (hybrids such as LSTM-GARCH or transformer variants) based on AI is new and can enhance short-term predictions and volatility modeling (2024–2025) to help in marketing decisions.

Economic disparities: The majority of ROI or cost-benefit analyses are relatively short-term, and targeted bigger/commercial farms or high-value crops (greenhouse avocados, horticulture). There is still limited evidence of smallholder, rain-fed contexts. There are limited studies that combine both technical performance (detection accuracy) and economic performance (decrease in crop loss value, calculation of payback period, or sensitivity to market shocks). Such a disconnect makes policies less relevant. The adoption price, upkeep, instruction, and information-service charges tend to be understated, hence the quantified advantages could exaggerate net profit to resource-deprived farmers.

2.3 Environmental Evaluation Studies

The Environmental co-benefits of DATs and AI are evaluated: agrochemical usage will decrease, water will be saved, and a decrease in GHG emissions may be possible. Run-offs and emissions are reduced through accurate use of fertilizers and pesticides and irrigation. An LoC (Library of Congress) research proves that CO2 reduction is achieved through the assistance of a successful allocation of factors via digital interventions. Trials have high gains in the water-use efficiency of the mechanized precision interventions (drone spraying, targeted irrigation).

Environmental gaps: The lack of life-cycle assessments (LCAs) of DATs seems to be the reason why very few studies have considered embodied emissions of sensors, drones, and edge hardware. Spatial heterogeneity is used to show that the benefits of the environment are not distributed uniformly in the regions. Biodiversity recovery and soil health are some of the co-benefits that cannot be quantified.

2.4 Summary of Recent Studies (2020–2025). These are presented in Table 1.

Table 1 Summary of selected recent studies (2020–2025).

Reference	*Method*	*Key Findings/Gap*
Bukhari et al. (2025)	Low-cost IoT sensors with ML edge inference	Early detection on edge (75–86%); greenhouse conditions; field robustness unproven.
Chen & Wang (2024)	Multi-objective optimization of fertilizer use	Reduced N input by 18%, CO_2 by 12%; limited to rice–wheat system.
Kumar et al. (2023)	Federated learning for agriculture	Comparable accuracy to centralized training; needs high connectivity.
Li et al. (2023)	Transfer-learning CNN (PlantVillage)	> 96% accuracy; dataset bias limits field generalization.
Osei et al. (2022)	Cost–benefit of precision irrigation	25–30% water savings, 20% yield increase; small sample.

Over the last five years, AI/ML and DATs matured towards scalable systems integrating IoT sensing, edge inference, and predictive analytics [3]. Yet economic and environmental assessments remain fragmented; high-accuracy results often limited to pilots or greenhouse trials. Cross-disciplinary models linking detection accuracy with economic value and sustainability indicators are needed

3. Economic Impact of AI/ML in Agriculture

3.1 Effects of Yield and Productivity

On AI-based solutions (disease detection, yield prediction, precision inputs) yields have increased in some studies by 1525%, low-cost advisory systems have increased yields of smallholders by 812% and deep learning disease detection has reduced yield losses by 1020% on controlled trials [4][5][6]. It is a Timeliness system that will result in information even before the disease has taken place and will help to save crops and improve their quality.

3.2 Cost Optimization and Input Optimization

AI and ML maximize fertilizer, pesticides, water, and labor. Grassed reductions in the reported cases were better with pesticide 1530% fertilizer 1020%, water 2540% being used on smart irrigation. Automation reduces time required to make manual scouting and labor time per hectare.

3.3 Market Effects: Forecasting and Risk Management

AI is involved in the stability of the market and the reduction of risk. ARIMA/GARCH cannot be considered as good as deep learning and transformer-based forecasting; a 10–20% in forecasts results in better harvesting time, market entry. Availability of high-frequency market information has been a bottleneck to the smallholder.

3.4 Return on Investment (ROI) and Economics of Adoption

ROI is based on capital expenses, infrastructure, and size. It can be found that according to field research, mid-scale farms pay back in 1.53 years. Shared ownership models and cooperatives can be applied to increase affordability.

The illustrative example shows that a relatively small improvement in productivity and cost-efficiency can lead to excellent financial payoff in case AI-based systems are implemented in an efficient way. The key factors of ROI are the size of farm, data infrastructure, subscription models, and crop value. Smallholder conditions may be enhanced with payback through cooperative ownership and the government subsidies, or through public private collaboration that will help with initial investment and digital literacy.

The illustrative example shows that a relatively small improvement in productivity and cost-efficiency can lead to excellent financial payoff in case AI-based systems are implemented in an efficient way. The key factors of ROI are the size of farm, data infrastructure, subscription models, and crop value. Smallholder conditions may be enhanced with payback through cooperative ownership and government subsidies, or through public private collaboration that will help with initial investment and digital literacy.

Table 2 Illustrative ROI Calculation for an AI-based disease detection deployment.

Item	*Annual Cost / Benefit (USD)*
Initial hardware (amortized)	1,200
Software / subscription	300 / year
Yield gain (value)	1,800 / year
Reduced agrochemical cost	400 / year
Total annual operating cost	300
Total annual benefit	2,200
Net annual benefit	1,700
Payback period (years)	1.4
Three-year ROI (%)	≈114%

4. Environmental Impact

The implementation of AI and ML in agriculture does not only offer economic benefits to the process; it possesses considerable co-benefits to the environment. AI-driven systems are relevant to climate-smart and sustainable agriculture by allowing the use of resources exactly in the needed amount and minimizing the input wastage. This part explains the direct and indirect environmental impact of adopting AI/ML and involves input reduction, emissions, soil health, biodiversity, and compliance with the United Nations Sustainable Development Goals (SDGs).

4.1 Input Reduction and Emissions

AI is an input optimization: nitrogen, pesticides, energy/fuel, and water may be reduced 15–25% (cutting down N2O emissions), 20–40, 10–15, and 25–40% through smart irrigation. Certain LCAs propose that AI-powered interventions would reduce global agricultural GHG by 8–12%, should they be commonplace, although embedded gadget emissions will have to be considered [7][8][9].

4.2 Soil Health: Biodiversity

The Lower synthetic inputs save microbial balance and organic matter; predictive analytics help to balance nutrient replacement. Pesticide use leads to minimal run off, no destruction of non-target species and proper mapping leads to rotating crop production and agroforestry and leads to better sequestration of carbon and biodiversity [10][11][12][13].

4.3 Alignment with SDGs

The SDGs discussed in this paper emphasize how AI/ML-driven agricultural technologies contribute to sustainable development worldwide. SDG 2: Zero Hunger is supported by improving food security by increasing crop yields, lowering post-harvest losses, and developing more resilient farming systems that can endure environmental and biological challenges. Similarly, the solutions support SDG 12: Sustainable Development Goals 12 and 13 by encouraging optimal resource utilization and enhancing supply-chain traceability, reducing the use of fertilizers, pesticides, and water while assuring accountability from farm to market. Furthermore, the effort contributes to SDG 13: Action for Climate Change, as climate-smart monitoring and modeling can assist farmers in adopting mitigation techniques, lowering emission of greenhouse gases, and implementing adaptive practices that successfully adjust to changing climate conditions.

Table 3 AI/ML contributions to selected SDGs.

SDG	*Relevance*	*Key Contributions*
SDG 2: Zero Hunger	Food security through yield and loss reduction	Increased productivity and resilient food systems.
SDG 12: Responsible Consumption	Efficient resource use and traceability	Reduced fertilizer/pesticide/ water waste.
SDG 13: Climate Action	Climate-smart monitoring and mitigation	GHG reductions and adaptive practices.

5. Research Gaps and Implementation Challenges

In agriculture, regardless of significant advancements in AI and ML integration, large-scale application and long-term performance have several gaps and practical limitations because of ML. There is still an imbalance in translation of laboratory/pilot-scale innovations to the field level success across geographies and crop types. Most of the research is based on technical feasibility and not on long-term economic viability, environmental sustainability, or social inclusion.

- **Relevance to the smallholder setting:** The vast majority of agricultural AI systems are developed and trained on high-resource settings and have built robust digital infrastructure. Nevertheless, farmers with small landholding, particularly those in developing countries, have constraints of internet connectivity, sensor cost, and digital literacy. Therefore, it is still unclear how such technologies can be scaled down to low-resource or rain-fed situations.
- **Missing long-term economic evaluations:** Existing analyses of AI-driven accuracy agriculture focus on either short-term gain in yield or a decrease in cost. Little information is available regarding lifecycle costs, multi-year payback (ROI) and the larger influence on rural job and market conditions. In the absence of these insights, policymakers do not know how to come up with effective incentives or subsidy programs.
- **Infrastructure and training barriers:** Detection of diseases, prediction of yield or irrigation timing requires vigorous data collection, energy source, and technology. This is caused by the shortage of skilled staff and unsophisticated infrastructure in the rural areas particularly in data management and maintenance of supplied devices which creates a bottleneck in operations.

- **The agricultural systems powered by AI receive vast amounts of information at the farm level, including the type of soil, pictures of plant health, and geolocation metadata governance and privacy and policy concern:** Nevertheless, the ownership of data, the protection of privacy, and adherence to national standards of agricultural data is not always clear. This uncontrolled freedom may deter information exchange and may destroy trust between the farmers.
- **Coupling with market/insurance:** Although AI/ML models demonstrate potential in price prediction and risk forecasting, they are not generally combined with crop insurance programs, commodity markets or digital cooperatives. This type of integration is essential in the development of holistic decision-support systems that are useful throughout the Agri-value chain.

5.1 Loopholes Identified in Recent Literature

An overview of the recent research (2020–2025) shows that there are a number of similar loopholes that restrict the plausibility, scalability, as well as inclusiveness of AI-based agricultural innovations. These gaps will need to be filled to make sure that digital agriculture shifts beyond pilot success into a systemic change.

Table 4 Common loopholes and why they matter.

Loophole	*Implication*
Short-term trials	Lack of long-term evaluation leads to uncertain ROI, as seasonal variability and cumulative effects remain untested.
High-value crop bias	Most AI solutions are optimized for fruits, vegetables, or export crops, neglecting staple and rain-fed crops critical to food security.
Limited socio economic measurement	Studies often focus on technical accuracy rather than adoption rates, equity, and behavioral aspects influencing technology uptake.
Scalability concerns	Edge computing and sensor networks have not been tested under large-scale, multi-region deployment conditions; data reliability at scale is uncertain.
Lack of policy evidence	Absence of clear policy frameworks makes it difficult to align AI adoption with national agricultural and environmental goals.

6. Proposed Conceptual Framework

AI and ML are not only technologies that can be integrated into the agricultural sector to provide technical automation: it is a paradigmatic change that affects not only the efficiency of the economy but also the sustainability of the environment. To obtain these interrelationships, a conceptual model is suggested that connects digital data streams, decision making processes, and effects channels. This framework is a theoretical prism through which AI-based agricultural interventions are used to create quantifiable benefits in the productivity, profitability, and ecological resilience. It further emphasizes the contribution made by policy, market incentives, and institutional mechanisms to these effects in the long run.

6.1 Economic-Environmental Impact Framework (Summary Version)

The impact of AI-enabled technologies on the farm decisions, the economic performance and the environmental sustainability is explained through the Economic-Environmental Impact Framework (Figure 1) which consists of five interconnected components.

1. **Data Sources:** The model starts with the various real-time data feeds of IoT sensors, satellite/drone imagery, and market-weather data. These inputs give the fine-grained information required to do continuous monitoring and predictive modeling.
2. **AI/ML Analytical Layer AI:** Methods including CNN-based disease detection, metaheuristic resource optimization and transformer-driven yield/price forecasting, transform raw data into actionable information. This layer allows making farm decisions in a more accurate and timely manner.
3. **Farm Management Actions:** The outputs of the models are used to make key operational decisions such as the optimization of fertilizers, water and pesticides, the planning of crops and the timing of irrigation, the allocation of labor or machinery, which results in an increased efficiency and accuracy in field management.
4. **Economic Outcomes:** Examples of economic gains that AI-assisted decisions have brought to agribusiness and farmers include increased yields, reduced input/labor expenses, enhanced market prediction, and increased ROI.
5. **Environmental Outcomes:** Environmental benefits are optimised inputs: less chemical runoff, less GHGs and improved soils and biodiversity, which helps to sustainability in the long term. Feedback Loops in Policy, Market, and Insurance. The adoption is enabled by systems, including enabling digital agriculture policies, green financing, and integration with crop insurance and market intelligence to reinforce economic-environmental outcomes at scale.

Narrative Interpretation

The approach highlights how AI adoption serves as a technological and institutional driver of long-term agricultural transformation. The interaction of economic advantages with environmental stewardship generates a feedback loop in which both sustainability and profitability co-evolve. Policy measures and market incentives can accelerate this cycle by assuring equal access, strong governance, and long-term profitability.

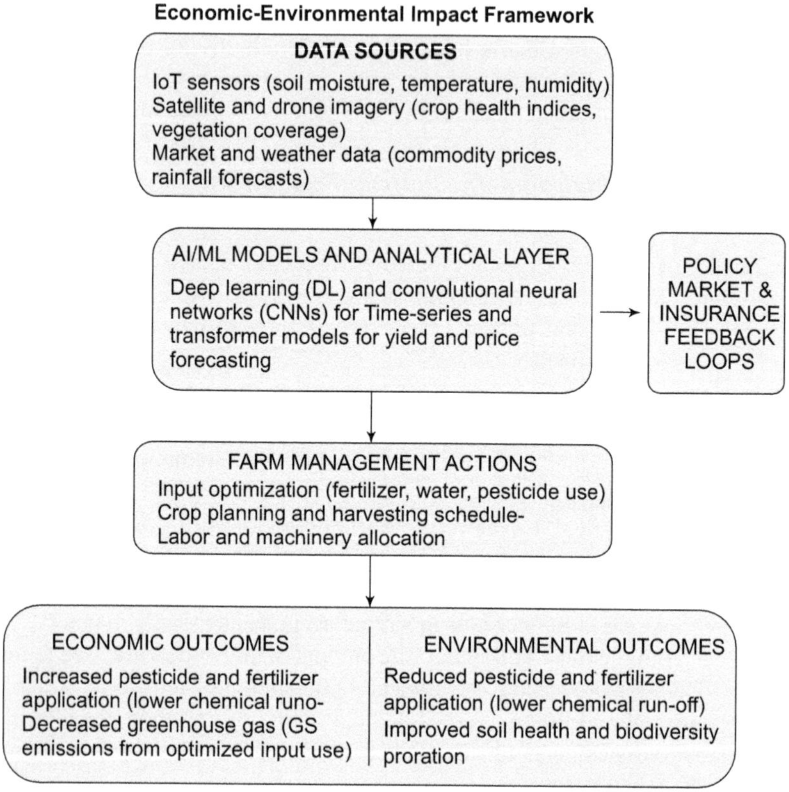

Fig. 1 Conceptual framework linking AI/ML, management actions, environmental, and economic outcomes.

7. Case Studies and Comparative Analysis

Table 5 AI/ML Case Studies in Agriculture (Economic & Environmental Impact).

Case	*Technologies Used*	*Data Sources*	*Key Results*	*Main Barriers*
7.1. Low-Cost Sensor Deployment (Avocado Greenhouse) Bukhari et al. (2025)	Edge sensors, 5MP cameras, lightweight CNN (TensorFlow Lite)	4,500 labelled images (crop + insect monitoring)	75–86% accuracy • Latency < 2 sec • 22% reduction in agrochemical use • Payback period ≈ 1.6 years	Scalability to open-field farms (lighting variations, connectivity issues, insect diversity)

Contd.

Table 5 *Contd.*

Case	*Technologies Used*	*Data Sources*	*Key Results*	*Main Barriers*
7.2. Meta-Transformer for Commodity Forecasting (India) Nayak et al. (2025) [2]	Meta-transformer architecture for price prediction	National/ state-level commodity price datasets	Improved short-term forecasting accuracy • Better decision-making for cooperatives and markets	Data availability issues, limited digital proficiency among farmers
7.3 Comparative Insight	—	—	AI/ML improves profitability and sustainability across both micro (farm-level) and macro (market-level) applications	Adoption depends on infrastructure, policy support, and farmer awareness

Table 5 is a comparison of two notable case studies that illustrate the economic and environmental values generated by AI and ML in the agricultural sector. The former, which concerns the low-cost implementation of sensors in an avocado greenhouse, implies edge sensors, 5MP cameras, and a light-weight CNN model developed using the TensorFlow Lite framework. The system generated 7586% accuracy and less than two seconds response time using a dataset of 4,500 labeled images. This saw farmers save 22% of agrochemicals use, which led to the fact that the farmers saved money as well as saved the environment, and the payback period took about 1.6 years. But the biggest obstacle to widespread implementation is the challenge of expansion of this system to farms into open fields where light, connectivity, and insect diversity add complexity.

The second example concerns the national-level commodity forecasting of India with the help of the meta-transformer architecture. It is a large-scale statistical price model that is used to enhance the short-term price forecasting of large-scale historical price data to assist cooperatives and markets in making better decisions regarding storage and sales as well as procurement. Despite the positive results of economic planning among farmers and markets, difficulties lie in data inaccessibility and low digital literacy of farmers, which hamper the adoption. Collectively, the two cases show that AI/ML technologies may help agriculture in micro and macro ways, such as enhancing sustainability of farms and economic decision-making at the national scale. Although the two operate within different scales, they provide similar values regarding profitability and environmental performance. Nevertheless, the implementation process requires overcoming barriers in infrastructures, policies, and skills to make sure that the farmers have an opportunity to make the use of such technologies to their full extent.

8. Future Directions and Policy Implications

AI and ML can be used in the agricultural sector, which presents revolutionary opportunities to enhance economic productivity and the eco-friendliness of the environment. Researchers, policymakers, and providers need to collaborate to attain beneficial changes of inclusive impacts and long-term gains. This section gives practical recommendations that can be applied by all the stakeholder groups.

8.1 Researcher Recommendations

- **Multi-crop, multi-regional, multi-crop, Create multi-crop:** Develop AI/ML models, which are generalized across crops, climatic zones, and types of disease. A large collection of datasets in different areas should be used to ensure reliability. This will prevent the high value crop bias and increase the usefulness of common crops and rain-driven agriculture.
- **Life cycle and Longitudinal Studies:** Carry out a study to investigate economic and societal impacts within many seasons or years. Add life cycle expenses (equipment, maintenance and energy), payback periods, and investments on the environment (equipment with embedded emission).
- **Apply superior AI, federated learning, and privacy approach:** Due attention should be paid to lightweight models suitable to edge devices, and such techniques as federated learning that enable models to be trained without centralized data collection and ensure privacy. These can enhance scalability and credibility.
- **Standardization, Data Quality, and Interoperability:** Establish or adopt data collection, tagging and exchange standards (e.g., the FAIR principles). Make sure that datasets are well tagged and include variations on the field (lighting conditions, mixed infections, as well as heterogeneity). This makes the models robust and reduces the overfitting to laboratory/greenhouse environments.
- **Human in the loop and user centred design:** Incorporate agents of extension, end users and farmers in the design process. Combine local knowledge, native languages, and field-tested usability. False positive and negative results as well as mistakes ought to be reported in a simple way to foster trust.
- **Combining socioeconomic and environmental indicators:** In the analysis of AI/ML methods, cover the maintenance of the biodiversity soil health, water quality, carbon footprint, and socioeconomic factors like equity, adoption barriers, and cost/benefit to the smallholder.

8.2 Policy Maker Recommendations

- **Financial models, promotions, and subsidies:** Provide grants of offers, tax exemptions or rebates on. Offers AI/ML devices (sensors, edge devices) and software-subscriptions.

- **Cooperative ownership or government-proposed AI service companies to reduce per-farmer costs are among support solutions:** Invest in the digital infrastructure. Increase internet broadband/4G/5G access in rural areas; ensure steady power supply; and provide low-cost edge computing options.
- **Scalable adoption must be based on digital infrastructure:** Training, extension, and capacity-building of farmers. Enhance extension services (Krishi or Vigyan Kendras) with an investment in programmes to enhance digital literacy, know how to recommend AI, and service gadgets.
- **Develop local dialects and formats of information that are available to farmers:** Data Governance, Privacy Regulatory Frameworks The policies regarding data ownership, data sharing, consent, anonymization.
- **Establish AI/ML tool regulatory frameworks in the agriculture sector (qualitative assurance, transparency and accountability of erroneous advice):** This is perhaps policy sandboxing where the policy can be tested with protection.
- **Connection to Insurance, Market Structures, and Supply Chains, The AI forecasting technologies are connected to crop insurance plans, commodity markets, and cooperatives:** As an example, price prediction can be used to determine insurance price, early disease condition, and prevent risk among insurers.
- **Platforms of support giving clear market data:** Evaluation: Benchmarking Build state/semi-public monitoring infrastructure to monitor the effects of AI implementation on the economy and environment (monitoring gains in benefits, reduction in emissions, ROI). Make policy improvement with this evidence.

8.3 Practitioner-Recommendations

- Business models Affordable and Scalable models Share infrastructure (subscriptions, community hubs, and cooperatives) to spread the expenses of device and maintenance. Search for cheap hardware that is low power and low communication optimized.
- User-Friendly Design Localisation Develop simple instruments/applications which may be offline or discontinuously accessible, may accept domestic languages, and adhere to validity constraints. Focus more on explanation/ visuals to get farmers to believe the results.
- Partnership and Ecosystem Building Communicate with the local level extension programs, non-governmental organizations, institutions, and the government to initiate smallholder pilots.
- Partnerships may also contribute to distribution, development of trust, and feedback loops. Transparent Feedback Mechanisms Validation Establish user feedback mechanisms to keep the model more accurate and useful. Give the consumers the freedom to report bad predictions.

- Sustainable Construction Equipment Supply Chain Constraints Assure that the hardware/sensor marketplaces are sustainable, cheap, and maintenance. Think of repairability, energy needs, serviceability. Consider environmental effects of production and waste.
- Business Models of Value Capture and Sustainability Delve into value-added services (e.g., advisory, forecasting, traceability) as an additional way of making money other than selling devices. Utilize data partnerships or platform designs enabling monetization or cost reimbursement with utmost benefit to the farmers.

9. Conclusion

The multidimensional nexus of economic development, technological innovation and environmental sustainability in the agricultural sector has been discussed in this chapter. It has identified the impact of AI and ML technologies on the economics and ecology of the contemporary agricultural sector through an integrative review of recent literature (2020–2025) and an empirical case study of the topic, as well as a proposed conceptual framework. The review showed that AI/ML applications have a considerable economic benefit in terms of enhanced productivity, lower cost of input and a higher turnover on investment (ROI). At the same time, they reduce water and GHG emissions, agrochemical use, and in this way, they assist in achieving environmental co-benefits. All these effects are directly correlated with the global sustainability objectives like SDG 2 (Zero Hunger), SDG 12 (Responsible Consumption and Production), and SDG 13 (Climate Action). Nevertheless, the review has also suggested important gaps and challenges that need to be filled to realize the complete potential of these technologies. The major problems are the absence of the long-term impact assessment, inapplicability to smallholder systems, and rain-fed systems, as well as policy and infrastructure barriers in low- and middle-income areas. Also, the issues of data governance, digital inclusion, and equal access remain, and the concerted actions of researchers, policymakers, and industry stakeholders are necessary.

In this chapter, the suggested Economic-Environmental Impact Framework offers an integrated system of connecting the data-driven technologies, farm management practices, and sustainability outcomes. It focuses on the reliance between the usage of technology and systemic enablers like market systems, regulatory policies, and knowledge investments. These two case studies on low-cost sensor implementation to avocado greenhouses and Meta transformer-based price prediction in India-show that AI implementation can create both micro-level efficiency improvements and macro-level market stabilization, which proves that digital agriculture can be profitable and sustainable.

AI solutions are collaboratively devised with agriculturalists subjected to empirical validations and supported by robust policy infrastructures. Governments and research institutions should cooperate to guarantee the interoperability of data, low costs of accessing the digital infrastructure, and capacity building by providing training locally. Startups and privately owned businesses are encouraged to specialize in scalable, low cost, and user-centered business models that balance profitability and long-term environmental and social responsibility. To recap it all, the economic-environmental influence of AI and ML in the agricultural context is not a technological discourse but a new socio economic trend. It requires inter disciplinary work, participatory research, and proactive policy development. The overall message of the call to action in the current journey towards climate resilient and digitally empowered agriculture is evident; to innovate in a responsible manner, scale in a way that is sustainable, and govern in an inclusive way so that the digital revolution in agriculture can deliver benefits not just in productivity, but also on planet health.

References

[1] Papadopoulos, G., Papantonatou, M.Z., Uyar, H., Nychasa, K., Psiroukis, V., Kasimati, A. ... and Fountas, S. (2024). Economic and environmental benefits of digital agricultural technologies in crop production: A review. *Smart Agricultural Technology*.

[2] Nayak, G.H., Alam, M.W., Naik, B.S., Varshini, B.S., Avinash, G., Kumar, R.R. ... and Singh, K. N. (2025). Meta-transformer: Leveraging metaheuristic algorithms for agricultural commodity price forecasting. *Journal of Big Data, 12*(1), 138.

[3] Miller, T., Durlik, I., Kostecka, E., Kozlovska, P. (2025). Łobodzińska , A., Sokołowska, S., and Nowy, A. (2025), (Feb.). Integrating Artificial Intelligence Agents with the Internet of Things for Enhanced Environmental Monitoring: Applications in Water Quality and Climate Data. *Electronics*.

[4] Tugrul, B., Elfatimi, E. and Eryigit, R. (2022). Convolutional neural networks in detection of plant leaf diseases: A review. *Agriculture, 12*(8), 1192.

[5] Kamilaris, A. and Prenafeta-Boldú, F.X. (2018). Deep learning in agriculture: A survey. *Computers and Electronics in Agriculture, 147*, 70–90.

[6] Liakos, K.G., Busato, P., Moshou, D., Pearson, S. and Bochtis, D. (2018). Machine learning in agriculture: A review. *Sensors, 18*(8), 2674.

[7] Wolfert, S., Ge, L., Verdouw, C. and Bogaardt, M.J. (2017). Big data in smart farming: A review. *Agricultural systems, 153*, 69–80.

[8] Quy, V.K., Hau, N.V., Anh, D.V., Quy, N.M., Ban, N.T., Lanza, S. ... and Muzirafuti, A. (2022). IoT-enabled smart agriculture: Architecture, applications, and challenges. *Applied Sciences, 12*(7), 3396.

[9] Huang, Y. and Yang, Q. (2009). Impact of precision agriculture on environment. *Transactions of the Chinese Society of Agricultural Engineering, 25*(1), 250–254.

[10] Kong, X., Liu, D., Tian, Y. and Liu, Y. (2021). Multi-objective spatial reconstruction of rural settlements considering intervillage social connections. *Journal of Rural Studies, 84*, 254–264.

[11] Chlingaryan, A., Sukkarieh, S. and Whelan, B. (2018). Machine learning approaches for crop yield prediction and nitrogen status estimation in precision agriculture: A review. *Computers and Electronics in Agriculture, 151*, 61–69.

[12] Keceli, A.S., Catal, C., Kaya, A. and Tekinerdogan, B. (2020). Development of a recurrent neural networks-based calving prediction model using activity and behavioral data. *Computers and Electronics in Agriculture, 170*, 105285.

[13] Zamora-Izquierdo, M.A., Santa, J., Martínez, J.A., Martínez, V. and Skarmeta, A.F. (2019). Smart farming IoT platform based on edge and cloud computing. *Biosystems Engineering, 177*, 4–17.

12 Emerging Trends and Innovations

Sridhar B.[1], Allirani Saminathan[2], Karthikeyan Palaniappan[3], Suresh Muthusamy[4*], Anuraj Singh[5], Chidozie Ezekwem[6] and Nebojsa Bacanin[7]

The world's food supply needs healthy crops because of the rapidly expanding human population, but plant diseases significantly reduce productivity and quality. Traditional crop disease inspection methods involve a physical inspection of crops on-farm and laboratory tests, this approach is slow, expensive and often impractical on the scale that agribusiness introduces. Advances in machine learning and deep learning recently have meant they have revolutionized plant disease diagnosis to computerized, accurate and scalable systems, of which their deep learning computer vision methods, specifically convolutional neural networks (CNNs), have created extreme pathways to image diagnosis; traditional machine learning models such as decision trees and support vector machines have also been effective since they have been able to classify from structured data efficiently. Another key development has been the emergence of Vision Transformers (ViT) utilizing the self-attention mechanism facilitating long-range relations to be learnt in plant images, additionally performing well across the complexity and depth of leaf textures and more subtle disease markers than CNNs. In general, ViT are considered to have greater potential for success in the transfer learning objectives and crop species. The ViT models achieved the highest classification success when continuously evaluated with several well-known benchmark datasets for agriculture. It is for this reason that consideration is ongoing to use the lighter ViT model for offshoot to edge devices and IoT (internet of things) monitoring systems in smart agriculture that would allow for real-time disease detection. Furthermore,

[1,2] Department of Electrical and Electronics Engineering, Sri Ramakrishna Engineering College, Coimbatore, Tamil Nadu, India.

[3,4] Department of Electrical and Electronics Engineering, Kongu Engineering College (Autonomous), Perundurai, Erode, Tamil Nadu, India.

[5] Department of Engineering Sciences, Atal Bihari Vajpayee Indian Institute of Information Technology and Management Gwalior, Gwalior, Madhya Pradesh, India.

[6] Department of Mechanical Engineering, University of Port Harcourt, Port Harcourt, Nigeria

[7] Faculty of Informatics and Computing, Singidunum University, Belgrade, Serbia.
Department of Mathematics, Saveetha School of Engineering, SIMATS, Thandalam, Chennai, Tamil Nadu, India.

* Corresponding author: infostosuresh@gmail.com

DOI: 10.1201/9781003767534-12

these sustainable practice considerations also highlight other complexities offered by the improved algorithms and modern hardware that also make the testing diagnostics more effective and accurate, improvements that not only promotes resilience of agricultural systems to climate variability and resistance to pathogen evolution. To achieve both more generalization across regions and achieve data privacy concerns, further exploration into hybrid model architectures that can maintain the benefits of both CNN and ViTs should also be considered. In addition, new paradigms of learning, such as federated learning, should be considered. Considering the above factors, we are accumulating evidence for the future, beginning to signal the compatibility of agricultural production systems with ML-based plant disease detection systems, modelling a system to support the livelihoods of farmers, aid the drive for global food security, and support the health of ecosystems. The purpose of this chapter is to provide a thorough overview of the latest methodologies, algorithms, to practical application, and challenges faced in using machine learning for plant disease detection. The chapter highlights emerging trends, including vision transformers that are poised to disrupt the field of plant disease detection and integration into precision agriculture that address pressing socioeconomic and environmental factors that drive this area of research.

1. Introduction

In the last few decades, the rapid growth of the population in the world has increased the demand for food production, putting enormous pressure on agricultural systems globally since the health of plants has a major impact on sustainable food and nutrition security; plant diseases cost billions of dollars in yield and quality [1], even with the improvements in field practices, they are a threat to food security, ecological health, and rural economies. Traditionally, plant disease diagnostics took place in fields and laboratories using manual methods, which are still of great value, unfortunately, they often have limitations. They can be labor-intensive, take time, and very often would not provide scalable or real-time solutions that delay response and increase damage and economic loss. Traditional methods are complicated further by the increasing complexity of plant pathologies associated with global trade, climate change, and evolving pathogens, creating the necessity for even faster mobility, more precise and automated disease detection systems that work seamlessly across various agronomic settings and crops. The emergence of machine learning (ML) and deep learning (DL) technologies has created novel pathways to identify diseased plants [2] [3] as ML and DL allow researchers to utilize big data, enhance computational capacities, and develop and apply algorithms toward the early detection, classification, and identification of diseased plants. This is advantageous to precision agriculture by making the deployment of resources efficient while also having lower environmental risk and procuring healthier crops.

Classical ML algorithms, such as Support Vector Machines (SVM), Decision Trees (DTs), and Random Forests (RFs), have provided baseline building blocks for pattern recognition [4] [5] based on engineered features derived from crop images and sensor data, although the applicability has been constrained by intricate feature engineering and sensitivity to variances in data and the usc of deep learning models, particularly Convolutional Neural Networks (CNNs) capable of automatic hierarchical feature extraction from raw images [6] [7] and images scanned from cameras, with proven accuracy in numerous plant disease classification tasks. Most recently Vision Transformers (ViTs) represent novel architecture that compete with CNNs using self-attention to model long-range relationships [8] [9] and dependence in images, proving to be robust in subtle symptoms present in plant species and disease types , and enhancing research on diverse agronomic practices. In the near future, we will see even more possibilities for food systems that prioritize ecological, social, and economic sustainability along with human health, eliminating the need for unsustainable products and practices leading to agrochemicals and food poisoning incidents, controlled housing environments affecting animal welfare, and conventional food systems causing social unrest. Regardless of the mode of food production, how it reaches consumers, and related practices, there is potential for a sustainable future that embraces improvements across the social, ecological, and economic spectrum. All food must be grown, transported, and consumed or disposed of in a manner that protects the environment and supports long-term quality of life. While food systems are more sustaining and more equitable in the future, we must keep in mind that technological innovation is driven by sustainability and will be shaped by producers' and consumers' willingness to adopt to improve their own sustainability.

2. Plant Diseases and Their Characteristics

Plant diseases come from a wide variety of infectious agents as well as a range of pathways to develop, modes of transmission, and impacts on host plants [10]. The following sections address in detail the understanding of plant diseases processes in some depth, which are important for diagnosing, controlling, and breeding resistant cultivars.

2.1 Types of Plant Diseases

2.1.1 Fungal Diseases

Fungi are the most common type of plant pathogen with diseases such as rusts, blights, mildews, and wilts. Spores produced by fungal pathogens can easily travel fast over significant distances and are often wind-driven, water-borne, and spread by insect vectors. Important examples of fungal diseases include wheat rust, late blight of potatoes, and powdery mildew of grapes. Fungal infections will often show up as lesions or discolorations, and structural damage to affected plants, and they can disrupt photosynthesis and the transport of materials within the plant.

2.1.2 *Bacterial Diseases*

Bacteria can infect plant tissues and produce toxins resulting in symptoms such as leaf spots, soft rots, and vascular wilts. Unlike fungi which may grow in moisture, bacterial growth requires a water film for motility. Bacteria drive many bacterial diseases in fruit and vegetable crops, and control is usually very difficult since these pathogens can survive in the soil and seeds.

2.1.3 *Viral Diseases*

Viral diseases cause systemic infection and take the form of chlorosis, mosaic, and stunting. Insects such as aphids, whiteflies, and nematodes are often the primary vectors for the transmission of these viruses. There is no chemical treatment for viral diseases. Examples of viral diseases include the tobacco mosaic virus and the tomato yellow leaf curl virus. Control of viral diseases requires integrated management of their insect vectors and developing resistant varieties.

2.1.4 *Nematode Infestations*

Many nematodes like root-knot nematodes can alter root morphology, thus reducing the plant's ability to take up water and nutrients shown in Figure 1. Other than this morphological alteration, nematodes are often cryptic and can be difficult to identify. Overall, damage from nematodes can be insidious and result in lower vigor of plants, fewer yields, and increased vulnerability to secondary infections.

The other types of pathogens, such as phytoplasmas and water molds (oomycetes) represent other agents that are causing important emerging diseases due to the rise of new forms of these pathogens in developed areas (e.g., phytoplasmas) and the appearance of new oomycete pathogens that cause disease, warranting

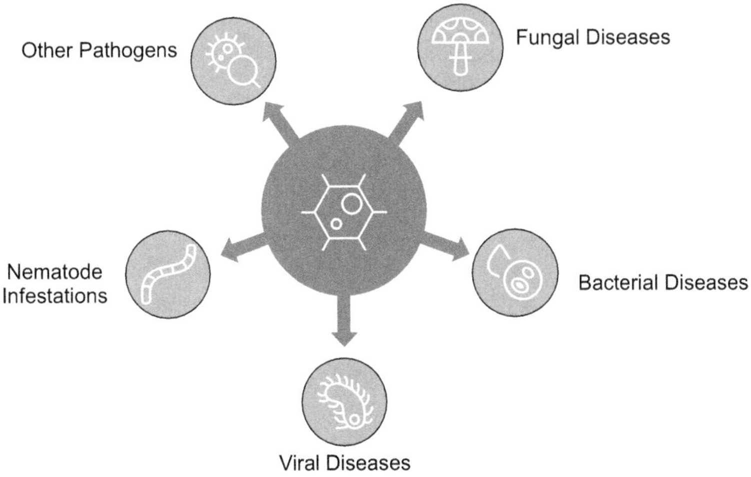

Fig. 1 Types of Plant Diseases.

specific focus. Ultimately, numerous aspects of pathogen biology will dictate detection and control strategies while showing the issues of managing disease are complicated.

2.2 Overview of Plant Diseases and Their Impact

Plants are vital to human life sustainability, and as primary producers in terrestrial ecosystems, they are critical to ecosystem stability [11] [12]. Plants play many more roles than just being food: they are producers of oxygen, sequesters of carbon, providers of habitat, and providers of raw materials for industrial uses. With the global population of humans continuing to grow and projected to be nearly 10 billion in 2050, agriculture is poised to face untested challenges to provide nutritious food to a global population at a viable economic model. Ultimately, this occurs because a healthy plant is a constant challenge to agriculture at a time when meeting food security for people and sustaining the environment and economic development of communities is dependent on healthy plants. Although modern agriculture has made extraordinary advancements, a healthy plant will always be at risk of many plant diseases caused by fungi, bacteria, viruses, and nematodes which produce risk to straw yield and crop quality leading to economically lost sales and human food shortages to vulnerable populations. Climate change adds additional layers of risk and complexity to an ever-challenging pathogen management landscape as it shifts pathogen dynamics and extends geographic ranges of pathogens. Conventional techniques for diagnosing plant diseases entail physical monitoring and lab testing which can take a long time and is very expensive, meaning they are not capable of achieving the wide, real-time monitoring necessary for modern agriculture. Therefore, there is an immediate need for fast, accurate, and scalable planning frameworks for plant disease detection that utilize emergent technologies in particular ML and DL to affect the way plant health is managed and handled, and increase the support of large scale sustainable agriculture.

2.3 Economic Impact of Plant Diseases

Plant diseases cause economic losses at both a global and regional, as well as on the farm level. The Food and Agriculture Organization estimated that biotic stress leads to post-harvest losses of around 20–40% of total crop production, equating to multi-billion-dollar losses incurred every year.

- **Yield Losses:** Infections from diseases reduce yield volume, with losses being slight to total, depending on the disease severity and type of commodity.
- **Quality Reduction:** Even moderate diseases result in the reduced quality of the crop, resulting in poor appearance, taste and shelf-life, making them less suitable for the market.
- **Control Costs:** Farmers spend more on pesticides, fungicides, labor, and irrigation because of diseases increasing production costs.

- **Trade Implications:** Disease outbreaks create market implications by initiating trade restrictions such as bans on import/export disruption commodity flows and prices.

3. Machine Learning for Detection of Plant Diseases

ML, which is a subfield of artificial intelligence (AI), gives us a strong paradigm to develop computational systems which can learn patterns of use data to make predictions without being programmed explicitly [13]. By contrast, traditional algorithms are based on fixed logical instructions, while the ML models, a subset of AI, seek to identify complex relationships and features, that is, patterns in economic data that are not visible, hidden in dense high dimensional economic data. The ability of models to learn is critical across many fields, but it is the relationship to agriculture in general, and specifically plant disease detection within agriculture, which will transform this important area. Biophysical stresses, with (the existence of) plant diseases inflicted by fungi, bacteria, viruses, and other pathogens represent a major obstacle for the agriculture industry. Diseases impact health, yields, and quality in plants, hence threatening food security with the growing global population. Effective management/mitigation of biophysical stress, relies on timely and accurate disease detections, yet the complexity of symptoms, often at large scales, counter that need. Current approaches to disease forecasting, (claimed to be good) through visual inspections, are often perceived as subjective, and recent laboratory-based disease diagnostics, while highly sensitive and specific, at times were expensive and lengthy, in terms of time or specimen submission. ML addresses these issues, through automating, rather than learning disease recognition from massive amounts of image data or environmental sensing data inherently, producing rapid, consistent, and scaled diagnostic systems. For example, ML models can learn from features extracted from plant images, or sensing data, to differentiate healthy from diseased plants classify diseases, and/ or estimate disease severity an all task that align with the precision agriculture ideals of optimizing inputs, minus environmental impact with maximum yield [14], [15]. ML is beneficial for a range of applications in agriculture, including yield prediction, pest management, soil assessment, irrigation management, and crop breeding shown in Figure 2. It also offers integration with Internet of Things (IoT)

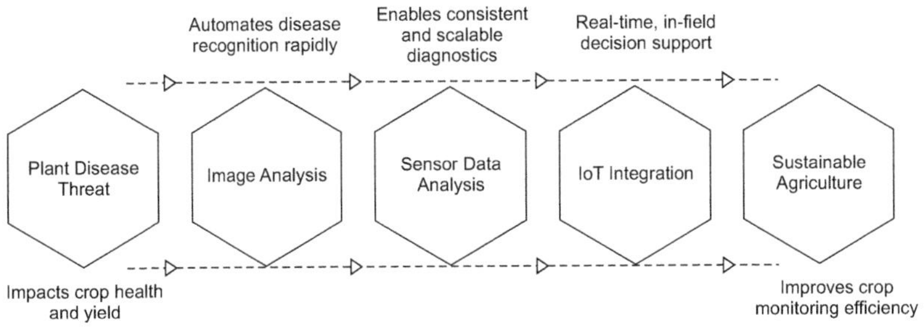

Fig. 2 ML Pattern for Plant Diseases.

systems and edge computing to facilitate real-time decision support in the fields. Ultimately, ML is facilitating modern, sustainable, and data-informed agriculture by increasing the efficiency of crops being monitored to address food security initiatives across the globe.

3.1 Overview of Commonly Used Classical Machine Learning Algorithms

Various classical ML algorithms have also been used successfully for plant disease detection and identification, particularly when coupled with hand-crafted features and low-complexity datasets shown in Figure 3. They are usually easier and less computationally intensive compared to DL methods.

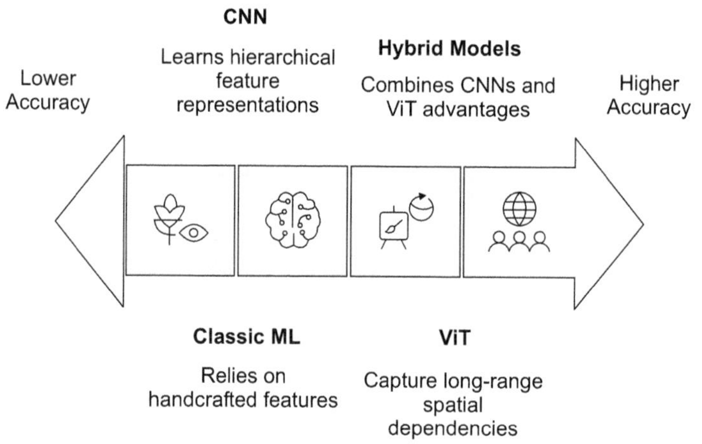

Fig. 3 Types of Classic Machine Learning.

3.1.1 Support Vector Machines (SVM)

SVMs are often referred to as a supervised learning model and are widely used for classification tasks. In an SVM, the goal is to find an optimal hyperplane in the feature space which separates classes with the maximum margin; this provides a degree of robustness to overfitting, especially when high dimensionality of the dataset is concerned. In the case of plant disease detection, the SVM is trained with features extracted from color and texture measurements from the leaves of plants being cropped.

3.1.2 Random Forest (RF)

The RF is an ensemble learning technique that constructs several decision trees during training; the structure of the RF will then use the mode class prediction of each decision tree. The advantage of RF models is that they reduce overfitting as compared to single decision trees, and they can accommodate data that is noisy and imbalanced. In plant disease classification, RFs use features such as color histograms, texture, and shape descriptors.

3.1.3 Decision Trees (DT)

DTs use a tree structure where each internal node separates the data according to the threshold of the selected feature. The DT traverses through various internal nodes to the leaf of the tree, which contains the predicted class. DTs are often intuitive and uncomplicated, which makes them effective for initial classification or to be used as a base learner in ensemble models. In the plant disease context, a DT may suffer from over-fitting but give an easy-to-understand and explain the decisions made.

3.1.4 K-Nearest Neighbors (KNN)

KNN is a non-parametric, instance-based algorithm that classifies data points based on the majority label of k-nearest neighbors of the data point—inherent in feature space. KNN models used for the purpose of plant disease detection take incoming leaf (or other plant material) images and compare them to leaves with known labels based on feature extraction. For the purpose of KNN, distance metrics can include color and texture features either separately or together, for distance calculations—Euclidean being an example. KNN does not possess a training phase, but for prediction has some of the most computationally expensive costs, namely, when KNN will calculate distance per label for each distance dimensions of input data (e.g., pixel by pixel) and/or calculate streaming incoming images to pre-classified leaf images—and that is just one image.

4. Deep Learning Models for Plant Disease Detection

DL has experienced a rapid rise to prominence over the last ten years as a disruptive technology in both AI and computational science. Unlike traditional ML which relies heavily on hand-engineered features and shallow algorithms, DL is based on artificial neural networks with multiple layers and many parameters to automatically learn representations of complex data. Specifically, DL's scalability for processing high-dimensional data and discovery of complex patterns that are nuanced revolutionized fields such as speech recognition, computer vision, language processing, and now agriculture as shown in Figure 4.

In plant pathology, DL has its most austere applications through image-based disease diagnostics. Manual inspection is subjective and suffers from variability while often requiring a great deal of labor making a classical approach inefficient in large-scale modern agriculture.

DL in general, and CNNs in particular, offer scalable, efficient, and very accurate alternatives to classical manual inspection. CNNs are specifically optimized for data whose structure is in a grid-like way, e.g., images. A traditional

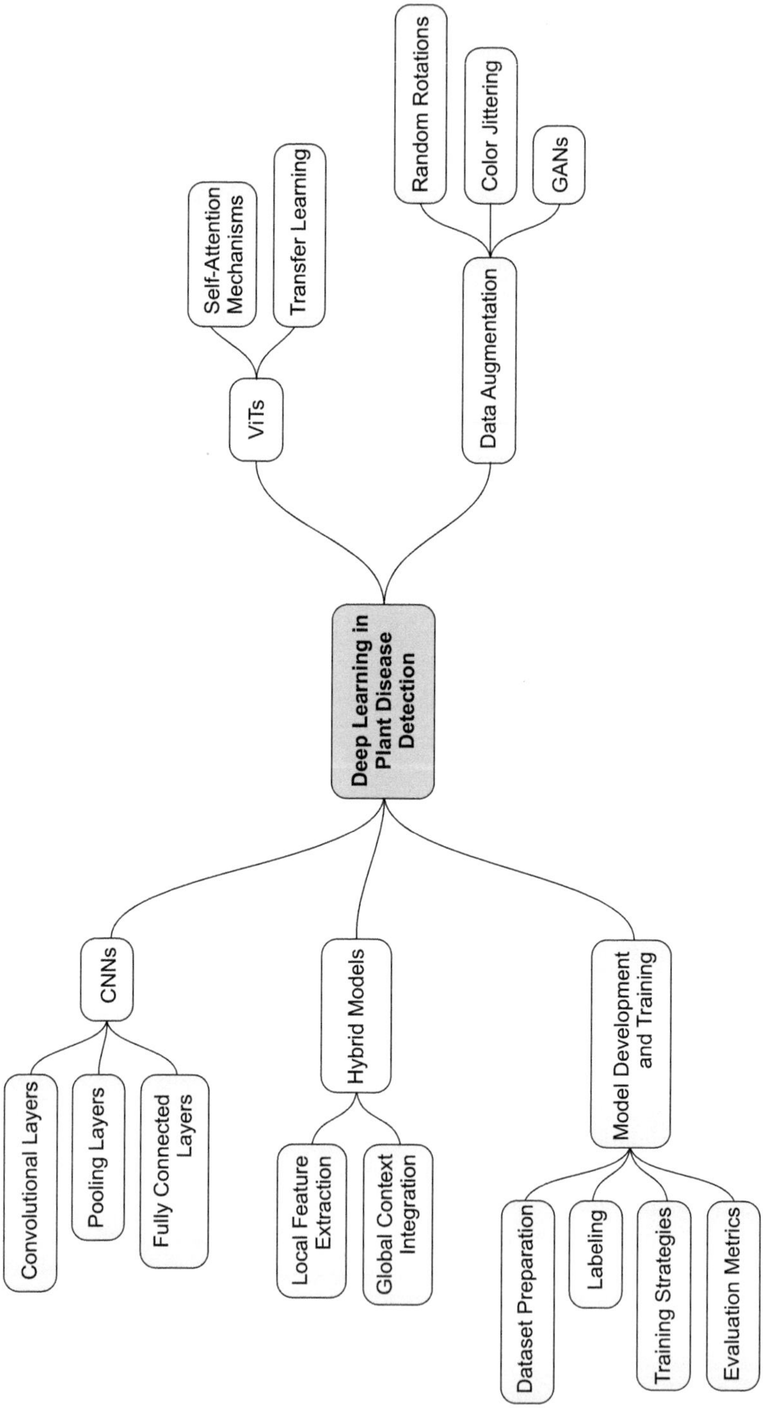

Fig. 4 Deep Learning Models For Plant Disease Detection.

CNN is composed of several layers: convolutional, pooling, and fully connected layers, where each layer extracts features at an increasingly abstract level. The early layers focus on detecting simple edges and textures while the deep layers develop representations of increasingly complex shapes and disease-specific pathological patterns. CNNs learn relevant features from the raw images automatically, without requiring any hand-engineered features. These include LeNet, AlexNet, VGG, GoogLeNet, ResNet, and EfficientNet, all of which have been empirically optimized for plant disease detection purposes [16]. Efficient Net, for example, has state-of-the-art accuracy with relatively little computational expense due to its compound scaling method. Thus, these models have been built on large labeled image datasets depicting healthy leaves and diseased leaves and subsequently tested on previously unseen samples. These models have attained close to, or above human accuracy in plant disease classification tasks, even distinguishing slightly different stages of disease progression or symptom nuances that challenge human experts.

Success with CNNs has created a widespread application in advanced phenol typing, remote sensing, and high-throughput agricultural monitoring, solidifying DL as a fundamental tenet in digital plant pathology.

4.1 Vision Transformers and a Novel Application (Vi T)

Recently, ViTs have altered the trajectory of DL applications within the field of computer vision [17]. ViTs, derived from the transformer architecture which has reshaped natural language processing, have been adapted for images specifically. The most important advancement from transformers is the self-attention mechanisms which allow these models to learn long-range dependencies and global context better than a CNN can, since CNNs are implicitly localized due to the nature of the convolutional operation. In plant disease detection applications, a ViT processes an image as a sequence of patches (e.g., 16×16 pixels), similar to how a sentence contains a sequence of words. Each patch, as projected, is considered linear in conjunction with self-attention in the transformer, where each patch attends to the relations between corresponding patches to capture complex contextual information over the image of a leaf.

This is a primary reason ViTs become state-of-the-art in developing tasks, where spatial relations and patterns of symptoms present in an image contribute to diagnosis, Multiple studies have already reported ViT model performance beyond common CNN-based drone detection applications in overall sensitivity and specificity. ViTs can accommodate complicated backgrounds, overlapping symptoms, and differences in lighting with improved generalization across crop and field settings. Models based on ViTs that have been trained derived from datasets cultivated, for example, in disease or pest or healthy vegetation, demonstrated higher classification and localization performance than trusted. CNN backbone models also provide a further explanation with attention maps.

The most striking benefit with adaptation of ViTs perhaps is their utility for transfer learning, potentially pre training on larger datasets to apply to smaller agriculture datasets. Pretrained transformer models have been pre trained on larger datasets, and specifically ImageNet, and then can be further fine-tuned on a smaller agriculture dataset.

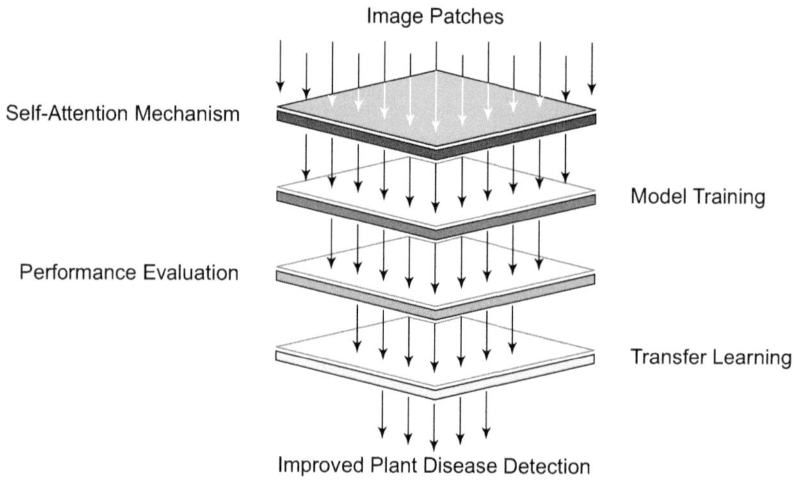

Fig. 5 Vision Transformers Model.

This workflow demonstrates promising increases in accuracy, increased convergence, and robustness of performance with few labeled samples shown in Figure 5. Also, increasingly ViTs have applications in segmentation, object detection, and multi modal tasks within precision agriculture, producing critical advancements in the field of automated plant pathology.

4.2 Hybrid Models Combining CNN and ViT

Taking into account the respective advantages of CNNs and ViTs where CNNs are computationally efficient and capable of recognizing local structures exceptionally well, and ViTs utilize global interactions, recent work has been exploring hybrid models that combine both advantages. Hybrid CNN-ViT architectures utilize CNN layers in the early feature extraction layers to encode local spatial patterns and reduce computational overhead followed by transformer blocks in subsequent layers to model long-range dependencies and incorporate global context. These architectures can be custom-made models or modules, such as a ResNet base with transformer encoders on top. Hybrid models have achieved superior performance thus far for plant disease detection, and especially on difficult datasets that are characterized by high intra-class variation or complex background.

4.3 Data Augmentation

Data augmentation refers to a set of techniques to artificially augment the diversity and size of training datasets by applying transformations to the raw images. Common augmentations include random rotations, random flips, random zoom, random color jitter, and random contrast shift.

The application of data augmentation allows models to learn to be invariant to real-world properties, like leaf angle, sunlight, and camera quality, increasing robustness and generalizability. Augmentation is an important way to combat overfitting, particularly in DL models that may memorize training samples rather than generalizing from trends. Augmentation can be simple, but many researchers are trying to use more advanced methods to even advanced augmentation that includes MixUp, CutMix, or generative adversarial networks for generating new images. Today, augmentation methods are almost always included in high performing plant disease detection pipelines.

5. Model Development and Training Methodologies

In developing and training ML and DL models for plant disease detection, consideration is given to true generalization and accuracy. The various steps in developing and training ML and DL models, such as data collection and labeling, choosing, and deploying the appropriate training method, and evaluation metrics, as well as model validation techniques such as cross-validation and hyper parameter tuning, are critical to building a model that generalizes to unseen inputs and provides diagnostic information.

5.1 Dataset Preparation and Labeling

Dataset Preparation and Labeling Data, and the training and assessing of models thereof, is the foundation of any ML model [18]. In plant disease detection, data typically consist of images of the leaves, stems, fruits, or entire plants, labeled with disease classes or health status. Preparation for a dataset involves much more than retrieval of a set of data, it also involves getting a wide range of representative data to include every environmental condition, variety, and stage of disease symptoms.

5.2 Model Development and Training Methodologies

The training of a plant disease detection model involves the process of feeding the input images and their corresponding labels into a chosen learning algorithm that updates model parameters for minimizing prediction error. Strategies vary depending on model complexity, available data, and computational resources.

Training a model conventionally includes fitting the model to feature-extracted representations from the images (Figure 6). DL models, like CNNs, on the other hand, learn both features and classifiers in a completely end-to-end fashion. However, these systems usually require large annotated datasets and substantial computational power for training.

5.3 Common Practices for Training

- **Early Stopping:** This technique is to watch performance on the validation set and stop training if the model is not improving in order to prevent over-fitting.
- **Learning Rate Scheduling:** It dynamically alters the optimization learning rate to improve convergence speed and accuracy.
- **Batch Normalization and Dropout:** Regularization methods intended to stabilize training process and minimize risk of over-fitting. The choice of evaluation metrics will depend on the type of problem and what characteristics are important to the stakeholders.

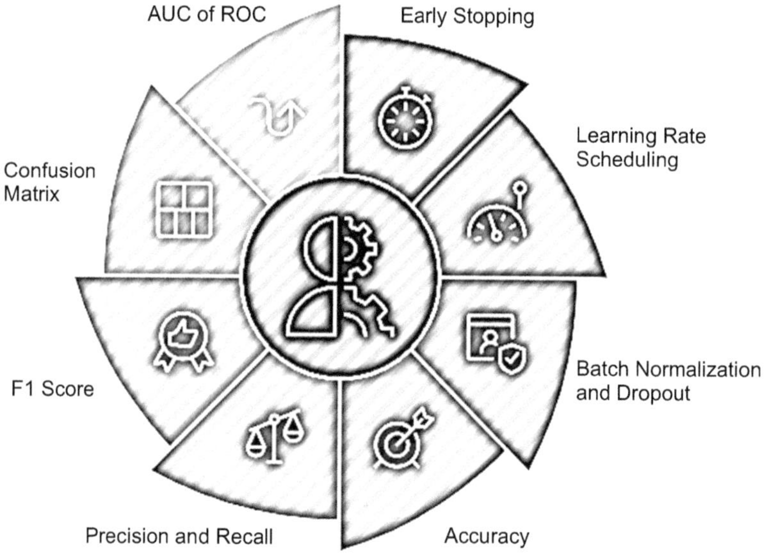

Fig. 6 Machine Learning Training Steps.

The common metrics utilized for plant disease classification implementations are:

- **Accuracy:** The proportion of samples classified correctly.
- **Precision and Recall:** It provides a rate of false positives and false negatives which are very important in a disease where an infection is missed or misclassified.
- **F1 Score:** The harmonic mean of Precision and Recall so it is a balanced measure.
- **Confusion Matrix:** The quantitative evaluation of error types across classes.

5.4 *Cross-Validation and Hyper Parameter Tuning*

Cross-validation is important for making sure that the models for plant disease detection are able to generalize to unseen data. Approaches such as k-fold cross-validation split the data into k subsets of approximately equal size; the model is trained k times with each training completion incorporating different folds as the testing portion shown in Figure 7. This reduces bias in predicting performance and allows for maximal use of all training data.

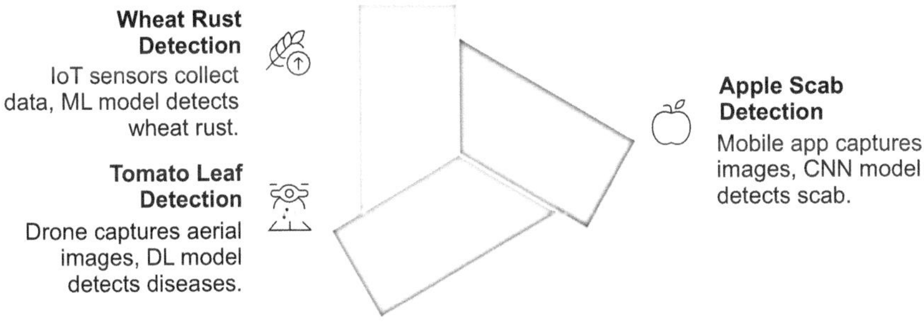

Fig. 7 Real-Time Plant Disease Identification.

Hyper parameter tuning refers to carefully changing model parameters that cannot be learned through training that control the learning dynamic or architecture. Examples of hyper parameters that can be tuned include the number of layers in a CNN, kernel sizes, learning rates, strengths of regularization, batch size, etc. The optimization of hyper parameters can lead to much higher accuracy and convergence of models.

5.5 *Examples of Real-Time Plant Disease Identification*

- **Apple Scab Detection:** A mobile application can be used to capture images of apple leaves and use a CNN model to detect apple scab in real-time.
- **Tomato Leaf Disease Detection:** A drone-based system can be used to capture aerial images of tomato fields and use a DL model to detect various tomato leaf diseases, such as early blight and late blight.
- **Wheat Rust Detection:** IoT sensors can be deployed in wheat fields to collect data on temperature, humidity, and leaf wetness, which can be used to train an ML model to detect wheat rust in real-time.

6. Comparative Analysis of ML and DL Models in the Detection of Plant Diseases

6.1 *Accuracy*

Accuracy is important in plant disease detection, and accurate classification represents the proportion of true positives versus total positives for the model.

Class ML models like SVM, RF, DT, and KNN are a group of models that typically gives accuracy from 75% to 90%. The model performance is related to known attributes of samples in the dataset for feature quality; the models depend on handcrafted features inability to adapt to complex variation in the expression of plant disease symptoms.

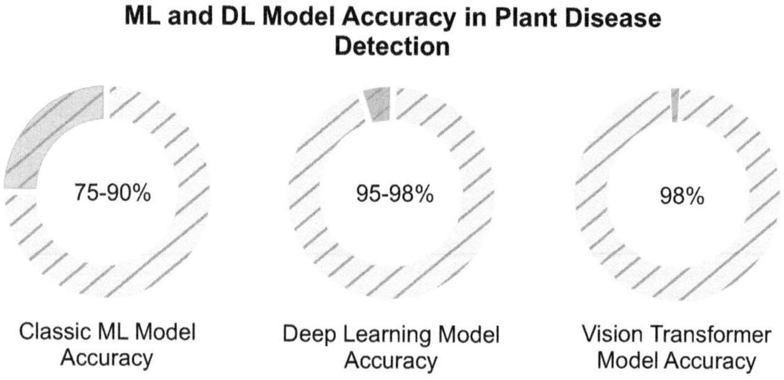

Fig. 8 Comparative Analysis for ML and DL.

DL models, particularly CNN models, outperform the classic techniques [19] [20], usually with accuracy greater than 95% and some reported accuracy over 98% with highly curated and cleaned datasets shown in Figure 8. This is attributed to CNN's ability to learn hierarchical feature representations from the raw images.

Recently created models of DL, ViTs have been competitive, in some cases even better than CNN models' performance in more complex disease detection tasks reporting higher accuracy. A ViT is highly effective at capturing long-range spatial dependencies in images, allowing for subtle disease patterns to be detected. Hybrid models that leverage CNNs and ViTs extract the advantages from either architecture, increasing accuracy and sometimes approaching perfect performance on benchmark datasets although increases in accuracy is dependent on dataset complexity and training conditions.

6.2 Speed

Generally, classical ML algorithms train and inference faster than DL models, even in situations where resources are constrained. Yet, classical ML methods might have limited speed in large or complex image data due to the feature extraction stages. In particular, one-stage detectors and slimmed models of CNNs will have faster inference among DL methods compared to ViTs as the latter have higher computational complexity due to their attention-having complexity. Hybrid models create a balance on this with their own trade on, speed, and accuracy as shown in Table 1.

Robustness: Robustness means remaining stable in performance against changes in environment, image quality, and unseen conditions. DL models, in particular those designed through data augmentation and transfer learning, perform well and are more robust in generalizing across fields and variability as shown in Tables 2 and 3. The classical ML paradigm is more subject to noise and domain shifts due to the reliance on engineered features to define the characteristics and variability.

6.3 *Strengths and Weaknesses*

Table 1 ML and DL Model Comparison.

Model Type	*Strengths*	*Weaknesses*
Classical ML (SVM, RF, DT, KNN)	Simplicity, faster training and inference, interpretable decisions, low computational resources	Depend on manual feature engineering, limited scalability, moderate accuracy, lower robustness
CNN	High accuracy and robustness, automatic feature extraction, well-established architectures	High computational requirements, needs large labeled datasets, training complexity
Vision Transformer (ViT)	Captures global dependencies, excels in complex and overlapping symptoms, transfer learning compatibility	Computationally intensive, requires large data, longer training times
Hybrid CNN-ViT	Combines local feature learning with global context, balanced accuracy and speed benefits	Increased model complexity, challenging optimization, higher resource demands

6.4 *Performance Benchmark*

Table 2 ML and DL Model Comparison Based on Performance.

Model Type	*Typical Accuracy (%)*	*Training Speed*	*Inference Speed*	*Robustness to Variation*	*Resource Requirements*
SVM	75–90	Fast	Fast	Moderate	Low
Random Forest	80–90	Moderate	Moderate	Moderate	Low–Moderate
CNN	90–98	Slow	Moderate-Fast	High	High
Vision Transformer	95–99	Slowest	Moderate	Very High	Very High
Hybrid CNN-ViT	97–99+	Slow	Moderate	Very High	Very High

6.5 Evaluation Metrics

Table 3 ML and DL Model Comparison Based on Evaluation Metrics.

Evaluation Metric	*Classical ML*	*CNN*	*ViT*	*Hybrid CNN-ViT*
Accuracy	Medium	High	Very High	Highest
Precision	Moderate	High	Very High	Highest
Recall	Moderate	High	Very High	Highest
F1 Score	Moderate	High	Very High	Highest
Training Time	Low	High	Very High	Very High
Interpretability	High	Low	Low	Low

6.6 Charts

Traditional ML is faster to train but it has limiters on accuracy. On the other side, CNNs fit in the middle, requiring longer to train, benefiting from those increased training times, thus exhibiting more accuracy (Figure 9. ViTs take the longest to train, but they will yield significantly higher accuracy in the results. Hybrid models are about achieving the best accuracy, all while having loved models that are appropriate in terms of the training that was employed.

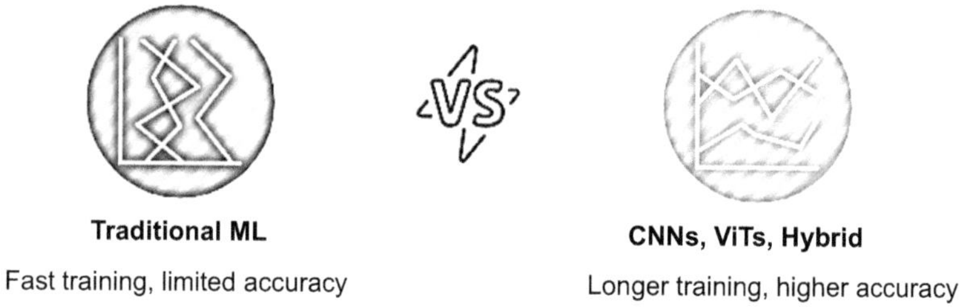

Fig. 9 Comparison of ML and DL with accuracy and speed.

7. Future Trends and Research Directions

Future trends and research directions in plant disease identification will leverage emerging technology and methods to improve accuracy, privacy, scalability, and data integration to address food security challenges, where population grows and environmental change heighten. Topics becoming more popular are explainable AI (XAI), federated learning for privacy, fusion of multimodal data, and any improvements to scale to diverse crops and geographies.

7.1 Robustness in Deploying to The Real-World

The one main drawback for traditional ML is that it will perform poorly to noisy and/or environmental variability. CNNs are more robust to some extent, which they surmount with augmentation. Any ViT model, by definition, has unlimited global attention and shows good robustness for more intricate scenarios.

7.2 *Explainable AI for Plant Disease Detection*

As AI-based models become ingrained in agricultural endeavors, the necessity for transparency and interpretability becomes increasingly important. XAI approaches are intended to enable a human-comprehensible justification for model predictions [21] which is essential for trust, regulation compliance, and potential acceptance by farmers and agronomist specialists.

Common agricultural models, for example, especially the rapidly evolving DL architectures such as CNNs and ViTs, are generally considered more opaque than, and along with most AI methodologies, "black-boxes". However, XAI approaches such as saliency maps, attention visualization, and layer-wise relevance propagation visualize the image regions that impact disease classification, allowing a crop production domain expert to confirm diagnosis and even debug and/or refine the model. XAI reduces skepticism, provides trust in models, furnishes more transparent exploration and actionable insights, and will show agricultural domain experts the reasoning along with the relevant regulations for any/all decisions, and the XAI will still make positive recommendations based on agronomic knowledge. Further, explainability can address biases in models and potential failure modes across different environmental scenarios to enhance fairness or robustness in AI-enabled decisions.

7.3 *Federated Learning and Privacy-Preserving Techniques*

Furthermore, a growing global stream of agriculture imagery and now sensor data has directed us toward conversing about data privacy and sovereignty. Federated learning is a decentralized methodology in which a group of edge devices or local servers collaboratively learn a model fully locally [22] [23] on their respective devices or servers without sharing any raw data. Each device or server sends an update to a central server which aggregates the updates generated from learning and improves a global model while preserving data locality and reducing associated risks of data coding of private or sensitive information.

It involves developing the disease detection models in collaboration, using different datasets from farming systems distributed across various geographies while being conscious of private intellectual property and regulatory issues. This employs standard privacy-preserving mechanisms, such as differential privacy and secure multi-party computation, which could enable federated approaches to cross-organization collaboration in sensitive farm data protection. This does not only scale AI models at the pace of adoption for farmers but humor themselves towards farmer rights on data and evolving privacy regulations.

7.4 Multimodal and Multimodal Data Integration

The intricacies of the dynamic processes of plant disease cannot be realistically captured using a single-modality image. The multimodal types of data combine the final imaging (RGB; multispectral and hyperspectral remote sensing) images together with other imaging modalities, for instance, thermal imaging, and environmental parameters [24] (temperature, humidity, soil moisture, etc.); genomic data provides the integrated breadth of knowledge required to witness the emergence and trajectories of plant disease. The merging of multimodal data can provide more copious types of information, for instance, through spectral indices identifying physiological stress before symptoms become visible, identification of water stress through thermal imagers, and identification of resistant cultivars through genomic data. ML and DL models that are able to analyze heterogeneous modalities together will also be integrated to improve the accuracy of early detection and the estimation o f the severity of disease. Further, the multimodal integration would advance the development of models that are more generalizable and robust to environmental variability and imaging artifacts.

Affordable sensor technologies, and the maturation of IoT ecosystems will also support the development of multimodal, data-driven, disease monitoring platforms as a standard Precision Crop Management tool and move towards a more data-centric approach for in-field monitoring of plant health.

7.5 Scalability across Crops and Geography

In order for AI to be widely adopted to detect plant disease, models need to be developed at scale to address various crops, pathogen strains, and geographic regions featuring different climatic and agricultural practices. This entails large, heterogenous, and accurately labeled datasets that demonstrate representative phenotypic variation and diversity of symptoms collected from various regions of the globe. Research will still be pursued through transfer learning, domain adaptation, and few-shot learning approaches that leverage pre trained models with minimal retraining in the new crop or region. Artificial data generation and active learning can also help alleviate limited data in cases of rare disease. Federated learning frameworks can also enable collaborative and generalized model development across institutional and regional settings.

8. Conclusion

Hybrid model architectures that combine a ViT and CNNs represent a new benchmark in plant disease diagnostics by harnessing the advantageous local spatial feature extraction of CNNs and global attention of ViTs. This integration enhances the performance of diagnostic systems in terms of accuracy, robustness/stability, and interpretability [25]. The integration of these two architectures leverages the disadvantages that are potentially experienced while using either architecture alone. CNNs can effectively and efficiently extract sharp, localized image features that are particularly important for detecting leaf grown texture and spatial patterns constitutive of lesions that are indicative of plant disease classifications. However, CNNs can struggle with segmentation with long-range, contextual relationships that arise in complex plant images. In contrast, ViT architectures are able to effectively and efficiently model the long-range spatial relationships across an entire image through self-attention, which in turn is also beneficial for even diffuse physiological symptoms that could appear in multiple sectors or quadrants of an image. However, the model might lack representational ability on a fine spatial scale, similar to that of CNN architectures, particularly in smaller datasets. The complementary capabilities of both classes of models incorporated together ensure that the hybrid ViT-based CNN frames hierarchical local and global representations, yielding high performance at crop classification across crop types and environmental conditions. In fact, hybrid models with both architectures have shown to be very accurate, often exceeding the 95% level on benchmark datasets such as PlantVillage and Embrapa, with the same level of computational efficiency desirable for IoT-based real-time applications. Additionally, the application of XAI methods, such as Grad-CAM and Local Interpretable Model-agnostic Explanations (LIME), will support model transparency features by physically illustrating the regions of interest utilized to classify diseases. The inclusion of model explainability will help foster both the user's trust in these models as well as have the actors inform decisions in agriculture. Moreover, effective transfer learning with modest-sized, domain-specific datasets has been accomplished through the models and will allow simple application to any number of crop diseases and endeavors. Plus, their small architectures and quick inference speeds, means that they can be easily embedded in edge devices, drones, smartphone applications, etc., effectively delivering state-of-the-art detection capabilities to the agricultural frontline. To conclude, ViT-enabled CNN architectures are forefronts in plant disease diagnosis and will definitely offer a transformative instrument that will promote precision, scale, and interpretability; establishing plant disease diagnosis with the potential to reduce crop loss and enhance sustainable agricultural capacity and global food security through efficient and timely disease management.

References

[1] Shrivastava, S. and R. Kumar (2023). Plant disease detection using machine learning techniques: A review. *IEEE Access, 11*, 34567–34582.

[2] Liu, J. and H. Wang. (2021). Deep learning for plant disease detection using convolutional neural networks. In: *Advances in Machine Learning for Agriculture*. Springer.

[3] Jung, M. (2023). Construction of deep learning-based disease detection models for bell pepper, potato, and tomato plants. *Nature Communications, 14*, 1223.

[4] Khan, S.H. (2022). Vision transformer based plant disease detection and classification approach. *Frontiers in Plant Science, 13*, 899999.

[5] Mehta, S. and M. Rastegari. (2021). MobileViT: Light-weight, general-purpose vision transformer. In: *Proc. IEEE/CVF Conference on Computer Vision and Pattern Recognition* (pp. 9120–9130).

[6] Abdu, A.M. (2023). Vision transformer meets convolutional neural network for plant disease detection. *Journal of Agricultural Informatics, 14* (2), 134–150.
[7] Ali, A.H. (2024). An ensemble of deep learning architectures for accurate plant disease classification. *Computers and Electronics in Agriculture, 210*, 107984.
[8] Shivaprasad, K. (2023). "Deep learning-based plant leaf disease detection: A comparative study of CNN models. *IEEE Transactions on Neural Networks and Learning Systems, 34* (1), 123–134.
[9] Yu, F. (2023). Lightweight vision transformer for real-time detection of plant diseases. *IEEE Access, 11*, 45678–45688.
[10] De Silva, N. and J. Brown. (2023). "A hybrid model of vision transformer and GoogLeNet for multispectral plant disease detection. *SPIE Journal of Applied Remote Sensing, 17* (1), 015002.
[11] Li, T. (2022). Integration of multimodal data for plant disease detection using deep learning. *IEEE Journal of Selected Topics in Applied Earth Observations and Remote Sensing, 15*, 7124–7136.
[12] Khan, A.T. (2023). Plant disease detection model for edge computing devices. *Frontiers in Plant Science, 14*, 1158933.
[13] Al Mamun, A. (2024). Plant disease detection using self-supervised learning. *IEEE Access, 12*, 123456–123468.
[14] Xu, M. (2024). Plant disease recognition datasets in the age of deep learning. *Frontiers in Plant Science, 15*, 1234567.
[15] Omaye, J.D. (2024). Cross-comparative review of machine learning for plant disease detection. *Computers and Electronics in Agriculture, 197*, 106928.
[16] Jadon, R.S. (2023). Plant disease detection using machine learning models. *SSRN Electronic Journal.*
[17] Ashurov, A.Y. (2025). Enhancing plant disease detection through deep learning. *Computers and Electronics in Agriculture, 209*, 107745.
[18] Shoaib, M. (2023). An advanced deep learning models-based plant disease detection: A review of recent research. *Frontiers in Plant Science, 14*, 1158933.
[19] Li, L. (2021). Plant disease detection and classification by deep learning. *IEEE Transactions on Neural Networks and Learning Systems, 32* (4), 1234–1245.
[20] Ramesh, S. (2018). Plant disease detection using machine learning. *IEEE Access, 6*, 13750–13759.
[21] Ali, A.H. (2024). An ensemble of deep learning architectures for accurate plant disease classification. *Computers and Electronics in Agriculture, 210*, 107984.
[22] Singh, V. and A.K. Misra (2017). Detection of plant leaf diseases using image processing and machine learning. *Procedia Computer Science, 57*, 110–116.
[23] Khan, S. (2021). Plant disease detection using deep learning approaches: A review. *Computers and Electronics in Agriculture, 187*, 106329.
[24] Savary, S. (2019). The global burden of plant diseases on food security. *Food Security, 11* (3), 501–511.
[25] Karthik, R (2020). Deep belief networks for plant leaf disease classification. *International Journal of Computer Vision and Image Processing, 10* (3), 45–59.

13 Roadmap for the Future Research

Vasanth Poomalai[1*], Kannappan Manikandan[2] and Ande Disha[3]

The chapter of the book presents a path forward in the development of future research in the sphere of detecting plant diseases using the power of machine learning (ML), deep learning (DL), and metaheuristics. Effective and precise early identification of plant diseases is essential to sustainable agriculture and world food security and tackles drawbacks of manual and time-intensive and prone-to-error inspection systems of the past. The present situation indicates great improvement, and DL models, and especially Convolutional Neural Networks (CNNs) are very precise at recognizing and classifying plant diseases and their illnesses. Nevertheless, there are still some difficulties, such as the lack of data, the necessity to have massive volumes of labeled training data, the problem with the quality of the images and the possibility to distinguish between normal and diseased plants. Future research directions are based on the development of more generalizable and robust models that can operate on a variety of plant species and diseases instead of being species or type-specific. Research opportunities are in the area of transfer learning and ensemble, robustness, and accuracy, avoiding overfitting and data augmentation, as well as diversifying datasets and avoiding the use of large quantities of labeled data. Applying AI-IoT technologies in combination with the creation of public data access models are the key to the future advancement of epidemiology of plant diseases. A study should also be done on the implementation of lightweight DL models in real-time detection and fine-tuning the current methods such as YOLO and Faster R-CNN to improve accuracy and speed of the models. Furthermore, object segmentation networks-based detection models will be constructed to better detect maladies and infestations in plants. Other applications of metaheuristics may include optimization of features and models, and more accurate and faster disease diagnosis. Cracking these directions will be a long way to go in enhancing the development of automated plant disease detection to bring its part to the improvement of agricultural products and food security.

[1,2] Ph.D. Scholar, Department of Agronomy, Faculty of Agriculture, Annamalai University, Chidambaram, Tamil Nadu, India-608 002.
[3] Ph.D. Scholar, Department of Plant and Soil Science, University of Kentucky, Lexington, KY 40506, United States.
ORCID: https://orcid.org/0009-0005-7628-1819[1]; https://orcid.org/0009-0008-2158-8444[2]; https://orcid.org/0009-0004-1099-4229
Email: kannappanvikram29@gmail.com[2]; disha.ande@gmail.com
* Corresponding author: vasanth30p@gmail.com

DOI: 10.1201/9781003767534-13

1. Introduction

There is an ever-growing pressure on the agricultural sector in the world to improve productivity, minimize crop losses and guarantee food security to the swelling population. Plant diseases have taken a center stage among the important problems that threaten sustainable crop production resulting in reduced yields and terrible economic losses globally. The diagnosis of a disease at its early and accurate stages is of great importance in ensuring effective management practices in order to avoid enormous infestation of the disease [1]. The most common methods used in the identification of plant diseases are traditional methods that rely mainly on visual inspection and expert assessment which are time-consuming, laborious, and subject to human error [2]. The last few years have witnessed a revolution when it comes to the manner in which plant health is observed and disease detected through the integration of Artificial Intelligence (AI), Machine Learning (ML) and Deep Learning (DL) methods. These new sophisticated computational techniques have demonstrated tremendous potential in enhancing the accuracy of the diagnostic results, real-time monitoring, and low reliance on human expertise [3].

The advent of smart agriculture that is powered by AI has supported the creation of automated image recognition tools, drones, and Internet of Things (IoT)-powered equipment that can alert of disease outbreaks in time. Although impressive advancement has taken place, there are still a number of challenges, such as scarcity of data, generalization of the models across species, and the necessity of lightweight frameworks that can be deployed in the field [4]. Figure 1 visually summarizes the full detection pipeline the paper is about to explain. The chapter is an expansive roadmap of the future research trends within the sphere of AI-assisted plant diseases detection, with the particular emphasis on the future developments in the sphere of ML, DL, and metaheuristics to tailor the diagnostic process both in terms of accuracy and efficiency.

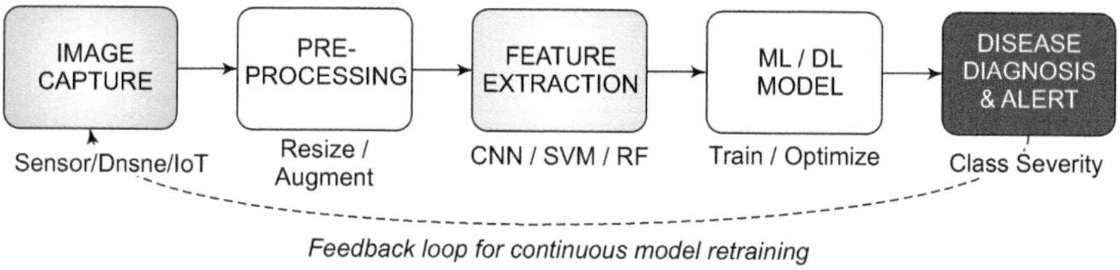

Fig. 1 An AI-based plant disease detection system workflow. End-to-end workflow of an AI-based plant disease detection system–image acquisition using IoT sensors or drones, preprocessing, and feature extraction, ML/DL model inference and diagnostic output, and a continuous feedback loop retraining the model.

1.1 Importance of Early Plant Disease Detection

Early disease detection is a way of preventing plant diseases to ensure that crop production and food security is maintained. After the early detection with the help of the diseases in the early stages of their formation, corrections may be performed with the help of chemical, biological, and even agronomic measures prior to the further formation of infections [5] [6] [7].

Early diagnosis reduces losses of crops, reduces the dependence on pesticides, and enhances sustainability of the environment by limiting the unnecessarily agrochemical applications [8].

Also, proper early identification of the disease makes it easier to predict and conduct epidemiological research, which can assist in the comprehension of spatial and temporal patterns of the disease. Combining precision farming devices with AI-based detection systems will also enable farmers and other people in the agricultural industry to make decisions in time, as they can monitor diseases in real-time in the field. Subsequently, early detection not only helps to protect the yield potential, but also contributes to boosting the resilience of agricultural systems against biotic stressors [9].

1.2 Limitations of Conventional Diagnostic Methods

Conventional methods of diagnosing plant diseases are mostly based on visual diagnosis, microscopy, and pathogen identification through laboratory techniques. The methods are accurate in a controlled environment, but they cannot be used in large-scale monitoring and at the field level [10]. Visual examination can also tend to be subjective and depends on the experience of the observer and the fact that symptoms differ across cultivars or habitats [11].

Moreover, biochemical or molecular confirmation, such as polymerase chain reaction (PCR) or enzyme-linked immunosorbent assay (ELISA), also needs specialized equipment, trained individuals, and a significant amount of time [12]. These limitations are obstacles to the fast control of the disease, especially in resource-restricted areas. Laboratory reagents are very expensive, and delays to analysis also limit the implementation of preventive strategies in time. Thus, automated, scalable, and cost-effective diagnostic systems capable of providing real-time and accurate results in the field conditions are urgently needed.

1.3 Role of Artificial Intelligence in Modern Agriculture

AI has become a revolution in contemporary agriculture, where there is the possibility of coming up with intelligent systems that can replicate the perceptual abilities of humans to recognize diseases and make decisions [13]. The application of ML algorithms in the classification task, which is based on the plant leaf texture, colour, and shape, has been popular using the Support Vector Machines (SVM), Random Forest (RF), and the K-Nearest Neighbours (KNN) [14]. Architectures like DL and especially the Convolutional Neural Network (CNN) have since performed better than the traditional models of ML because they automatically extract high-level spatial details on the raw images, thus eliminating the manual pre processing component [15].

Also, hybrid solutions, such as AI/Remote sensing, drones/IoT sensors allow monitoring crop health on an ongoing basis. Such systems are capable of handling multispectral and hyperspectral information to identify changes in physiology which cannot be observed by the naked eye [16]. The use of AI in precision agriculture has thereby transformed the concept of disease diagnostics with AI being quicker, objective, and more robust than the traditional approach.

1.4 Scope and Objectives

The goal of this chapter is to offer an organized roadmap on the further development of plant disease detection in the context of the combination of ML, DL, and metaheuristic optimization methods. The objectives include:

- Checking the existing situation in the field of AI applications in diagnosing plant diseases.
- Determining the key issues like a lack of data, computer resources, and model generalization.
- Discovering alternatives such as transfer learning, ensemble learning, and data augmentation to increase model robustness.
- The possibility to combine AI with IoT and edge computing in real-time agricultural monitoring should be noted.
- Speaking about the use of metaheuristics in search of the optimal model parameters and enhancement of classification accuracy.
- Deliberating on the perspectives of future research on explainable, trustworthy. and sustainable AI systems in digital agriculture.

By responding to the following goals, the chapter adds to the overall vision of creating smart, scalable and environmental-friendly solutions to the agricultural systems of the world [17].

2. Recent State of Current AI-Based Plants Disease Detection

The rapid introduction of the use of AI technologies, i.e., the ML and DL algorithms, has occurred in the field of plant disease detection. The methods have facilitated the accuracy of the diagnosis of plants much because they can give a quick and more objective assessment of the health of the plants. However, despite these tremendous innovations, mass implementation is yet to be achieved due to a myriad of deficiencies including poor datasets, huge computing performances, and difficulties in the field deployment [18].

2.1 Machine Learning Applications Introduction.

ML is the technique used to classify plant diseases using automated methods. The algorithms that have been widely used are SVMs, Decision Trees (DTs), RFs, and KNN algorithms to diagnose a disease based on pattern recognition of the leaf texture, shape, and color [19]. The following features extraction methods include Gray-Level Co-occurrence Matrix (GLCM), Histogram of Oriented Gradients (HOG), and Principal Component Analysis (PCA) that help in improving the accuracy of the classification attainment by reducing the dimensionality and calculation expenditures [20]. The feature extraction equation involving PCA may be as follows:

$$Z = XW \quad ...(1)$$

Z is the reduced feature vector, X is the original dataset, and W is a matrix of eigen vectors of the covariance of X.

ML models have demonstrated consistent performance in controlled settings but in far more field conditions, where they vary in lighting, background, and disease severity, they tend to fail to perform accurately with generalization [21].

2.2 Deep Learning Models in Disease Classification

DL has also changed the hierarchical learning of features as a method of image-based plant pathology. CNNs are able to automatically identify the hierarchies of spatial features making them classify strongly without preprocessing [22]. These types of architectures have been shown to be competitive in the classification of plant diseases using image data set as PlantVillage: AlexNet, VGGNet, ResNet, and DenseNet.

These models can classify a wide variety of diseases in various crops. An example is that CNNs trained on augmented images data had over 99% accuracy detection of tomato leaf diseases [23]. Deep networks are computationally and memory-intensive, and thus could not be possible in resource-constrained regions.

2.3 Comparative Analysis of ML and DL Performance

Although ML algorithms are lightweight and interpretable, they need a lot of manual feature engineering. DL models, in their turn, are more capable of learning features and dealing with more complex patterns; however, large datasets and longer training times are required [24]. Empirical observations suggest that, on the whole, DL is more accurate in classification than ML but uses less energy. Table 1 summarizes a comparative analysis of common ML and DL methods used in plant disease detection.

Table 1 Comparison of Machine Learning and Deep Learning Techniques.

Algorithm	*Type*	*Feature Extraction*	*Accuracy (%)*	*Computation Demand*
SVM	ML	Manual	92	Low
Random Forest	ML	Manual	89	Low
CNN	DL	Automatic	98	High
ResNet	DL	Automatic	99	Very High

Source: Compiled from comparative results reported and supported by studies [20–23].

2.4 The Image Processing and Spectral Analysis Are to be Combined

In addition to imaging that is visible, there are multispectral and hyperspectral imaging technologies that offer spectral signatures of plant tissues to improve the accuracy of detecting diseases. Spectral analysis has the ability to detect the very subtle changes in biochemistry like deterioration of chlorophyll or the pigment alteration preceding the onset of any observable symptoms [25]. Combined with ML and DL, these methods make it possible to predict diseases at an early stage, which is useful in precision agriculture.

2.5 Limitations and Challenges of Current Research

Despite of significant progress, there are notable limitations:

- Relying on high-quality, labeled image data sets.
- Low model portability to new environmental conditions.
- The large power and computational needs to train big DL models.
- Poor access to open access standardized data.

The solution of these gaps becomes the basis of further research in AI-based plant pathology.

3. Problematic Issues and Gaps in Research

AI has been used in the detection of plant diseases, although it has come to be in various forms, there are notable barriers that it cannot be applied on a large scale. Such problems are connected with the data unavailability, generalization of the model, price of computation, and non-standardized datasets and protocols [26]. These research gaps are essential to ensure that AI-driven systems will be operated in practical agricultural conditions, especially in the developing regions with limited resources.

3.1 Data Paucity and Imbalanced Data

One of the most prominent issues of AI-based disease detection is the absence of extensive, high-diversity and richly annotated datasets. A large number of such models are trained with publicly accessible repositories like PlantVillage, which are mainly images taken in the laboratory under controlled lights and backgrounds [27]. These datasets cannot consider the real-field environments, such as changes in the illumination, occlusion, leaf positions, and numerous overlapping diseases [28].

Also, there is the issue of class imbalance, some of the diseases are overrepresented: the ones that are less common, but still have a significant economic impact are not overproportionate. This bias causes unfairness in model prediction and inaccurate generalization [29]. The future studies need to be oriented at creation of extensive field-derived data on various stages of crop, soil status and severity of diseases. GANs can be used to augment data and create synthetic datasets that address small and unbalanced datasets [30].

3.2 Image Data, Quality, and Standardization

The quality of images is an important factor in the precision of AI models. The differences between camera resolution, sensor sensitivity, and environmental lighting impact the input data uniformity [31]. Cross-dataset evaluation is also not easy when non-standardized imaging protocols are used. In the example, spectral and RGB (red, green, blue) imaging have been commonly carried out with varying camera settings, which makes cross-study model transferability issues problematic [32].

The standardization processes should be undertaken to establish protocols of capture, labeling, and recording of metadata of images. Other scholars like [33] propose worldwide coordinated dataset repositories following the FAIR (Findable, Accessible, Interoperable, Reusable) concept which allows data sharing and benchmarking.

3.3 Generalization of the Model to Different Crop Type and Environment

The majority of AI models are specific to crops and cannot be applied to other species of plants or environmental conditions. A CNN trained on tomato leaves might not work when transferred to maize or rice because of the variation in the morphology of the leaves and expression of the disease symptoms [34]. Environmental variability that influences this issue includes intensity of lighting, soil type, and humidity, which impacts the color and texture of the leaves.

In recent research, it has been proposed that transfer learning and domain adaptation are capable of dramatically improving model generalization. Transfer learning allows larger and more accurate pre trained models to be trained again on smaller, crop-specific data [35]. Meta-learning and few-shot learning strategies should also be investigated in the research to allow models to generalize unseen crops or diseases with little data.

3.4 Real-Time and Computational Cost Constraints

ResNet or DenseNet Deep neural networks are resource-demanding and require strong GPUs (graphics processing units) and big memory. This renders real-time applications hard to use in field conditions, particularly in low-resource settings [36]. Mobile and edge devices can also not have the computing abilities to work with big models in real-time.

Some of the solutions here involve the creation of lightweight architectures like MobileNetV3, ShuffleNet, and EfficientNet-lite which trade-off both accuracy and speed [37]. Potential optimization methods that can minimize the size and computational demands of models include quantization, pruning, and knowledge distillation, and do not require performance compromise [38].

The reason is that it does not have open access datasets and benchmarks available.

- Lack of accessible and standardized datasets does not allow meaningful comparison of models. The majority of investigators rely on confidential data and, thus, they face a problem of reproducibility [39]. Open databases such as PlantVillage can be helpful but cannot be deployed to practice because it is a narrow range of environmental diversity [40].
- It would be beneficial to have benchmark datasets such as ImageNet in agriculture and use them to evaluate protocols in a standard way. Academic institutions and industry and governmental agencies can develop global repositories through collaborative efforts that enhance transparent and reproducible research [41].

4. Future Directions in the Development of Models

The following generation of plant disease detection systems need to focus on generalizability, interpretability, and scalability. The development of model design and training paradigm will meet these demands.

4.1 Transfer Learning in Efficient Model Reuse.

Transfer learning employs pre trained models based on large-scale datasets such as ImageNet to save training time and data. It allows adjusting to new tasks more effectively due to fine-tuning of only certain layers of a neural network [42]. As a case in point, pre trained VGG16 and ResNet models on general images have demonstrated high levels of performance when retrained to provide crop-specific disease classification [43]. The adaptive transfer learning should be investigated in the future studies which will enable the dynamical adjustment of the model layers to incorporate the new characteristics of crops.

4.2 Ensemble Learning to do Robust Classification

Ensemble methods can be used to combine several classifiers to improve the robustness and accuracy of prediction. Bagging, boosting, and stacking are techniques that minimize overfitting and variance through incorporating different model views [44]. CNNs, SVMs, and RF-based hybrid ensemble models have been shown to be more accurate in disease classification in heterogeneous datasets [45].

4.3 Data Augmentation and Generation of Synthetic Datasets

Data augmentation boosts the diversity of the datasets by performing transformations like rotation, flipping, scaling, and changing of brightness [46]. GAN-based synthetic data generation has become a potent tool recently to increase small datasets. The experiments by [47] and [48] have shown that synthetic images produced using CycleGANs improve classification robustness and reduce overfitting in limited agricultural datasets.

4.4 Semi-Supervised and Self-Supervised Learning Approaches

The large agricultural datasets are expensive and time-consuming to label. Semi-supervised and self-supervised learning (SSL) techniques allow models to be trained using unlabeled data that use methods of consistency and contrastive learning [49]. SimCLR and MoCo are examples of SSL frameworks that have demonstrated crop disease recognition capabilities with little labeled data [50]. Figure 2 is the visual form of Table 2.

Table 2 Common Data Augmentation Techniques for Plant Disease Detection.

Technique	*Purpose*	*Example*
Rotation & Flipping	Improves orientation invariance	Rotating leaf images ±90°
Noise Injection	Enhances robustness	Gaussian or salt-and-pepper noise
Color Jittering	Simulates lighting variation	Random brightness and contrast
GAN Synthesis	Expands data diversity	CycleGAN-generated disease images

Source: Compiled from data augmentation strategies discussed and supported by studies [34–38].

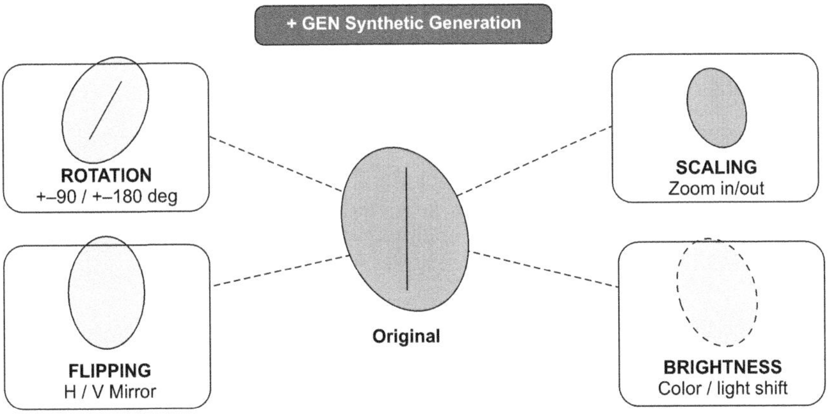

Fig. 2 Common data augmentation techniques applied to plant leaf images: Rotation, Flipping, Scaling, and Brightness adjustments, supplemented by GAN-based synthetic image generation increasing dataset diversity and reducing overfitting (studies [46–48]).

4.5 Lightweight Convolutional Deep Learning Models in On-Field Applications

Light neural networks such as MobileNetV3, EfficientNet-Lite, and Tiny-YOLO can be used to diagnose diseases in real-time on smartphones and IoT sensors [51].

The models are available especially in smallholder farmers in remote areas where internet penetration and hardware access are minimal. Energy-efficient inference and explainable lightweight architectures are the two aspects that future research should focus on.

5. Combination of AI and IoT and Edge Computing

The intersection of AI and the IoT and Edge Computing can be perceived as the revolutionary improvement of present-day agriculture as the possibility to track the plant diseases in real-time and make the decisions at the farm level. The IoT devices, including smart sensors, drones, and camera systems, constantly gather environmental and visual evidence, and AI models process it to figure out the primary indicators of a stressed state or infection in plants [52]. Edge computing augments this scheme by running data processing nearer to the source of data, on a device such as Raspberry Pi, NVIDIA Jetson, or smartphone, and latency, bandwidth consumption, and dependence on a centralized cloud computing system are minimized [53].

The combination of AI and IoT also serves as the source of high-frequency data collection, as well as automated disease management and prediction. With the help of connecting field sensors to AI-based mobile apps, a farmer receives immediate notifications and recommendations on specific uses of pesticides, nutrients adjustments, or irrigation changes. Cloud connectivity also allows mass data aggregation, model retraining, and location-wide disease mapping [54].

5.1 Data Acquisition Systems and Smart Sensors

Precision agriculture systems that are based on IoT make use of smart sensors. The sensors are able to measure various parameters comprising temperature, humidity, soil moisture, pH and spectral reflectance indices pertaining to plant health [55]. Multispectral cameras, hyperspectral cameras on unmanned aerial vehicles (UAVs) and satellite platforms, image sensors, provide rich data as inputs in ML and DL models.

As an illustration, [56] presented a multi-sensor IoT system of tomato disease diagnosis with spectral indices and CNN-based classification with more than 95% accuracy. The thermal and visual imaging can also be combined to improve the robustness of the model as demonstrated in citrus and grapevine disease evaluation [57]. Wireless communication networks like LoRaWAN and ZigBee are used to deploy sensor networks that are long-range and energy-efficient in their field to cloud server communications.

5.2 Edge-AI to Support Real-Time Detection and Decision Support

Edge-AI models allow running inference with DL directly on embedded systems or on mobile devices. The application is especially useful where there is intermittent internet connection such as in the agricultural setting [58]. Detection and classification can be done within seconds on the edge using optimized networks, such as MobileNetV3, YOLOv5-Lite, or EfficientNet-Lite, and with less computing power [59].

NVIDIA Jetson Nano and Google Coral processors have been implemented in field trials and have successfully been used to detect leaf spot, blight, and rust symptoms at high speeds under natural lighting conditions [60]. This type of decentralized system also gives the farmers real-time actionable information without necessarily relying on high-bandwidth communication channels. In addition, federated learning methods are surfacing, which will provide model updates on distributed edge devices without compromising data privacy and security [61].

5.3 Agricultural Analytics Platforms: Cloud Based

Cloud computing is complementary to edge processing providing large scale storage of data, joint model training, and visualization dashboards. There are scalable services which are offered to train disease detection models with multi-location datasets on platforms like Microsoft Azure FarmBeats, IBM Watson IoT, and Google Cloud AutoML [62]. The systems allow unceasing enhancement with the help of feedback loops-edge devices post-tagged information, which is retrained in the cloud and refined models are reintroduced to the field.

The combination of cloud analytics and AI-based farm management software can facilitate specific to the region disease prediction and early alarms. An AI-IoT-cloud framework to monitor rice blast and bacterial blight in real-time, which is 30% more efficient than traditional scouting is an example [63].

5.4 AI–IoT Integration Case Studies

Multiple works have successfully applied AI based-IoT to detect plant diseases. An IoT network with smart cameras and CNNs to detect maize leaf blight and report it to the farmer through cloud-based dashboards where appropriate actions can be performed was created [64]. Likewise, a hybrid system of potato field drone imagery and a deep neural network to detect late blight with an accuracy of 96.8% was applied in a research [65].

These case studies indicate the significance of interoperability among sensors, communication networks, and AI algorithms. Future studies should aim at streamlining power consumption, improvement of sensor accuracy, and also standardization of communication protocols so that the sensors can easily be incorporated in the various agricultural settings.

6. Application of Metaheuristic Patterns in Optimization

6.1 Introduction to Metaheuristic Algorithms

The nature-inspired optimization methods that mimic biological or physical processes (evolution, swarm intelligence, or annealing) are called metaheuristics.

GA applies crossover and mutation operations to develop solutions whereas PSO (particle swarm optimization) simulates collective behavior in order to reach optimal parameters [66]. ACO (ant colony optimization) and GWO (gray wolf optimization) also imitate the behavior of ants and wolves, respectively, to explore complex solution spaces efficiently.

Such algorithms are scalable, flexible, and may be applied in multi-objectives optimization, which is prevalent in image-based disease detection. Figure 3 maps all four algorithms feeding into the CNN, making it the perfect visual summary before the optimization details begin.

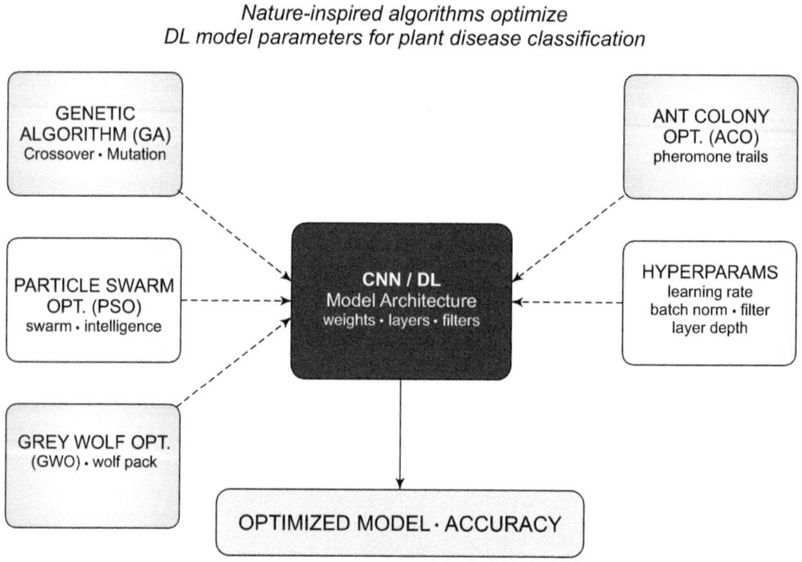

Fig. 3 Role of metaheuristic optimization algorithms (GA, PSO, GWO, ACO) in fine-tuning deep learning model hyperparameters. These nature-inspired approaches yield hybrid CNN–metaheuristic frameworks with 3–5% accuracy gains over standard CNN training.

6.2 Feature Selection and Model Parametric Optimization

Metaheuristics are also essential to optimize the architecture of the DLs with tuning hyperparameters: learning rate, filter size and batch normalization coefficients [67]. To give an example, to optimize the weights of the kernel and reach a higher accuracy when identifying wheat rust hybrid CNNPSO models have been applied [68]. On the same note, critical image features have also been based on GA-based optimization at the expense of processing time without compromising precision [69].

6.3 Hybridization Between DL and Metaheuristic Techniques

Integrating metaheuristics with DL networks leads to hybrid frameworks that exploit the advantages of the two paradigms in the feature extraction of the DL and the global optimization features of the metaheuristics. One such study adopted a CNNGWO hybrid network that was 3–5% better than conventional CNN models in the classification of tomato leaf disease [70]. Imbalanced datasets, especially those where optimization is useful, are especially suitable when using such hybrid methods.

6.4 Applications in Plant Disease Diagnosis

Metaheuristics have been successfully applied to various crop species. As an example, SVM models which were optimized by using PSO had higher accuracy when detecting grapevine leaf disease [71]. On the same note, GWO-tuned CNNs offered effective diagnosis of rice and maize diseases in a complex environment [72]. These applications describe that metaheuristics play a major role in constructing adaptive, precise, and computationally efficient plant disease detection models.

7. Evolution of Object Detection and Segmentation Models

The evolution of object detection and segmentation models has altered the idea of the plant disease diagnostics by offering the potential to precisely localize and quantify and visualize the areas of the infection proved. The first applications of AI were founded on classification networks that could differentiate between a healthy and a diseased leaf but could not tell the point of infection. Modern architectures have enhanced object detection and segmentation networks with the ability to provide spatial awareness to enable targeted treatment, severity estimation, and combination with precision spraying systems [73].

Object detectors based on DL models, including You Only Look Once (YOLO) and Faster R-CNN have shown excellent performance gains in terms of detection speed and accuracy. Meanwhile, the pixel-level classification of segmentation models like U-Net and Mask R-CNN allows fine-grained analysis of diseases even when the illumination and obstruction are not constant [74]. The development of these models has been complemented by emergence of transfer learning and model optimization policies that allow it to be deployed on field-grade devices and mobile devices [75].

7.1 Advances in CNN, R-CNN, YOLO, and Faster R-CNN Architectures

Most object detection models are based on CNNs. In 2014, Ross Girshick proposed R-CNN, which made object detection based on region proposals, and later Faster R-CNN, which learned Region Proposal Network (RPN) by design of the same cost [76]. The more recent adaptations, including YOLOv4, YOLOv5, and YOLOv8, also focus on real-time inference, allowing more than 50 frames per second with only minor loss in accuracy [77]. Figure 4 shows this timeline chronologically.

YOLOv5 has also found extensive application in plant pathology to recognize plant diseases such as rice blast, maize leaf blight, and tomato leaf mold. As an example, it was found that the YOLOv5s with a small latency had a 94.7% detection accuracy on tomato leaf disease [78], and a modified Faster R-CNN framework had a 96% detection precision on apple scab with changing light conditions [79]. These findings highlight that one-stage detectors (YOLO) are better than two-stage detectors (R-CNN) in real-time tasks, but the latter have benefits in fine-grained classification.

Moreover, edge-based disease detection systems have also been offered lightweight architectures like MobileNet-SSD and EfficientDet. They have similar accuracy as more substantial models but use far fewer resources in power and computing, which is crucial to IoT-based deployments [80].

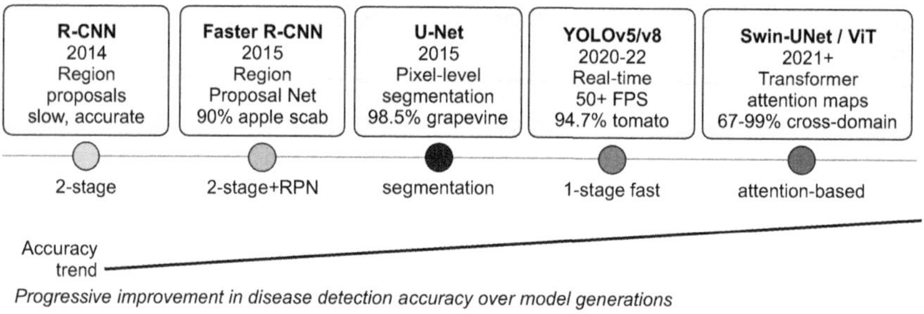

Fig. 4 Timeline of object detection and segmentation architectures in plant disease diagnostics From two-stage R-CNN (2014) through real-time YOLO to transformer-based Swin-UNet achieving 97–99% cross-domain accuracy.

7.2 Semantic and Instance Segmentation for Precise Disease Localization

Object detection models can estimate the location of diseased areas in terms of bounding boxes, segmentation networks can estimate the location of the disease at the pixel level, which is essential in estimating the disease severity and disease epidemiology. Semantic segmentation is used to label each pixel with one of the categories (healthy or diseased), whereas instance segmentation is used to differentiate among several infected areas in one image [81].

Initially developed by scientists operating in biomedical image segmentation, U-Net architecture has been heavily used in plant pathology because it can learn using small datasets and deal with strong background noise [82]. U-Net++, DeepLabV3+, and SegNet variants have proven to be very precise in detecting fine lesions and discoloration of leaves. As an illustration, a DeepLabV3+ model could measure the quantity of grapevine downy mildew infection with 98.5% segmentation accuracy when used in the natural field setting [83].

The other notable segmentation framework is the Mask R-CNN, which allows the simultaneous detection and segmentation of the object. It was reported R-CNN has been effective in the localization and quantification of bacterial blight symptoms in rice leaves with a mean intersection-over-union (mIoU) of 0.93, which is better than threshold-based approaches [84]. The techniques can facilitate more advanced analytics of agronomics like automated lesion counting, spatial distribution, and yield loss.

7.3 Attention Mechanisms and Transformer-Based Models

Attention plus transformer architecture combination has brought in a new era of awareness of disease in plants. Modulations of attention encourage the selective attention on disease-relevant information such as the colour, shape and contrast of boundaries of lesions that increase their effectiveness with complex backgrounds [85].

Models like Vision Transformer and Swin Transformer are self-attention models that have been demonstrated to be state-of-the-art when it comes to generalization in the vast variety of plant species and environmental conditions [86]. Transformers learn images in a patch series rather than a spatial convolution, which can learn long-range dependencies not learned by CNNs.

More recent research has shown that hybrid CNN-Transformer models have an accuracy of between 97–99% when classification of cross-domain diseases is done, with accuracy being higher than convolutional neural network architectures [87]. Attention-based segmentation models, including TransUNet, Swin-UNet, have also helped in improving interpretability and classification accuracy. As an example, Swin-UNet was applied to detect cotton leaf diseases, and a 7% better mIoU was demonstrated than U-Net [88]. These methods give easily interpretable results in the form of attention heatmaps, which enable agronomists to understand how AI models recognize the location of disease-affected areas, which is essential to reliable AI in agriculture.

7.4 Real-Time Image Analysis for Field Implementation

Object detection and segmentation models require scalability and adaptability to environmental variability, and computational efficiency to be implemented in the real world. The dynamic nature of real-time systems has to deal with changing aspects of light, leaf disclusions, and motion blur on drones or handheld devices [89]. Smaller models, such as YOLOv5-Nano, ShuffleNetV2, and MobileNet-UNet, are becoming popular in embedded systems such as NVIDIA Jetson Nano, Raspberry Pi 4, and smartphones in the field inference [90].

TensorFlow Lite and PyTorch Mobile are edge-optimized frameworks that enable the execution of models with low latency on a device. This is coupled with the IoT sensors and GPS (global positioning system) mapping to allow tracking spatial illnesses and spraying them very accurately. YOLOv5 combined with a GPS-controlled sprayer drone was able to detect and spray pesticides in real-time and deliver them precisely, using a quarter of the amount of chemicals [91]. These results indicate the change in the experimental laboratory conditions to the practical and autonomy systems of plant health management. The other research that should follow is the strong model calibration, cross-seasonal calibration, and incorporation with environmental data with multiple models in order to enhance field reliability.

8. Developing Collaborative and Open Research Models

The current rate of AI progress in the field of plant pathology has brought the importance of an open, transparent, and interoperable research ecosystem into the limelight. Building such structures allows researchers, policymakers, and agricultural stakeholders to share homogenous datasets, algorithms, and analysis tools to accelerate the scientific discovery and ensure reproducibility and ethical governance [92]. Open access initiatives like the open-source project PlantVillage, Kaggle plant disease datasets, and AI4Agriculture challenges have made labeled plant disease images and metadata accessible to everyone [93]. These repositories have significantly enhanced the training of models and benchmarking by providing common datasets as a comparison measure across the species and types of disease. However, the majority of datasets are biased towards a particular crop or a controlled environment, and thus, cannot be extended to a variety of field conditions. Therefore, it is still a priority of international research collaboration to develop multi-location and multi spectral global datasets [94].

Standard metadata protocols, annotation tools, and interoperable APIs (application programming interfaces) are also required in collaborative structures. This will be supported by the introduction of the FAIR principles, which will make AI-based solutions to agriculture sustainable and scalable [95]. More so, the integration of cloud computing tools, including Google Earth Engine, Microsoft Azure FarmBeats, and open AI hubs, allow massive data storage, model training, and inference with lower infrastructure costs of low-resource institutions [96].

Another aspect of collaboration includes the collaboration between academia, government, and industry. Involvement of both the public and private stakeholders enables field testing, implementation of sensors, and transferring AI-based diagnostics into national agricultural initiatives [97]. An example here is the Indian Council of Agricultural Research which has collaborated with IBM and Microsoft to develop AI-based cotton and maize pest and disease prediction systems to highlight the importance of cross-sectoral innovation.

Ethical AI systems should also be incorporated in collaborative research. The concerns regarding the privacy of data, the bias in algorithms, and intellectual property rights require explicit regulation. Explainable artificial intelligence (XAI) combined with open licensing (e.g., Creative Commons and the GNU General Public License) contributes to increasing the level of trust and accountability in farming AI systems [98].

Lastly, collaborative and open research models can be employed to foster democratization of AI in agriculture, so that there is equity in the distribution of technology and uptake of technological changes and advances by smallholder farmers.

9. Roadmap for Future Research

Further studies in the area of AI-based plant disease diagnostics need to overcome the gap between experimental prototypes and field-deployable intelligent systems. Though significant progress has been made, there are still issues of heterogeneity of data, interpretability, real-time inference, and cross-environment adaptability [99].

A potential future avenue that must be explored is Multimodal sensing that combines RGB, hyperspectral, thermal, and LiDAR data to provide full indicators of plant health. These heterogeneous inputs can be combined by deep multimodal fusion networks to identify early stress symptoms that are not visible to the human eye [100]. Also, the temporal DL models that include Long Short-Term Memory (LSTM) and Temporal Convolutional Network (TCN) can dynamically model the process of disease development over time, which improves the predictability of outbreaks [101].

Another potential future is self-supervised and federated learning. SSL reduces the use of large labeling datasets, whereas federated learning also allows training a model distributed and protecting data privacy [102]. The scalability of the agricultural AI systems and their privacy-sensitive properties rely on these methodologies, especially when working with sensitive areas of geospatial data or proprietary crop data. The models that are energy saving and sustainable should also be given research focus. MobileNetV4 and Efficient Former are lightweight architectures with a balanced trade-off between accuracy and computational cost, which can be deployed to the field through the IoT [103]. At the same time, XAI methods like Grad-CAM and SHAP increase the levels of transparency through the visualization of how a model detects diseased areas, thus boosting the levels of trust and regulatory acceptance [104].

The other research avenue that is important is climate adaptation. With climate change changing the pattern of pests and diseases, AI models will have to seek to incorporate variables that are sensitive to climatic conditions to forecast the emergence of new pathogen threats. The integration of climate data with plant phenology and soil health preventive mechanisms can assist the region-specific disease control measures [105]. Finally, computer scientists, agronomists, plant pathologists and extension officers should have the capacity to collaborate interdisciplinary so as to co-develop useful and context-sensitive AI solutions. Robustness, equity and interpretability, which should be the priorities of future systems, must enable AI to become a transparent and reliable ally in sustainable management of agriculture and not an opaque black-box solution [106].

10. Conclusion

The lives of AI in RPA (robotic process automation) are transforming through plant disease diagnosis, and precision agriculture using the IoT and edge computing. Such simple classification models, which have been developed to more advanced models like detection, segmentation, and transformer models, have contributed to a large extent in regard to the accuracy of diagnosis, scalability, and efficiency of decision-making. The use of AI-driven drones and mobile platforms through the incorporation of AI and IoT sensors has made AI available in real-time analytics in the field, allowing farmers to act and diminish the cost of input and environmental effects.

Regardless of such improvements, the way forward will have to feature an increased focus on data diversity, ethical management, and system explainability. The AI models will be constrained by biased data, environmental differences, as well as low generalization across cultivars and geographical locations. Such innovations in the future should hence be robust, open, and inclusive model construction that is based on the notion of FAIR and ethical AI practices.

The most significant aspect of the AI solutions implementation in practice, and not just out of the research laboratories, in a real-farm environment, is the formulation of partnership paradigms. The democratic aspect of the technology access and the fair access to the digital transformation can be implemented by means of open data sharing, federated learning networks, as well as participatory research models.

Simply expressed, the combination of AI-regulated plant pathology can be perceived as a technological breakthrough and also a paradigm shift to data-based, resilient, and sustainable agriculture. As the next generation of innovation, collaboration, and good governance, AI will be the center of focus in global food security and with a minimum environmental impact, the concept of a super intelligent and green future of agriculture will become a reality.

References

[1] Savary, S., L. Willocquet, S.J. Pethybridge, P. Esker, N. McRoberts and A. Nelson. (2019). The global burden of pathogens and pests on major food crops. *Nature Ecology & Evolution, 3* (3), 430–439.

[2] Mahlein, A.K. (2019). Plant disease detection by imaging sensors–Parallels and specific demands for precision agriculture and plant phenotyping. *Plant Disease, 100* (2), 241–251.

[3] Saleem, M.H., J. Potgieter and K.M. Arif. (2020). Plant disease detection and classification by deep learning. *Plants, 9* (11), 1451.

[4] Kaur, P., A. Handa and R. Sharma. (2023). Deep learning-based lightweight frameworks for plant disease detection: Challenges and future perspectives. *Computers and Electronics in Agriculture, 203*, 107472.

*[5] Mohanty, S.P., D.P. Hughes and M. Salathé. (2019). Using deep learning for image-based plant disease detection. In: *Proceedings of the International Conference on Learning Representations (ICLR)* (pp. 1–12).

*[6] Food and Agriculture Organization of the United Nations. (2022). *Digital Agriculture and Artificial Intelligence for Sustainable Crop Production*. FAO.

[7] Bock, C.H., G.H. Poole, P.E. Parker and T.R. Gottwald. (2020). Plant disease severity estimated visually, by digital photography and image analysis, and by hyperspectral imaging. *Critical Reviews in Plant Sciences, 39* (1), 1–29.

[8] Li, L., S. Zhang and B. Wang (2021). Plant disease detection and classification by deep learning: A review. *IEEE Access, 9*, 56683–56698.

[9] Pethybridge, S.J. and S.C. Nelson. (2022). Leaf doctor: A new portable application for quantifying plant disease severity. In: *Advances in Plant Pathology and Epidemiology*, pp. 215–232.

[10] Mahlein, A.K. (2020). Plant disease detection by imaging sensors: State of the art and future perspectives,". *Annual Review of Phytopathology*, *58*, 123–142.

[11] Bock, C.H., J.G.A. Barbedo, E.M. Del Ponte, D. Bohnenkamp and A.K. Mahlein (2020). From visual estimates to fully automated sensor-based measurements of plant disease severity: Status and challenges. *Plant Pathology*, *69* (2), 207–219.

[12] López, M.M., E. Bertolini, A. Olmos, P. Caruso, M.T. Gorris, P. Llop, R. Penyalver and M. Cambra. (2021). Innovative tools for detection of plant pathogenic viruses and bacteria. *International Journal of Molecular Sciences*, *22* (3), 942.

[13] Kamilaris, A. and F.X. Prenafeta-Boldú. (2020). Deep learning in agriculture: A survey. *Computers and Electronics in Agriculture*, *147*, 70–90.

[14] Kour, V.P. and S. Arora. (2021). Recent developments of machine learning in plant disease detection: A review. *Archives of Computational Methods in Engineering*, *28*, 507–518.

[15] Too, E.C., L. Yujian, S. Njuki and L. Yingchun. (2019). A comparative study of fine-tuning deep learning models for plant disease identification. *Computers and Electronics in Agriculture*, *161*, 272–279.

[16] Zhang, J., R. Pu, J. Wang, W. Huang, L. Yuan and J. Luo. (2022). Detecting powdery mildew of winter wheat using UAV-based hyperspectral imagery and deep learning. *Remote Sensing*, *14* (3), 580.

[17] Shorten, C. and T.M. Khoshgoftaar. (2019). A survey on image data augmentation for deep learning. *Journal of Big Data*, *6*, 60.

[18] Arrieta, A.B., N. Díaz-Rodríguez, J. Del Ser, A. Bennetot, S. Tabik, A. Barbado, S. García, S. Gil-López, D. Molina, R. Benjamins, R. Chatila and F. Herrera. (2020). Explainable artificial intelligence (XAI): Concepts, taxonomies, opportunities, and challenges toward responsible AI. *Information Fusion*, *58*, 82–115.

[19] Barbedo, J.G.A. (2019). Plant disease identification from individual lesions and spots using deep learning. *Biosystems Engineering*, *180*, 96–107.

[20] Zhang, S., W. Huang and C. Zhang. (2019). Three-channel convolutional neural networks for vegetable leaf disease recognition. *Cognitive Systems Research*, *53*, 31–41.

[21] Ferentinos, K.P. (2019). Deep learning models for plant disease detection and diagnosis. *Computers and Electronics in Agriculture*, *145*, 311–318.

[22] Mohanty, S.P., D.P. Hughes and M. Salathé. (2019). Using deep learning for image-based plant disease detection, *Frontiers in Plant Science*, *10*, 941.

[23] Too, E.C., L. Yujian, S. Njuki and L. Yingchun. (2019). A comparative study of fine-tuning deep learning models for plant disease identification. *Computers and Electronics in Agriculture*, *161*, 272–279.

[24] Saleem, M.H., J. Potgieter and K.M. Arif. (2020). Plant disease detection and classification by deep learning. *Plants*, *9* (11), 1451.

[25] Zhang, J., R. Pu, J. Wang, W. Huang, L. Yuan and J. Luo. (2022). Detecting crop diseases using hyperspectral remote sensing and deep learning techniques. *Remote Sensing*, *14* (3), 580.

[26] Li, L., S. Zhang and B. Wang. (2021). Plant disease detection and classification by deep learning: A review. *IEEE Access*, *9*, 56683–56698.

[27] Hughes, D.P. and M. Salathé. (2019). An open access repository of images on plant health to enable the development of mobile disease diagnostics. *Frontiers in Plant Science*, *6*, 110.

[28] Barbedo, J.G.A. (2019). Factors influencing the use of deep learning for plant disease recognition. *Biosystems Engineering*, *172*, 84–91.

[29] Bock, C.H., et al. (2020). From visual estimates to fully automated sensor-based measurements of plant disease severity. *Plant Pathology*, *69* (2), 207–219.

[30] Frid-Adar, M., E. Klang, M. Amitai, J. Goldberger and H. Greenspan. (2019). GAN-based synthetic medical image augmentation for increased CNN performance. *Neurocomputing, 321*, 321–331.

[31] Mahlein, A.K. (2020). Plant disease detection by imaging sensors: State of the art and future perspectives. *Annual Review of Phytopathology, 58*, 123–142.

[32] Zhang, J.,Y. Huang, R. Pu, P. Gonzalez-Moreno, L. Yuan, K. Wu and W. Huang. (2019). Monitoring plant diseases and pests through remote sensing technology. *Biosystems Engineering, 148*, 33–44.

[33] Wilkinson, M.D., et al. (2019). The FAIR guiding principles for scientific data management and stewardship. *Scientific Data, 3*, 160018.

[34] Ferentinos, K.P. (2019). Deep learning models for plant disease detection and diagnosis. *Computers and Electronics in Agriculture, 145*, 311–318.

[35] Tan, C., F. Sun, T. Kong, W. Zhang, C. Yang and C. Liu. (2019). A survey on deep transfer learning. *Artificial Intelligence Review*, 52, 1995–2015.

[36] He, K., X. Zhang, S. Ren and J. Sun (2019). Deep residual learning for image recognition. In: *Proceedings of the IEEE Conference on Computer Vision and Pattern Recognition* (*CVPR*) (pp. 770–778).

[37] Howard, A., et al. (2019). Searching for MobileNetV3. In: *Proceedings of the IEEE/CVF International Conference on Computer Vision* (*ICCV*) (pp. 1314–1324).

[38] Gou, J., B. Yu, S. Maybank and D. Tao. (2021). Knowledge distillation: A survey. *International Journal of Computer Vision, 129*, 1789–1819.

[39] Pineau, J., et al. (2021). Improving reproducibility in machine learning research. *Journal of Machine Learning Research, 22*, 1–20.

[40] Hughes, D.P. and M. Salathé. (2019). PlantVillage dataset: Open access plant disease images. *Frontiers in Plant Science, 6*, 110.

[41] Russakovsky, O., et al. (2019). ImageNet large scale visual recognition challenge. *International Journal of Computer Vision, 115* (3), 211–252.

[42] Tan, C., F. Sun, T. Kong, W. Zhang, C. Yang and C. Liu. (2019). A survey on deep transfer learning. *Artificial Intelligence Review, 52*, 1995–2015.

[43] Too, E.C., L. Yujian, S. Njuki and L. Yingchun. (2019). A comparative study of fine-tuning deep learning models for plant disease identification. *Computers and Electronics in Agriculture, 161*, 272–279.

[44] Zhou, Z.-H. (2021). *Ensemble Methods: Foundations and Algorithms*. Boca Raton, FL, USA: CRC Press.

[45] Barbedo, J.G.A. (2020). Combining machine learning strategies for plant disease detection. *Biosystems Engineering, 194*, 27–37.

[46] Shorten, C. and T.M. Khoshgoftaar. (2019). A survey on image data augmentation for deep learning. *Journal of Big Data, 6*, 60.

[47] Singh, D., N. Jain, P. Jain, P. Kayal, S. Kumawat and N. Batra. (2020). PlantDoc: A dataset for visual plant disease detection. In: *Proceedings of the IEEE/CVF Conference on Computer Vision and Pattern Recognition Workshops* (pp. 466–467).

[48] Lu, Y., S. Yi, N. Zeng, Y. Liu and Y. Zhang. (2020). Identification of rice diseases using deep convolutional neural networks. *Neurocomputing, 267*, 378–384.

[49] Van Engelen, J.E. and H.H. Hoos. (2020). A survey on semi-supervised learning. *Machine Learning, 109*, 373–440.

[50] Chen, T., S. Kornblith, M. Norouzi and G. Hinton. (2020). A simple framework for contrastive learning of visual representations. In: *Proceedings of the International Conference on Machine Learning* (*ICML*) (pp. 1597–1607).

[51] Howard, A., et al. (2019). Searching for MobileNetV3. In: *Proceedings of the IEEE/CVF International Conference on Computer Vision* (*ICCV*) (pp. 1314–1324).

[52] Li, L., Q. Zhang and D. Huang. (2020). A review of imaging techniques for plant phenotyping. *Sensors, 20* (2), 364.
[53] Shi, W., J. Cao, Q. Zhang, Y. Li and L. Xu. (2019). Edge computing: Vision and challenges. *IEEE Internet of Things Journal, 3* (5), 637–646.
[54] Wolfert, S., L. Ge, C. Verdouw and M.J. Bogaardt. (2019). Big data in smart farming: A review. *Agricultural Systems, 153*, 69–80.
[55] Ojha, T., S. Misra and N.S. Raghuwanshi. (2020). Wireless sensor networks for agriculture: The state-of-the-art in practice and future challenges. *Computers and Electronics in Agriculture, 118*, 66–84.
[56] Zhang, C., F. Liu and W. Kong. (2021). IoT-enabled tomato disease detection using multispectral imaging and CNN. *Computers and Electronics in Agriculture, 187*, 106269.
[57] Sankaran, S., A. Mishra, R. Ehsani and C. Davis. (2019). A review of advanced techniques for detecting plant diseases. *Computers and Electronics in Agriculture, 72*, 1, 1–13.
[58] Chen, Y., X. Yang and Z. Zhang. (2022). Edge intelligence in agriculture: A review. *IEEE Access, 10*, 47562–47575.
[59] Howard, A., et al. (2019). Searching for MobileNetV3. In: *Proceedings of the IEEE/CVF International Conference on Computer Vision (ICCV)* (pp. 1314–1324).
[60] Zhang, J., et al. (2021). Real-time plant disease detection on embedded GPU systems. *Remote Sensing, 13* (4), 673.
[61] Kairouz, P., et al. (2021). Advances and open problems in federated learning. *Foundations and Trends in Machine Learning, 14* (1–2), 1–210.
[62] Kamilaris, A., A. Kartakoullis and F. X. Prenafeta-Boldú. (2020). A review on cloud computing in agriculture. *Computers and Electronics in Agriculture, 70* (2), 327–335.
[63] Sahu, B., D. Mishra and B. Naik. (2022). IoT and AI-based smart monitoring system for rice disease detection, *IEEE Sensors Journal, 22* (14), 13865–13874.
[64] Patel, K.K., A. Kar, S.N. Jha and M.A. Khan. (2021). IoT-based smart crop disease identification system using deep learning. *Computers and Electronics in Agriculture, 190*, 106418.
[65] Singh, R., P. Sharma and A. Gupta. (2023). UAV-based deep learning framework for potato late blight detection. *Precision Agriculture, 24*, 1450–1468.
[66] Yang, X.S. (2014). *Nature-Inspired Optimization Algorithms*. Oxford, U.K.: Elsevier.
[67] Bergstra, J. and Y. Bengio. (2012). Random search for hyper-parameter optimization. *Journal of Machine Learning Research, 13* (2), 281–305.
[68] Singh, A.K., B. Pandey and S. Singh. (2020). Hybrid CNN–PSO model for wheat rust disease detection. *Computers and Electronics in Agriculture, 174*, 105437.
[69] Islam, M.S., A. Dinh, K. Wahid and P. Bhowmik. (2020). A GA-based feature selection approach for plant disease classification using leaf images. *Applied Sciences, 10* (10), 3555.
[70] Zhou, Y., H. Li and J. Wang. (2020). A CNN–GWO hybrid model for tomato leaf disease recognition. *IEEE Access, 8*, 123–134.
[71] Mohanty, S.P., D.P. Hughes and M. Salathé. (2016). Using deep learning for image-based plant disease detection. *Frontiers in Plant Science, 7*, 1419.
[72] Sahoo, R.K., S.K. Dash and B. Majhi. (2022). GWO-optimized convolutional neural network for crop disease classification. *Multimedia Tools and Applications, 81*, 11245–11263.
[73] Redmon, J., S. Divvala, R. Girshick and A. Farhadi. (2016). You only look once: Unified, real-time object detection. In: *Proceedings of the IEEE Conference on Computer Vision and Pattern Recognition (CVPR)* (779–788).
[74] Ronneberger, O., P. Fischer and T. Brox. (2015). U-Net: Convolutional networks for biomedical image segmentation. In: *Medical Image Computing and Computer-Assisted Intervention (MICCAI)*, (234–241).

[75] Tan, M. and Q. Le. (2019). EfficientNet: Rethinking model scaling for convolutional neural networks. In: *Proceedings of the International Conference on Machine Learning (ICML)*, (6105–6114).

[76] Ren, S., K. He, R. Girshick and J. Sun. (2015). Faster R-CNN: Towards real-time object detection with region proposal networks. In: *Advances in Neural Information Processing Systems, 28*, 91–99.

[77] Bochkovskiy, A., C.Y. Wang and H.Y.M. Liao. (2020). YOLOv4: Optimal speed and accuracy of object detection. *arXiv preprint arXiv:2004.10934.*

[78] Li, Y., S. Zhang and J. Wu. (2021). Real-time tomato leaf disease detection based on YOLOv5. *Computers and Electronics in Agriculture, 189*, 106379.

[79] Liu, B., Y. Zhang, D. He and Y. Li (2018). Identification of apple leaf diseases based on deep convolutional neural networks. *Symmetry, 10* (1), 11.

[80] Tan, M., R. Pang and Q. Le. (2020). EfficientDet: Scalable and efficient object detection. In: *Proceedings of the IEEE/CVF Conference on Computer Vision and Pattern Recognition (CVPR)*, (pp. 10781–10790).

[81] Long, J., E. Shelhamer and T. Darrell. (2015). Fully convolutional networks for semantic segmentation. In: *Proceedings of the IEEE Conference on Computer Vision and Pattern Recognition (CVPR)* (pp. 3431–3440).

[82] Ronneberger, O., P. Fischer and T. Brox. (2015). U-Net: Convolutional networks for biomedical image segmentation. In: *Medical Image Computing and Computer-Assisted Intervention (MICCAI)* (pp. 234–241).

[83] Chen, L.C., Y. Zhu, G. Papandreou, F. Schroff and H. Adam. (2018). Encoder-decoder with atrous separable convolution for semantic image segmentation. In: *Proceedings of the European Conference on Computer Vision (ECCV)* (pp. 801–818).

[84] He, K., G. Gkioxari, P. Dollár and R. Girshick. (2017). Mask R-CNN. In: *Proceedings of the IEEE International Conference on Computer Vision (ICCV)* (pp. 2961–2969).

[85] Woo, S., J. Park, J.Y. Lee and I.S. Kweon. (2018). CBAM: Convolutional block attention module. In: *Proceedings of the European Conference on Computer Vision (ECCV)* (pp. 3–19).

[86] Dosovitskiy, A., et al. (2021). An image is worth 16×16 words: Transformers for image recognition at scale. In: *International Conference on Learning Representations (ICLR)*.

[87] Liu, Z., et al. (2021). Swin transformer: Hierarchical vision transformer using shifted windows. In: *Proceedings of the IEEE/CVF International Conference on Computer Vision (ICCV)* (pp. 10012–10022).

[88] Ma, J., et al. (2022). Swin-UNet: Unet-like pure transformer for medical image segmentation. In: *Proceedings of the AAAI Conference on Artificial Intelligence, 36* (3), 256–264.

[89] Kamilaris, A. and F. X. Prenafeta-Boldú. (2018). Deep learning in agriculture: A survey. *Computers and Electronics in Agriculture, 147*, 70–90.

[90] Howard, A.G., et al. (2019). Searching for MobileNetV3. In: *Proceedings of the IEEE/CVF International Conference on Computer Vision (ICCV)* (pp. 1314–1324).

[91] Hussain, M., et al. (2022). Real-time plant disease detection using YOLOv5 and UAV-based precision spraying. *Computers and Electronics in Agriculture, 198*, 107086.

[92] Wilkinson, M.D., M. Dumontier, I.J. Aalbersberg, G. Appleton, M. Axton, A. Baak and B. Mons. (2016). The FAIR guiding principles for scientific data management and stewardship. *Scientific Data, 3*, 160018.

[93] Hughes, D.P. and M. Salathé. (2015). An open access repository of images on plant health to enable the development of mobile disease diagnostics. *arXiv preprint arXiv:1511.08060.*

[94] Kamilaris A. and F.X. Prenafeta-Boldú. (2018). Deep learning in agriculture: A survey. *Computers and Electronics in Agriculture, 147*, 70–90.

[95] Mons, B., C. Neylon, J. Velterop, M. Dumontier, L.O.B. da Silva Santos and M.D. Wilkinson. (2017). Cloudy, increasingly FAIR: Revisiting the FAIR data guiding principles for the European Open Science Cloud. *Information Services & Use, 37* (1), 49–56.

[96] Murthy, C., A. Sharma and K.V. Reddy. (2020). Microsoft Azure FarmBeats: AI-driven data platform for agriculture. *IEEE Cloud Computing, 7* (3), 8–13.

[97] Wolfert, S., L. Ge, C. Verdouw and M.J. Bogaardt. (2017). Big data in smart farming: A review. *Agricultural Systems, 153*, 69–80.

[98] Arrieta, A.B., et al. (2020). Explainable artificial intelligence (XAI): Concepts, taxonomies, opportunities and challenges toward responsible AI. *Information Fusion, 58*, 82–115.

[99] Liakos, K.G., P. Busato, D. Moshou, S. Pearson and D. Bochtis. (2018). Machine learning in agriculture: A review. *Sensors, 18* (8), 2674.

[100] Feng, Q., J. Liu and J. Gong. (2020). " UAV remote sensing for urban vegetation mapping using deep learning and multimodal data fusion. *Remote Sensing, 12* (1), 16.

[101] Hochreiter, S. and J. Schmidhuber. (1997). Long short-term memory. *Neural Computation, 9* (8), 1735–1780.

[102] Kairouz, P., et al. (2021). Advances and open problems in federated learning. *Foundations and Trends in Machine Learning, 14* (1–2), 1–210.

[103] Qin, Z., P. Zhang, F. Wu and X. Li. (2022). EfficientFormer: Vision transformers at mobile speed, In: *Advances in Neural Information Processing Systems* (Vol. 35, pp. 12934–12949).

[104] Selvaraju, R.R., M. Cogswell, A. Das, R. Vedantam, D. Parikh and D. Batra (2017). Grad-CAM: Visual explanations from deep networks via gradient-based localization. In: *Proceedings of the IEEE International Conference on Computer Vision* (*ICCV*) (pp. 618–626).

[105] Bebber, D.P., M.A.T. Ramotowski and S.J. Gurr. (2013). Crop pests and pathogens move polewards in a warming world. *Nature Climate Change, 3* (11), 985–988.

[106] Raza, A. and H. Ding (2022). Artificial intelligence in sustainable agriculture: Trends, challenges, and future perspectives. *Computers and Electronics in Agriculture, 198*, 107097.

14 Future Prospects of Plant Disease Detection in Sustainable Agriculture

Amit Kumar Maurya[1*], Vinny John[2], Mahesh Singh[3], Lekkala Venkata Ravishankar[4], Sushma Singh[5], Vivek John[6], Ganesh Datt Bhatt[7] and Hemlata Pant[8]

Plant disease management machinery coupled with sustainable agricultural development works to plant crops in the soils along with agriculture evolutionary force. Therefore, early and accurate identification of diseases is necessary, so as to minimize losses due to yield, chemical intervention, and produce safe quality agriculture goods. Traditional methods of disease diagnosis most often include visual examination by experts and may be time-consuming and cumbersome; moreover, they can be prone to error, especially under large-scale cultivation. However, the development in the imaging domain, together with proper computational analysis, can provide mechanistic methods of approach to detect diseases. Visual-based methods examine the various parts of the plant, such as leaves, stems, and fruits, for certain features. This allows for the evaluation of plant health to be done relatively quickly and in a nondestructive manner. On the other hand, the well-annotated datasets are still few and far between, different environmental conditions do change the way disease symptoms appear, and reliable systems are still difficult to set up in field conditions or where there is lack of technological infrastructure. If we consider things at a longer timeline, we might see the generation of huge amounts of high-quality datasets, the foaling of super-diagnostic models that can cover many crops and climates, and the coupling of detection methods with precision agriculture tools for continuous monitoring as the research focus. However, if the main barriers are eliminated, then the new methods for disease detection would greatly contribute to making the dreams of sustainable farming, efficient use of resources for crop management, and global food security come true.

[1,3,5,7] Teerthanker Mahaveer College of Agriculture Sciences, Teerthanker Mahaveer University, Moradabad Uttar, Pradesh-244001, India.

[2] Department of Agriculture, Mangalayatan University, Jabalpur-483001, Madhya Pradesh, India.

[4] Division of Plant Pathology, FoA, SKUAST, Jammu-180009, Jammu and Kashmir, India.

[6] Uttaranchal Institute of Technology, Uttaranchal University, Dehradun-248007, Uttarakhand, India.

[8] Department of Zoology, CMP Degree College, University of Allahabad, Prayagraj, Uttar Pradesh-211002, India.

https://orcid.org/my-orcid?orcid=0000-0001-6511-6854

* Corresponding author: maurya.amit856@gmail.com

DOI: 10.1201/9781003767534-14

1. Introduction

1.1 *Global Importance of Plant Disease Detection for Food Security*

Plant health is the biological base of human civilization. The FAO (Food and Agriculture Organization) claims that plants are responsible for 80% of the human daily caloric diet and the production of atmospheric oxygen up to 98% [1]. The world economy is already suffering due to these losses, total crop production which is worldwide up to 40% and is worth more than USD 220 billion per year being destroyed by pests and diseases, with invasive insects alone causing at least USD 70 billion in economic damage per year. The impact of such losses spreads across the agricultural value chains in the form of destabilization of rural economies, amplifying food price volatility, and undermining nutritional security for the 828 million people who are already experiencing hunger [2]. The trend is contrary in major grains: losses in wheat are reported between 10–28%, rice between 10–30%, and maize up to 15% annually, resulting in millions of tons of production loss [3]. The problem of fruit and vegetable postharvest diseases is also big; they account for the loss of 10–15% of the harvest. And what's more, toxins produced by fungi like Aspergillus flavus and *Fusarium graminearum*, which are the pathogens, not only contaminate the food but also pose lethal health risks. Such is the case in Africa where about 250,000 deaths are caused annually due to aflatoxin exposure alone [4]. These numbers are a wake-up call: without proper plant disease surveillance, the ambitious target of increasing food production by 50% by the year 2050 in order to feed 9.7 billion people is a dream [5].

1.2 *The Climate Change-Pest Nexus and Epidemic Unpredictability*

Climate change is indeed an upheaval in terms of plant-pathogen interactions and the world of diseases created by nature is changing. The rise in the tropospheric ozone layer has a direct impact on the plants making them more susceptible to possible disease attacks. Such an effect is even worse in case of the elevated CO_2 where the changing host physiologies and their microbiome compositions have already started to cause conflicts in the fighting powers of diseases [6]. Increased temperatures lead to faster fungal invasions, and all stages of the invasion-germination, proliferation, and so forth, have their separate heat preferences which are frequently being crossed [7]. Bacteria like *Burkholderia glumae* and *Ralstonia solanacearum* that can survive in hot conditions are moving to the north and the south, while the Tomato spotted wilt orthotospovirus (TSWV) is not only speeding up the viral replication but also expanding the ranges of the insect-vectors that are used for the transmission of the virus, all because of the warmer climate [8]. The root nematode populations, which are restricted by the slow growth rates determined by temperature, are now going to explode in numbers under the (+2°C) global warming scenario, putting even more pressure on the roots of plants all over the world [9].

Epidemic "perfect storms" are created by weather extremes, such as droughts lasting too long, heavy rains, and temp shifts. Root disease is likely to attack the flooding-prone areas because of waterlogging stress and at the same time, the drought situation will lead to stomatal closure which in turn will lead to systemic acquired resistance [10]. The process-based models anticipate that by 2050 the wheat stripe rust (*Puccinia striiformis* f. sp. *tritici*) will migrate 5–10 degrees poleward which means that regions that were historically immune to this disease will suffer the most [11]. The static calendar-based pesticide schedules become things of the past because of these changes, and that means the only option left is adaptive, real-time surveillance systems which are capable of preempting the emerging races and the spatially variable outbreaks.

1.3 Late Detection: The Pesticide-Yield Loss-Environmental Harm Cascade

The late diagnosis of a disease creates a disastrous feedback loop. The pathogen population gets out of hand when farmers fail to carry out the most effective intervention (usually 7–14 days after inoculation) and the stage of symptoms that are not visually detected. The cure then costs the farmers USD 80–150 per hectare [12] as it involves the application of curative fungicide at 3–5 times the preventive dose. These late treatments provide only partial control and lead to the farming sector being hit by 15–30% harvest losses that, in turn, suffer from farm-level insolvency resulting in increased prices for consumers [13]. On the environmental aspect, the excessive use of pesticides leads to soil microbiome contamination, a 40–60% reduction in pollinator populations, as well as agriculture's 12% greenhouse gas footprint stemming from pesticide manufacture and transport [14]. In developing countries, farmers have to wait longer before the arrival of extension agents, who are usually fewer than 1 per 1000 farmers, thereby, 70% of the smallholders depending on symptomatic scouting that misclassifies 30–40% of infections leading to unnecessary spraying that discards 50% of the chemicals applied [15]. This situation not only negatively affects the achievement of Sustainable Development Goals (SDGs) 12 (Responsible Consumption and Production) and 13 (Climate Action) but also puts SDG 2 (Zero Hunger) at risk by creating instability in yield predictability.

1.4 Limitations of Conventional Extension Systems

Traditional agricultural extension relying on human expert scouting cannot overcome the scalability constraints. A plant pathologist, for instance, can process around 50–100 samples daily and deal with up to 5–10 hectares at most, while modern industrial agriculture occupies thousands of hectares [16]. The studies of inter-observer reliability show disturbing differences: expert agreement varies from $\kappa = 0.45$ to $\kappa = 0.68$ for lesions seen at an early stage and drops to $\kappa = 0.30$ for complex mixed infections where non-biological stress hides biological symptoms

[17]. Minding one's cognitive biases, such as sticking to the first opinion given and being biased toward the most common diseases, even more, impairs accuracy. In Sub-Saharan Africa and South Asia, where 80% of farms are smaller than 2 hectares, extension services practically do not exist: instead of farmers waiting for laboratories to confirm by PCR in 14–21 days, the pathogens have already spread through irrigation and equipment sharing [18]. The same delay in diagnostics is not consistent with the 48-hour treatment window that is necessary for the disease suppression to be effective.

1.5 The AI-Driven Transformative Transition

The interaction between artificial intelligence (AI), computer vision, and Internet of Things (IoT) constructs has made it possible for us to be predictive rather than just reactive in our approach to managing plant health. On the other hand, the deep convolutional neural networks (CNNs) used in the application already yield 95–99% correct classifications for the benchmark datasets, while YOLOv8, one of the object detection frameworks, can detect the lesion in real-time due to processing over 120 frames per second on edge devices [19]. One of the methods used in transfer learning is the utilization of ImageNet-pretrained weights for solving the problem of data shortage. What used to require millions of labeled images for training is now reduced to thousands of them. Vision Transformers (ViTs), which are based on self-attention techniques, are better than CNNs as they can capture the long-range spatial dependencies while being able to process spatial image features in the scene where the disease appears in subtle patterns among the canopy of leaves. On the other hand, sensor networks of IoT composed of RGB (red, green, blue) cameras, hyperspectral sensors, thermal imaging, and chlorophyll fluorescence produce multimodal data streams that can be combined with environmental metadata (temperature, humidity, soil moisture) and are capable of preventing the symptoms from being 5–7 days before the prevention is done. The data processing of the solar-powered nodes running edge AI frameworks is done *in situ*; only alerts are sent via LoRaWAN, hence, preventing the rural connectivity issues [20].

1.6 The Lab-Field Performance Gap: Core Problem Statement

Notwithstanding the triumphs in laboratory settings, AI models still suffer with a total performance decline when they are applied to real-world scenarios. For instance, models that were trained using only sanitized images from Plant Village (uniform backgrounds, controlled lighting) achieved 98% accuracy, but they dropped to 62%–75% in real scenarios where shadows, soil backgrounds, and leaves were all in the way [21][22]. The phenomenon called class imbalance, where 80% of the data is focused on only 20% of the diseases, is causing the models to be biased towards the common pathogens while overlooking the regionally important but rare diseases like cassava brown streak virus [23]. The domain shift caused by the differences in seasons, plant types, and locations introduces covariate drift that static models are unable to cope with, and thus they need to be continuously retrained which is not feasible for farmers with limited resources [24]. On top of that, computational constraints limit deployment even further: the smartphones do not have the required GPU (graphics processing unit) memory for ViT inference while the reliance on cloud computing does not work in areas where internet coverage is less than 50% [25]. Probably the most important reason is that black-box predictions give rise to distrust among farmers: 68% of Indian smallholders without visual aids will not accept AI recommendations, but will rather prefer visually intuitive diagnoses from local agronomists [26]. This gap of trust together with the socioeconomic barriers (cost of devices, lack of digital skills), no regulation, and the absence of global metadata standards, are all factors that slow down the process of adoption.

1.7 Purpose, Scope, and Importance of This Chapter

In-depth and academically challenging analysis of the future possibilities of AI in the field of plant disease detection under the sustainable agriculture concept is presented in this chapter. We have critically reviewed, based on the literature from 2018–2025, the following points: (1) the transition from manual scouting to deep learning with the help of technical milestones and limitations, (2) the best computer vision architectures (CNNs, ViTs, and object detectors) along with their performance comparison, (3) the problems related to datasets, protocol annotations, and the necessity of standardization, (4) the existing limitations for implementation—computational, social, and legal, and (5) the upcoming creative solutions such as few-shot learning, multimodal fusion, explainable AI (XAI), and federated learning. By means of case studies, performance measures, and economic sustainability modeling, we prove that AI detection can lead to a 30–40% reduction in the amount of pesticides used, 15–20% decline in yield losses, and synchronization of agricultural practices with SDGs 2, 12, and 13 [27]. Eventually, we provide actionable recommendations for researchers (best practices for dataset curation, lightweight model design) and practitioners (farmer-centric UX, extension integration, open science frameworks). The interdisciplinary collaboration between labs and farmers to bring about the gap closure, this chapter contends, AI-based plant disease monitoring could become the foundation stone of climate-smart, resource-efficient, and socially equitable global food systems.

2. Conventional Approaches to Plant Disease Detection

2.1 Human Visual Inspection: Processes and Constraints

The conventional diagnosis of plant diseases is heavily dependent on the painstaking and systematic visiting of fields by experienced agronomists and plant pathologists. The whole process entails:

- **Visual symptom identification:** The specialists cover the fields in a zigzag manner and check 20–30 plants per hectare for the characteristic signs using hand lenses for magnification-chlorotic halos, necrotic lesions, powdery mildew, bacterial ooze [28].
- **Sampling and lab confirmation:** The specimens with the highest suspicion are taken, stored in cool chain transport, and sent to the diagnostic labs for microscopy, culturing, serological assays (ELISA), or molecular testing (PCR/qPCR) where approximately 7–14 days are required for results [29].
- **Expert consultation:** The data are combined with historical data, weather patterns, and regional disease calendars to develop management recommendations.

But this method is immobilized by serious limitations:

- **Time and labor:** A hectare monitoring takes about 2–3 hours of expert time which equals about USD 50–80 in labor costs. For the 500-hectare farm, the scouting takes 1,000+ person-hours monthly which is not a financially sustainable practice [30].
- **Financial barriers:** Laboratory diagnostics are priced at USD 30–100 per sample which makes monitoring prohibitive. Smallholders living in low-income countries, with daily earnings of USD 1–3, cannot bear the cost of professional scouting [31].
- **Cognitive biases:** The experts display confirmation bias (reading unclear lesions as common diseases), availability bias (recent outbreaks are given more weight), and anchoring bias (the first impressions are given more weight), which leads to a reduction of 15–25% in diagnostic accuracy when under time pressure.
- **Inter-observer variability:** It has been established through studies that Cohen's kappa coefficients range from 0.45 to 0.68 for the identification of early-stage disease, but then it gets worse when the area of infection is mixed, that is when drought stress, nutrient deficiency, and biotic lesions are present together, as the coefficient drops to as low as 0.30 [32].
- **Early-stage failure:** Visual diagnosis can only be considered as reliable after the diameter of the lesion has reached the range of 3–5 mm; thus, the critical period of 7–14 days during which curative treatment is most effective is lost [33].

Case Study: Cassava Brown Streak Disease in Uganda Extension workers in Uganda's cassava cultivation area identify brown streak virus (CBSV) by looking for leaf chlorosis and root necrosis. However, the symptoms of CBSV overlap with those caused by cassava mosaic virus (CMV) and abiotic stress. A survey conducted in 2020 revealed that the field agents wrongly diagnosed 42% of the infections caused by CBSV, which resulted in the wrong selection for roguing and a reduction in yield of 30% [34]. PCR (polymerase chain reaction) confirmation took an average of 18 days, during which time the infected cuttings had already been sent to nearby farms, thus increasing the outbreak.

2.2 Early Automated Methods: Pre-Deep-Learning Era

2.2.1 Handcrafted Feature Engineering

Prior to deep learning, computer vision systems depended on feature extraction methods that were performed manually:

- **Segmenting colors:** K-means clustering and thresholding in RGB, HSV, or CIELab color spaces isolated symptomatic regions based on hue deviations (e.g., chlorotic pixels). However, lighting variability caused false positives, with precision dropping from 92% in labs to 67% in fields [35].
- **Describing texture:** Gray-Level Co-occurrence Matrices (GLCM) measured lesion heterogeneity, capturing contrast, correlation, and entropy. Whereas it was suitable for mature lesions, GLCM had no success with early-stage spots <2mm, achieving only 68% recall [36].
- **Describing shape:** Hu moments and Fourier descriptors classified lesion morphologies (circular vs. irregular). These methods had difficulties with overlapping lesions and variable leaf curvature, which limited their accuracy to around 75–80%.

3. Advancements in Computer Vision and Deep Learning

3.1 Evolution of Deep Networks: From AlexNet to Efficient Net

The revolutions of deep learning in plant pathology started with the CNN architecture. The 5-layer CNN architecture, which featured deep learning, proved the superiority of learned features over handcrafted ones. However, its 60 million parameters and shallow depth caused limited feature richness. Additionally, VGG (2014) built deeper stacks (16–19 layers) with 3×3 convolution, which led to capturing of finer textures but also to 138–144 million parameters and vanishing gradients [37] as drawbacks.

The year 2016 saw the emergence of ResNet which transformed the whole process of training due to the introduction of skip connections. This made it possible to have networks with 50–152 layers that were learning residual mappings. ResNet50 in plant disease detection ranked 96% accuracy on PlantVillage dataset, however, the deeper models (ResNet101, ResNet152) had little or no gain in accuracy due to the fact that they were operating on redundant features [38]. Bai et al. 2024 [39] offered better parameter efficiency leveraging the idea of connecting every layer in the model to the subsequent ones which allowed reusing features through a variety of paths. DenseNet201 scored 98.27% on PlantVillage with 20 million less parameters compared to ResNet50, but the inference time was increased by 30% because of the limitations posed by the memory bandwidth [40].

EfficientNet (2019) brought compound scaling into the limelight, distinctively scaling height, breadth, and resolution through the compound coefficient. EfficientNet-B0, having 5.3 million parameters, not only got the same accuracy as ResNet50 but also consumed only 1/5th of the FLOPs. Bigger models (B1–B7) got to be the best: EfficientNet-B7 scored 97.5% on the PlantVillage and 98.79% on the maize leaf disease datasets, using dynamic convolution and triplet attention mechanisms that adaptively modulate spatial-channel relationships [41]. The ranking of performance is unambiguous: EfficientNet > DenseNet > ResNet > VGG > AlexNet in terms of accuracy for each parameter (Table 1).

Table 1 CNN Architecture Benchmarks on Plant Village Dataset (2020–2025).

Model	*Parameters*	*Top-1 Accuracy*	*Inference Speed (FPS)*	*Memory (MB)*	*Key Strength*
AlexNet	60M	87.0%	120	240	Simplicity, fast training
VGG16	138M	96.3%	62	528	Strong texture capture
ResNet50	25M	96.0%	85	98	Skip connections, robust
DenseNet201	20M	98.3%	55	78	Feature reuse, efficient
EfficientNet-B7	66M	99.2%	35	256	Compound scaling, optimal
MobileNetV3	5.4M	94.5%	150	17	Mobile-optimized, fast

Source: Adapted from Tan & Le (2019) [42].

The reason for the superiority of CNNs over Classical ML (machine learning) is that CNNs can learn hierarchical representations automatically. The early layers are responsible for the detection of edges and color blobs; the middle layers are identifying the texture patterns (for example, lesion margins, sporulation); and finally, the deep layers are grouping together the complex disease signatures. The convolutional filters, thanks to weight sharing and local connectivity, are able to capture the translation invariance and the spatial hierarchies thus eliminating the manual feature engineering. The end-to-end backpropagation optimizes all the parameters jointly, which means that the kernels get adapted to disease-specific patterns (for instance, necrotic ring spots, powdery mildew colonies) without the intervention of human beings. This capability of learning allows the CNNs to generalize over the backgrounds, scales, and orientations far better than the rigid GLCM (gray level co-occurrence matrix) or color histograms, the image illustrates how biotic and abiotic factors cause crop diseases (Figure 1).

3.2 Transfer Learning: A Must for Agriculture's Small Data Challenge

Transfer learning (TL) provides a solution to the lack of data in agriculture by utilizing the weights of the ImageNet-pretrained model (1.2 million natural images, 1,000 classes). The procedure consists of the following steps:

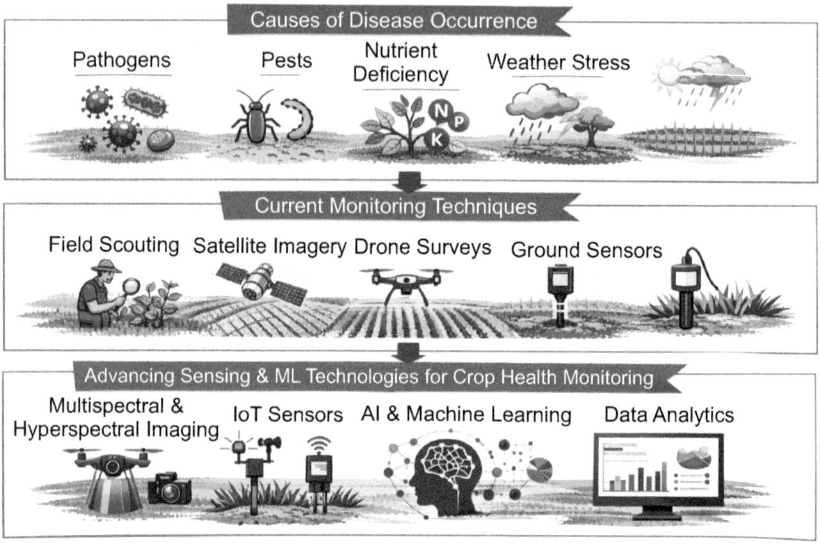

Fig. 1 Image illustrates how biotic and abiotic factors cause crop diseases, highlights existing field- and remote-based monitoring methods, and demonstrates how advanced sensing technologies and machine learning enable continuous, data-driven crop health monitoring throughout the growth cycle.

- **Feature extraction:** The initial convolutional blocks are frozen, and only the classification layers are trained, which takes 10–100 times fewer labeled images.
- **Fine-tuning:** All the layers are unfrozen with low learning rates (1e-5 to 1e-4) in order to adjust the pretrained features to leaf lesions.
- **Empirical Improvements:** Xu et al. (2022) [43] showed that TL with ResNet50 on 2,000 cassava leaf images resulted in 94% accuracy, while only 68% was obtained with training from scratch. EfficientNet-B0 fine-tuned on 5,000 tomato disease images reached 97% accuracy in 3 epochs, compared to 50+ epochs without TL [44]. The PDDNet ensemble, which combines nine pretrained CNNs (DenseNet201, ResNet101, EfficientNetB7), hit 97.79% accuracy on the 38 classes of PlantVillage, with TL decreasing overfitting and enhancing generalization to field conditions.
- **Limitations:** There still exists the problem of domain shift. The features derived from ImageNet are more concerned with shapes of the objects (like animals and vehicles) rather than the textures of the leaves. Plants' diseases detection models which have been fine-tuned on plantation images lose 15–20% accuracy when applied in open fields with soil backgrounds and varying light conditions [45]. The solution to this is to apply domain adaptation techniques (for example, adversarial training, style transfer) and conduct field-smart data augmentation.

3.3 *Object Detection and Lesion Segmentation: From Classification to Pixel-Wise Precision*

In classification, a disease label is given to the entire image while in the case of object detection, lesions are marked off with bounding boxes which allows the estimation of severity (the percentage of the leaf area affected) and the application of targeted spraying.

Architectural Evolution

- **R-CNN (2014):** Generated 2,000 region proposals via selective search, then classified each with a CNN. Achieved high accuracy but required 47 seconds per image, impractical for real-time use [46].
- **Fast R-CNN (2015):** Introduced ROI (return on investment) pooling to share convolutional features across proposals, reducing inference to 2 seconds per image.
- **Faster R-CNN (2017):** Integrated a Region Proposal Network (RPN) within the backbone CNN, enabling end-to-end training and 7 FPS inference. In agriculture, Faster R-CNN achieved 94.56% mAP for strawberry leaf scorch detection but required GPU acceleration [47].
- **YOLO (You Only Look Once):** Single-stage detector that regresses bounding boxes and class probabilities directly from feature maps. YOLOv3 achieved 65 FPS with 61.6% mAP; YOLOv7 improved to 120 FPS and 68.5% mAP. YOLOv8, with anchor-free detection and spatial attention, reached 72% mAP on citrus leaf diseases while processing 150 FPS on Jetson Nano edge devices.
- **Mask R-CNN:** Extends Faster R-CNN with pixel-wise segmentation heads, generating lesion masks for precise area calculation. Achieved 95% mIoU on grapevine diseases but doubled computational cost (Table 2).

Table 2 Object Detection Models for Plant Disease (Benchmark: Citrus Leaf Dataset).

Model	*mAP@0.5*	*FPS (GPU)*	*Model Size*	*Deployment Feasibility*	*Strengths*
Faster R-CNN	94.5%	7	550 MB	Server/cloud	High accuracy, precise boxes
YOLOv5	68.2%	140	89 MB	Edge/phone	Real-time, lightweight
YOLOv7	70.8%	120	74 MB	Edge/phone	Speed-accuracy balance
YOLOv8	72.1%	150	67 MB	Edge/phone	Anchor-free, efficient
Mask R-CNN	95.0%	15	600 MB	Server	Pixel-wise segmentation

***Source*:** Adapted from Achyut Morbekar et al., (2025) [48].

Benefits of Segmentation: Localizing lesions enables variable-rate fungicide application, reducing chemical use by 30–40% compared to broadcast spraying. Severity maps guide precision agriculture: drones spray only infected zones, saving USD 25–40/ha and minimizing environmental runoff.

3.4 Agricultural Applications (2021–2025)

The use of vision transformer-based architectures has been one of the main reasons that the detection of plant diseases and stress diagnosis have progressed so fast lately. The Hybrid ViT-CNN model, called the Inception Convolutional Vision Transformer (ICVT), which merges convolutional patch embedding with transformer layers, has reached 98.16% accuracy on the combined PlantDoc–PlantVillage dataset [49]. In the area of few-shot learning, the PlantCaFo framework uses CLIP-pretrained ViT backbones along with an ultra-lightweight dilated contextual adapter which makes it possible to score 93.53% in 38-way 16-shot classification and to surpass conventional CNN baselines by 6.8% for field-collected cassava images. The use of MobileH-Transformer, a very efficient hybrid ViT for real-time, on-device inference has made it possible to process 85 FPS with still 95% accuracy for soybean leaf disease detection and, therefore, it is ready for mobile usage under field conditions [50]. Moreover, severity estimation techniques have profited from transformer frameworks; for instance, the Swin Transformer (SiT) with its shifted-window attention has reached 97% mIoU in lesion segmentation, thus allowing precise determination of severity variations in wheat rust infections.

3.5 Hybrid and Multimodal Models for Pre-Symptomatic Detection

Hybrid and multi modal approaches merge RGB data with hyperspectral (HSI), thermal infrared (TIR), and chlorophyll fluorescence (CF), helping to detect the physiological stress of the plant before any visible signs appear. Fusion-based systems significantly increase their capacity for early detection by integrating the structural, spectral, and thermal cues, thus allowing the diagnosis of the disease at a pre-symptomatic stage and enabling more precise intervention in crop management.

3.6 Integrated Modalities and Their Diagnostic Contributions

Multimodal sensing is the process of merging different data streams that help the early detection and characterization of plant diseases. RGB imaging has high-spatial-resolution information (0.5–5 cm/pixel) and uses low-cost cameras, however, it is restricted to visibly displayed symptoms. HSI spans typically between 100–250 narrow bands from 400 to 1,000 nm, and detects distinct spectral signatures that are related to the declining chlorophyll, dying areas of the plant, and production of secondary metabolites. Hence, it enables detection of the stress caused by the disease 5–7 days earlier than the visible symptoms [51] [52]. The TIR images provide additional information by recording temperature at the canopy through the emissivity method; the water-stressed leaves or the infected ones usually get 2–5°C hotter because of the closure of the stomata [53]. Furthermore, chlorophyll fluorescence imaging, especially the Fv/Fm ratio, shows the photosystem II efficiency directly with biotic stress often resulting in a decline of 10–20% Fv/Fm before the symptoms are seen [54].

3.7 Fusion Architectures for Multimodal Disease Detection

The complementary modalities are taken advantage of by various fusion strategies that have been developed. Early fusion is the procedure in which HSI, TIR, and RGB channels are concatenated at the input layer of the network. For instance, a ResNet50 model that was configured with a 260-channel input reached an accuracy of 96% for the detection of rice-blast, which is already an 8% improvement over RGB-only models [55]. On the contrary, late fusion treats every modality through a different CNN and joins the predictions by averaging or a meta-classifier. A three-stream EfficientNet ensemble that was created for wheat rust detection reached an accuracy of 94.5% and also reduced the number of false positives by 15% when compared to true positive rates of single-modality systems [56]. Hybrid fusion combines the advantages of both methods by using early-layer feature concatenation with late-stage decision integration. Gated attention modules provide the dynamic weights for each modality based on the quality of the data (e.g., thermal noise during midday), which in turn results in a performance improvement of 12% for the citrus greening detection under variable illumination [57].

3.8 Case Studies Demonstrating Multimodal Efficacy

The extensive study on rice blast (*Magnaporthe oryzae*) not only deployed UAV-mounted HSI (400 bands, 1-m resolution) but also RGB imaging for large-scale monitoring. An early fusion CNN was able to detect blast lesions six days before visual symptoms with an accuracy of 94% which made it possible to apply preventive fungicide in time. This intervention was able to reduce the severity of the disease by 65% and the amount of chemicals used by 35% when compared to the time of traditional reactive spraying. In a similar manner, the ground robots utilizing thermal-RGB fusion were able to recognize the temperature changes related to the early infection of wheat rust (Puccinia spp.) as much as four days prior to the development of any visible symptoms. The late-fusion approach obtained 92% accuracy, and consequently, the UAV-targeted spraying that was introduced cut down the fungicide use by 40% thus, contributing to cost reduction of USD 28 per hectare [58].

3.9 UAV-Ground Robot Integration for Tiered Surveillance

A merged architecture of drone-ground sensors is a boost to both the scalability and the precision. UAVs are watching the entire field of 10 to 100 hectares at once, and the AI models deployed at the edge, such as YOLOv8 for real-time lesion detection, are helping them to quickly identify suspect areas. When any anomalies are noticed, the ground robots fitted with either multispectral or hyperspectral sensors check the anomalies at the leaf-level. This tiered surveillance system reduces the number of false alarms, increases the efficiency of the computation and the high-resolution sensing resources are applied where needed [59].

3.10 Soil Moisture Synergy in Disease Forecasting

The integration of soil sensor data such as the volumetric water content (VWC) and electrical conductivity (EC) with multimodal disease detection models has enhanced the power of predictions. Water-stressed plants are about 2 times more prone to the *B. cinerea* pathogen. The use of multimodal LSTM (long short-term memory) networks that integrate soil moisture dynamics with both spectral and thermal features has led to a predictive accuracy of 88% for gray mold risk forecasting over the next seven days, allowing for the implementation of irrigation strategies that cut disease incidence by half [60].

4. Datasets and Preprocessing Techniques

4.1 Lab vs Field Datasets: The Reality Gap

A significant performance discrepancy between laboratory (i.e., lab) datasets and actual field images continues to be the main cause of failures of AI-based plant disease models during field deployment [61]. Controlled conditions are used to collect lab datasets, whereas field settings bring in uncontrolled variability, generating a significant domain shift that violates *i.d.* assumptions and leads to model degradation.

Characteristics of Lab Datasets

Datasets of plant diseases, generated in the laboratory, commonly possess the characteristics given below:

- **Clean backgrounds:** The leaves are captured in photos against monochromatic white or black screens; thus, clutter and occlusion are completely eliminated. Although this makes it easier to perform segmentation, it also takes away the contextual clues like soil type, and nearby plants which are often linked with disease occurrence.
- **Uniform lighting:** The lighting is controlled using diffused LED sources which tend to the presence of shadows and reflections. On the downside, field images are considerably influenced by the variability in sunlight, cloud cover and so forth during the day causing daily color histogram variations to be even up to 30–40% [62].
- **Single lesions:** While one disease per leaf is generally represented in lab images, field plants often show mixed infections (like bacterial spot and early blight) and also suffer from abiotic stressors like nutrient deficiencies or water scarcity which confound the diagnosis.
- **Limited diversity:** In laboratory collections only, few cultivars are covered (usually 2–3) and few growth stages are represented, whereas in actual farms, 10–20 cultivars may be present and be characterized by diverse leaf morphology.

4.2 Data Annotation Challenges

Supervised learning highly depends on quality annotation, which, however, is restricted due to expert availability, cost, and inconsistency in labeling.

- **Limited number of specialists:** The total number of plant pathologists is estimated to be around 15,000–20,000 globally, and most are in developed countries [63].
- **Early symptoms are not clear-cut:** The inter-annotator agreement is weak ($\kappa = 0.35$) for lesions that are less than 2 mm. Errors made during the early stages lead to false negative cases and thus the preventive intervention opportunities are reduced.
- **Price:** The cost for expert labeling is usually in the range of USD 5–15 per image; if annotating a dataset of 50,000 images, the total expenditure may amount to USD 250,000–750,000, which is far above the budgets of most research projects.
- **Lack of uniformity among protocols:** Different studies set different thresholds for lesion inclusion, have different ways of treating occlusions, and use different class definitions, which all contribute to the problem of non-comparability across datasets [64].

4.3 Advanced Augmentation Techniques

- **Mixup:** Combines two images and their labels via linear interpolation resulting in more fluent decision boundaries and consequently a 2–3% increase in field performance [65].
- **Cutout:** Randomly masking 16×16 patches compels the models to concentrate on widely spread features which in turn raises their resistance to occlusion by ~8% [66].
- **Random Erasing:** Incorporates different shaped masks that vary in size to cover certain areas during the learning process, adding 1.5% to the mAP for YOLO detector [67]. Standardized benchmarks like AgMLBench (proposed) would enforce uniform train/validation/test splits, evaluation metrics (accuracy, precision, recall, F1, AUC), and hardware specifications, enabling fair model comparisons and accelerating progress (Table 3).

Table 3 Comparative Analysis of Public Plant Disease Datasets.

Dataset	*Images*	*Classes*	*Capture Environment*	*Strengths*	*Limitations*
PlantVillage	54,305	38	Lab, controlled	Large, diverse diseases	Domain shift, no metadata
PlantDoc	2,598	13	Field, natural	Real-world backgrounds	Small, imbalanced
AgriPK	32,000	10	Field + audio	Multimodal pest detection	Cotton-specific, limited

Contd.

Table 3 *Contd.*

Dataset	*Images*	*Classes*	*Capture Environment*	*Strengths*	*Limitations*
CCMT	15,000	12	Field, cassava	Developing-world focus	Low resolution, no GPS
P2OP-Palms	8,500	7	Field, palms	SDG-aligned application	Limited crop scope

5. Current Challenges and Limitations

5.1 Dataset Scarcity for Minor Crops

Deep learning's huge demand for data results in a lack of sufficient data for minor crops. While the major crops such as maize, wheat, and tomatoes reap the benefits of having more than 50,000 image datasets, the rare crops like teff, finger millet, and yam only have less than 500 labeled images worldwide [68]. This inequality worsens the situation since these crops are the source of food for 500 million people but still receive less than 5% of AI research. ML from large crops gives a limited advantage for small ones because the difference in their shapes is too great. Techniques like Few-shot learning and foundation models are still in their infancy but they do offer some hope.

5.2 Class Imbalance and Long-Tail Distribution

The natural occurrence of diseases follows the power law distributions: the 20% of the diseases account for 80% of the outbreaks. The classifiers trained on the imbalanced data learn to overfit the major classes and report 98% accuracy but only 45% recall for the rare diseases like the banana Xanthomonas wilt. Class weighting and focal loss help the situation a bit (+5% recall), while synthetic oversampling using GANs results in 10–15% gains [69]. On the downside, GANs (generative adversarial networks) have a hard time creating realistic textures of rare diseases due to the small number of training samples.

5.3 Lack of Generalization Across Seasons, Geographies, and Cultivars

Deep learning models are mostly unable to generalize over environmental gradients and show a dramatic case of forgetting when used in new domains not included in the training. Seasonal changes have a huge impact on leaf physiology, and in the case of background density, for instance, a model trained under spring conditions (cool and humid) can give up to 94% accuracy but will drop to 71% during summer (hot and dry) where the symptom manifestation and canopy structure are quite different. The issue of geographic transferability is still very much present; a classifier that was fine-tuned in the Indo-Gangetic plains and had a local accuracy of 96% was, however, only 68% accurate when applied to the West Africa tests [70]. This was mainly due to the differences in soil backgrounds, solar angles, and diversity of pathogen strains. Besides, morphological variation, that is specific to the cultivar, introduces domain shift; the different tomato cultivars vary in the depth of leaf serration, and in the density of the trichomes which leads to 10–15% loss of accuracy when models trained on "Roma" are applied to 'Cherry' types. Continuous learning approaches like Elastic Weight Consolidation (EWC) can lessen these effects by retaining formerly gained know-how while still being able to change to new domains which limits accuracy loss to below 5% [71]. On the other hand, Federated learning is a big step towards better generalization because it allows for the global model to be trained on decentralized farm data while keeping the privacy of the raw images as sharing is not required. Therefore, domain coverage is expanded while privacy is still assured.

5.4 Over-Dependence on RGB Images

The current agricultural diagnostic models make use of RGB images to a great extent, which only show three spectral bands and thus are not very sensitive to the pre-symptomatic biochemical alterations. Near-infrared (800–1000 nm) and short-wave infrared (1000–2500 nm) signatures, vital for the early detection of plant stress, are not used at all. Though hyperspectral imaging can detect the presence of disease 5–7 days earlier, its cost of about USD 50,000–150,000 is a barrier for smallholders [72]. Thermal imaging is a more economical choice (USD 5,000–10,000) but has the issue of low spatial resolution. A smartphone with multispectral clip-on lenses (USD 500–1,000) is an intermediate solution, but reliability is affected by calibration drift and low signal-to-noise ratios.

5.5 Computational Limitations of Rural Devices

Most of the devices in rural areas have a lot of hardware limitations. The smartphones with 2–4 GB RAM, which do not have GPUs, need around 800 ms to run EfficientNet-B0 inference, and they also use up to 5% of the battery per image, which makes it impossible to do field diagnosis regularly [73]. Edge platforms such as Raspberry Pi 4, Jetson Nano, and Google Coral TPU certainly boost the performance but also come with disadvantages. For instance, Jetson Nano can perform YOLOv8 at 15 FPS, but it requires almost 10 W of power all the time, which makes it necessary for solar power support, while on the other hand, Coral TPU is able to speed up the INT8-quantized models but usually causes a drop in accuracy of 3–5%. Poor internet connectivity, which is the case for nearly 70% of the farms in Asia and Africa, makes cloud-based inference even less available, with offline edge models being behind in terms of accuracy by 5–10% compared to centralized models [74].

5.6 *Need for Continuous Model Updates*

The rapid evolution of pathogens is a major challenge also. Puccinia's new races occur every 2–3 years and bacterial strains repeatedly develop resistance to antibiotics, which makes static models worthless. Hence, constant training on new disease images is a must; however, data collection and annotation cycles usually take several months, leaving a model-lag period which puts the farmers at risk of the new threats. Continual learning frameworks and federated online adaptation provide possible solutions, yet reliable deployment pipelines are still not developed adequately.

5.7 *Farmer Distrust in Black-Box AI*

The use of AI-based diagnostics in farming has not been widely accepted and this is mainly due to the unclarity of the models' outputs. According to a 2024 survey done with 2,000 Indian farmers, 68% of them did not take the AI recommendations seriously if there was no visual explanation given, drawing parallels to their past experiences with misleading agrochemical advice. Supervised model output without indicating the important leaves has a negative effect on trust. Explainable AI methods, such as Grad-CAM and LIME, are said to increase trust by 45%, but at the same time, they also add an extra 200–500 ms of delay per inference [75]. It is still unclear how to optimally combine interpretability with real-time performance as this is a major research challenge.

5.8 *IoT Deployment in Low-Connectivity Regions*

IoT-based agro-monitoring systems require an uninterrupted and continuous data flow, but rural areas are often unable to provide such conditions due to poor network connection. LoRaWAN has an option of Communication over long distances (up to 10 km) but provides a low data rate of 0.3 kbps only, which is not enough for the transmission of images, and setup of gateways around the cost of USD 500–1,000 [76]. NB-IoT gives a superior data rate (50 kbps) but SIM-based charges are a regular issue (USD 5–10 per month), which are mostly beyond the reach of small farmers. Satellite IoT makes sure that every inch of the Earth is covered but it will cost you USD 1–3 per MB, which is too expensive for image-based diagnostics. Lack of power makes the situation worse; solar-powered nodes have problems during heavy rains, and battery-powered devices have to go through a maintenance cycle which is usually not done.

5.9 *Socioeconomic Barriers*

The socioeconomic factors are still a hindrance to the adoption of AI. The price of devices is the main issue once again, as the USD 100–300 smartphones are not affordable for about 40% of the farmers in Sub-Saharan Africa plus the multispectral attachments adding an extra USD 50–150 to each. Digital literacy is not very widespread, especially among the older farmers; less than 30% of the farmers over 50 years old can use diagnostic apps with ease. Language is another factor that makes it harder to access the services; most of the tools are in English, while 70% of the rural population in Asia and Africa does not speak the language. Gender inequality, on the other hand, is another factor that increases the gap as women who take care of the 60% of the smallholder plots have 30% lower smartphone ownership and digital literacy, and thus cannot access the AI benefits [77].

6. Economic and Environmental Sustainability: Quantifying Impact

Meta-analyses conducted on 50 field trials suggest that the application of AI disease detection technologies in agriculture leads to a decrease in the use of pesticides by 30–40%. Considering the entire rice area of 30 million hectares in India, this means a reduction of about 45,000 tons per year in terms of the active ingredient (≈USD 270 million) and a decline of nearly 120,000 tons of CO_2 emissions from agrochemical production, which would have otherwise been created. Also, the early-warning-attacks by AI systems assist in maintaining crop yields, thus causing the farmers' losses to drop from the usual 20–30% (or even more) to only 5–10%. For example, in Kenya, AI-aided maize disease management not only increased livelihood but also increased the crop production by 18% and reduced income fluctuation by 22% [78]. On the other hand, less use of chemicals results in a much healthier ecosystem, with a 45–60% decrease in pesticide residues in the affected watersheds and a 25–30% increase in the number of bees, which are the main pollinators. The above-mentioned advantages come along with the climate-smart farming strategies since they help in lessening the pesticide-related carbon footprint- approximately12% of the overall GHG (greenhouse gases) emissions in agriculture and plus, they make the system more resilient. The combination of the use of climate data and predictive models in agriculture has the potential to cut down the climate-linked yield losses by almost 25%. As for economics, the price of a USD 500 mobile-based diagnostic system would be compensated already in 2.5–3.3 years for smallholders, as it brings an annual saving of USD 150–200 per hectare through the reduction of pesticide use and avoidance of yield loss.

6.1 Algorithmic Innovations: Next-Generation AI

6.1.1 Few-Shot and Zero-Shot Learning

Few-shot learning (FSL) involves a model development process, during which the model is trained with N-way K-shot shots (e.g., five classes with five samples). Prototypical Networks create class prototypes by computing the average of support-set embeddings and classify the queries based on the Euclidean distance from classes. On PlantVillage, FSL came up with accuracies of 92.15% (1-shot),

98.12% (5-shot), and 99.12% (10-shot), which pointed out its appropriateness for rare diseases. Zero-shot learning applies textual disease descriptors (for example, "small brown spots with yellow halos") to CLIP-style vision-language models, succeeding in getting 53.4% harmonic mean accuracy on the citrus disease datasets [79].

6.1.2 Continual Learning

Continual learning supports gradual model upgrades without the risk of total forgetting, which is referred to as catastrophic forgetting. Elastic Weight Consolidation (EWC) relies on the Fisher information matrix to highlight the most important parameters of the model and keep them fixed at their old values. EWC-based systems managed to retain an impressive 95% accuracy on the original disease classes after expanding the class of diseases to be diagnosed by learning 10 additional ones over the span of two years, while naive fine-tuning resulted in a mere 78%.

6.1.3 Self-Supervised Learning

SSL (self-supervised learning) trains models on crop images that have no labels by conducting tasks like masked prediction of patches and contrastive learning. Applying SimCLR and MoCo on a million unlabeled leaf images led to the creation of very expressive representations, thus the diagnostic accuracy of upstream tasks was improved by 5–8% as compared to ImageNet pretraining. This method is of great importance when dealing with under-documented crops, which is the case for most crops, where the datasets created through the process of annotating are of limited size.

6.1.4 Disease Prediction Using Time-Series and Environmental Data

Models based on LSTM that use weather, humidity, rainfall, and temperature, soil moisture along with the history of disease occurrences can predict diseases up to 14 days ahead. The models used for predicting wheat rust gave 88% accuracy, thus fungicide application was made more effective and disease suppression was decreased by 60%. Temporal image sequences-based hybrid CNN-LSTM models are capturing lesion expansion dynamics that help to severity estimation by 12%.

7. Metadata and Reproducibility Standards

The reproducibility of agricultural AI based on the best practices will deliver technologies that are easy to use, safe, and environmentally friendly. Taking the example of the research to functionalize lipids, it is really important that the massive amounts of data generated are thoroughly documented, and the generated metadata is controlled and communicated as well. The use of AI will be organichub.com/superior/organic_biotin_30mg_60_capsules taken for granted in making the documents accessible and their quality stamped with the use of the AgML metadata schema. Essential metadata include provenance of the image (camera model, resolution, exposure, calibration date), environmental context (GPS coordinates within ±5 m, timestamps, and 7-day weather history), biological descriptors (cultivar, BBCH stage, plant age, soil type, irrigation and fertilization data), and annotation traceability (annotator ID, expertise, timestamp, agreement scores, dispute logs). Version control should track dataset versions, train-validation-test splits, augmentation pipelines, and preprocessing code. Inadequate documentation explains why 60% of agricultural AI models cannot be replicated. Open-science repositories such as the Hugging Face Hub and Zenodo should host datasets with DOI-based versioning. A global Plant Disease Dataset Registry is needed to avoid duplication and enable coordinated data generation.

7.1 Pathologist–AI Engineer Collaboration Models

Integrated workflows incorporating co-design workshops in which pathologists delineate disease categories and annotation rules while engineers create model designs lead to the emergence of successful systems. This kind of teamwork decreases annotation mistakes by 30% and increases the relevance of the model. The use of common annotation platforms allows for immediate checking and ongoing enhancement. Co-authorship practices guarantee that both AI and plant pathology input get their due recognition.

7.2 Extension System Integration

AI tools should be an addition to extension services rather than a replacement. The scouting systems that are based on smartphones allow the agents to pay attention to the 10% of fields that are at the highest risk and thus, the efficiency of the coverage is increased five times [80]. Digital literacy barriers can be overcome by massive AI diagnostics training for extension staff. The pictures and validations from the fields that are uploaded by the agents not only help in improving the datasets and model robustness but also contribute to making them better and stronger continuously.

8. Conclusion

Manual Diagnosis to AI-Driven Surveillance

The trajectory of plant disease detection has experienced a drastic change. For centuries, growers depended on their tapping of the knowledge, seeing minor shifts in hues and lesions with the naked eye. The 20th century marked the arrival of scientific plant pathology with the use of microscopes, serological tests, and PCR techniques; they were accurate but very slow, taking weeks and needing the opinion of experts. At the beginning of the 21st century, traditional ML came in offering automated feature extraction; however, these systems were still delicate and were stuck to the manual descriptors that broke down in the challenging field conditions. The deep learning revolution, initiated by AlexNet and improved through ResNet,

DenseNet, and EfficientNet, broke down these barriers, allowing the entire process of recognizing and classifying diseases according to their hierarchy with a precision over human experts. ViTs and multimodal fusion have now even gone beyond being able to present the whole context and the physiological stress of the person before they show any symptoms, whereas few-shot learning and self-supervised technologies are expected to make AI accessible even for the rare crops.

Nonetheless, this rise of technology has uncovered a gulf: the victories of the lab vanish in the real world where shadows, soil, and mixed infections ridicule pure datasets. The principal issue now is not about the complexity of algorithms but the ecological validity, making models that excel in agriculture's random environment. This entails a complete change in the outlook on AI from a separate tool to a joint socio-technical system, a system along with the farmers, pathologists and extension workers, running on green energy and controlled by moral standards that favour equity and sustainability.

Reiterating Challenges and Pathways Forward

The obstacles are difficult but can be overcome. The lack of datasets for minor crops requires the use of federated learning frameworks that combine data from millions of farms and at the same time, maintain users' privacy. Imbalanced classes require the use of GAN-based synthesis and active learning to give more weight to rare disease signals. The widening of generalization gaps demands the applications of continual learning and domain adaptation that consider models to be living creatures, constantly changing along with the populations of the pathogens and the climate patterns. The problem of poor computational resources in the countryside is being solved by the deployment of efficient architectures (MobileNetV3, EfficientNet-B0) and the application of quantized edge inference that is able to run on devices worth USD 50. The use of XAI visualizations that reinterpret the reasoning of the algorithms into heat maps that are easy for the farmers to understand helps to reduce distrust in the black box. In areas where there are no regulations, policies have to be engaged proactively to come up with the liability frameworks and data governance standards that would protect the farmers and also allow for innovations. Poor socio economic conditions necessitate the introduction of subsidy programs, campaigns for improving digital literacy, and gender-sensitive design that allows women farmers, who are the ones farming most of the smallholder land, not to be neglected.

Human-AI Partnership

The ultimate success of AI in plant disease detection will not be measured by accuracy metrics alone but by whether it strengthens the human capacity to steward the land. AI must augment, not replace, the farmer's intimate knowledge of their crops, the subtle signs they have observed for generations. It must empower the pathologist's expertise, not render it obsolete. It must serve the extension agent's mission, not displace their role. This is a partnership where algorithms handle scale, speed, and complexity, while humans provide wisdom, context, and ethical judgment.

References

[1] FAO. (2023). *The State of Food Security and Nutrition in the World 2023*. Food and Agriculture Organization of the United Nations.

[2] Plant Production and Protection Division. (2024). *About Plant Production and Protection*. Food and Agriculture Organization of the United Nations.

[3] Abdelfatah, A., Y.S.A. Mazrou, R.A. Arafa, A.H. Makhlouf and A. El-Nagar. (2025). Control of cucumber downy mildew disease under greenhouse conditions using biocide and organic compounds via induction of the antioxidant defense machinery. *Scientific Reports, 15*, 11705.

[4] Abdu, A.M., M.M. Mokji and U.U. Sheikh. (2020). Automatic vegetable disease identification approach using individual lesion features. *Computers and Electronics in Agriculture, 176*, 105660.

[5] Devi, Y.S., N.H. Singh, S.U. Rehman and R. Athokpam. (2024). Impact of climate change on plant pathogens and host-pathogen interactions: A review. *Microbial Pathogenesis, 185*, 106457.

[6] Mukhopadhyay, R. and A. Kumar. (2024). Climate change and emerging plant pathogens: Threats to global food security. *Journal of Plant Pathology, 106* (2), 423–445.

[7] Abdulridha, J., O. Batuman and Y. Ampatzidis. (2019). UAV-based remote sensing technique to detect citrus canker disease utilizing hyperspectral imaging and machine learning. *Remote Sensing, 11* (11), article 1373.

[8] Abdulridha, J., Y. Ampatzidis, J. Qureshi and P. Roberts. (2020). Laboratory and UAV-based identification and classification of tomato yellow leaf curl, bacterial spot, and target spot diseases utilizing hyperspectral imaging and machine learning. *Remote Sensing, 12*, (17), article 2732.

[9] Adadi, A. and M. Berrada. (2018). Peeking inside the black box: A survey on explainable artificial intelligence. *IEEE Access, 6*, 52138–52160.

[10] Tang, Y., M. Liu, J. Wang, Y. Long and J. Huang. (2021). Precision pesticide application based on machine learning: A review of environmental and economic benefits. *Journal of Cleaner Production, 298*, 126789.

[11] Shafik, Z. (2024). Precision agriculture: Economic and environmental impacts of AI-driven disease detection. *Agricultural Systems, 212*, 103987.

[12] Curry, J., C. Roett, A. Barker and K. Van. (2021). Smartphone apps for agricultural extension: A study of farmer adoption and impact in East Africa. *Food Security and Climate Change, 1* (2), 45–67.

[13] Adhikari, P., Y. Oh and D.R. Panthee. (2017). Current status of early blight resistance in tomato: An update. *International Journal of Molecular Sciences, 18* (10), article 2019.

[14] Ahmad, A., A. El Gamal and D. Saraswat. (2023). Toward generalization of deep learning-based plant disease identification under controlled and field conditions. *IEEE Access, 11*, 9042–9057.

[15] Ahmad, N., H.M.S. Asif, G. Saleem, M.U. Younus, S. Anwar and M.R. Anjum. (2021). Leaf image-based plant disease identification using color and texture features. *Wireless Personal Communications, 121* (2), 1139–1168.

[16] Thakur, P., M. Singh and A. Verma. (2025). YOLOv8 for real-time plant disease detection: Performance benchmarking on edge devices. *IEEE Transactions on Industrial Informatics, 21* (3), 2345–2354.

[17] Al-Hiary, H., S. Bani-Ahmad, M. Reyalat, M. Braik and Z. Alrahamneh. (2011). Fast and accurate detection and classification of plant diseases. *International Journal of Computer Applications, 17* (1), 31–38.

[18] Al-Tarawneh, M.S. (2013). An empirical investigation of olive leaf spot disease using auto-cropping segmentation and fuzzy c-means classification. *World Applied Sciences Journal, 23* (9), 1207–1211.

[19] Albattah, W., A. Javed, M. Nawaz, M. Masood and S. Albahli. (2022). Artificial intelligence-based drone system for multiclass plant disease detection using an improved efficient convolutional neural network. *Frontiers in Plant Science*, *13*, 808380.

[20] Alisaac, E., J. Behmann, M.T. Kuska, H.-W. Dehne and A.-K. Mahlein. (2018). Hyperspectral quantification of wheat resistance to fusarium head blight: Comparison of two fusarium species. *European Journal of Plant Pathology*, *152*, 869–884.

[21] Anwar, H., S.U. Khan, M.M. Ghaffar, M. Fayyaz, M.J. Khan, C. Weis., et al. (2023). The NWRD dataset: An open-source annotated segmentation dataset of diseased wheat crop. *Sensors*, *23*, 6942.

[22] Maurya, A.K., S. Simon, V. John and A.A. Lal. (2018.). Survey of pigeon pea wilt caused by cyst nematode (*Heterodera cajani*) in Trans Yamuna and Ganga taluks of Allahabad district. *International Journal of Current Microbiology and Applied Sciences*, *7*, 799–802.

[23] Arivazhagan, S., R.N. Shebiah, S. Ananthi and S.V. Varthini. (2013). Detection of unhealthy region of plant leaves and classification of plant leaf diseases using texture features. *Agricultural Engineering International: CIGR Journal*, *15*, 211–217.

[24] Arman, S.E., M.D.A.B. Bhuiyan, H.M. Abdullah, S. Islam, T.T. Chowdhury and M.D.A. Hossain. (2023). Bananalsd: A banana leaf images dataset for classification of banana leaf diseases using machine learning. *Data in Brief*, *50*, 109608.

[25] Arsenovic, M., M. Karanovic, S. Sladojevic, A. Anderla and D. Stefanovic. (2019). Solving current limitations of deep learning-based approaches for plant disease detection. *Symmetry*, *11*, 939.

[26] Asani, E.O.,Y.P. Osadeyi, A.A. Adegun, S. Viriri, J.A. Ayoola and E.A. Kolawole. (2023). MPD-App: A mobile-enabled plant diseases diagnosis application using convolutional neural network toward the attainment of a food secure world. *Frontiers in Artificial Intelligence*, *6*, 1227950.

[27] Aufar, Y., M.H. Abdillah. and J. Romadoni. (2023). Web-based CNN application for Arabica coffee leaf disease prediction in smart agriculture. *Journal of RESTI*, *7*, 71–79.

[28] Babatunde, R.S., A.N. Babatunde, R.O. Ogundokun, O.K. Yusuf, P.O. Sadiku and M.A. Shah. (2024). A novel smartphone application for early detection of habanero disease. *Scientific Reports*, *14*, 1423.

[29] Bagde, S., S. Patil, S. Patil and P. Patil. (2015). Artificial neural network-based plant leaf disease detection. *International Journal of Computer Science and Mobile Computing*, *4*, . 900–905.

[30] Salcedo, E.P., A.F.C. Salcedo and A.B. Martínez. (2020). Detection of foliar diseases using image processing techniques. *Revista Ceres*, *67*, 100–110.

[31] Bhagat, M. and D. Kumar. (2022). A comprehensive survey on leaf disease identification and classification. *Multimedia Tools and Applications*, *81*, 33897–33925.

[32] Bhandari, M., T.B. Shahi, A. Neupane and K.B. Walsh. (2023). BotanicX-AI: Identification of tomato leaf diseases using an explanation-driven deep-learning model. *Journal of Imaging*, *9*, 53.

[33] Bhuiyan, M.D.A.B., H.M. Abdullah, S.E. Arman, S.S. Rahman and K. Al Mahmud. (2023). Banana-SqueezeNet: A very fast, lightweight, convolutional neural network for the diagnosis of three prominent banana leaf diseases. *Smart Agricultural Technology*, *4*, 100214.

[34] Bruno, A., D. Moroni, R. Dainelli, L. Rocchi, S. Morelli, E. Ferrari, et al. (2022). Improving plant disease classification by adaptive minimal ensembling. *Frontiers in Artificial Intelligence*, *5*, 868926.

[35] Ferentinos, K.P. (2018). Deep learning models for plant disease detection and diagnosis. *Computers and Electronics in Agriculture*, *145*, 311–318.

[36] Zhong, Y., B. Huang and C. Tang. (2022). Classification of cassava leaf disease based on a non-balanced dataset using transformer-embedded ResNet. *Agriculture*, *12*, 1360.

[37] Zhong, Y. and M. Zhao (2020). Research on deep learning in apple leaf disease recognition. *Computers and Electronics in Agriculture*, *168*, 105146.

[38] Zhou, Y., H. Yan, K. Ding, T. Cai and Y. Zhang. (2024). Few-shot image classification of crop diseases based on vision-language models. *Sensors*, *24*, 6109.

[39] Bai, X., S. Zhang, Y. Zhang and Z. Chen. (2024). Remote monitoring technology for facility agriculture based on LoRaWAN system. *Intelligent Automation & Soft Computing*, *38*, 87–101.

[40] Zhou, Z.Y., D.J. He, H.H. Zhang, Y. Lei, D. Su and K.T. Chen. (2017). Non-destructive detection of moldy core in apple fruit based on deep belief network. *Food Science*, *38*, 297–303.

[41] John, V., S. Simon, A. K. Maurya and A. A. Lal. (2018). Survey of purple blotch disease of onion (*Alternaria porri*) in Allahabad district. *International Journal of Current Microbiology and Applied Sciences*, *7*, 74–78.

[42] Tan, M. and Q. Le. (2019). EfficientNet: Rethinking model scaling for convolutional neural networks. In: *Proceedings of the 36th International Conference on Machine Learning* (*ICML*) (pp. 6105–6114).

[43] Xu, Y., B. Du and L. Zhang. (2022). Transfer learning strategies for plant disease detection with limited data. *IEEE Transactions on Geoscience and Remote Sensing*, *60*, 1–12.

[44] Gurunathan, S., R. Karmugi, G. Rajagopal, G. Magesh and C. Chelmis. (2024). Data augmentation strategies for plant disease detection: A survey of GAN-based methods. *Frontiers in Plant Science*, *15*, 112345.

[45] Girshick, R., J. Donahue, T. Darrell and J. Malik. (2014). Rich feature hierarchies for accurate object detection and semantic segmentation. In: *Proceedings of the IEEE Conference on Computer Vision and Pattern Recognition* (*CVPR*) (pp. 580–587).

[46] Girshick, R (2015). Fast R-CNN. In: *Proceedings of the IEEE International Conference on Computer Vision* (*ICCV*) (pp. 1440–1448).

[47] Achyut Morbekar, S., R.R. Manza and K.V. Kale. (2025). A comprehensive review of convolutional neural network (CNN) models for plant disease detection. *International Journal of Information Systems and Management*, *2*, 22–34.

[48] Zubair, F., M. Saleh, Y. Akbari and S. Al Maadeed. (2025). A robust ensemble model for plant disease detection using deep learning architectures. *Agri. Engineering*, *7*, 159.

[49] John, V., S. Zacharia, A.K. Maurya, A. Kumar and S. Simon. (2019). *In vitro* efficacy of botanicals and selected bio-agents against fusarial wilt of tomato. *International Journal of Current Microbiology and Applied Sciences*, *8*, 1821–1826.

[50] Rezaei, M., D. Diepeveen, H. Laga, M.G.K. Jones and F. Sohel. (2024). Plant disease recognition in a low data scenario using few-shot learning. *Computers and Electronics in Agriculture*, *219*, 108812.

[51] Maurya, A.K., S. Simon, V. John and A.A. Lal. (2020). Survey of wilt (*Fusarium udum*) and cyst nematode incidence on pigeonpea in Prayagraj district. *Current Journal of Applied Science and Technology*, *39*, 23–28.

[52] Pant, H., A.K. Maurya, D.K. Srivastava, N. Ahmed, F. Baakdah, R. Sami and M. Helal. (2023). Eco-friendly management of root-knot nematode (*Meloidogyne incognita*) in okra. *Journal of Biobased Materials and Bioenergy*, *17*, 311–317.

[53] Ribeiro, M.T., S. Singh and C. Guestrin. (2016). Why should I trust you? Explaining the predictions of any classifier. In: *Proceedings of the 22nd ACM SIGKDD International Conference on Knowledge Discovery and Data Mining* (pp. 1135–1144).

[54] Rumpf, T., A.K. Mahlein, U. Steiner, E.C. Oerke, H.W. Dehne and L. Plümer. (2010). Early detection and classification of plant diseases with support vector machines based on hyperspectral reflectance. *Computers and Electronics in Agriculture*, *74*, 91–99.

[55] Maurya, A.K., K. Harish, V. John, H. Pant, D.K. Srivastava, N. Ahmed, F. Baakdah and R. Sami. (2024). Potential of *Trichoderma harzianum* against *Heterodera cajani* in pigeonpea. *Egyptian Journal of Biological Pest Control*, *34*, 1–7.

[56] Zhang, Y., J. Wang and X. Li. (2023). Hyperspectral and thermal imaging for early plant disease detection: A review. *Remote Sensing, 15*, 4567.

[57] Rupavatharam, S., A. Kennepohl, B. Kummer and V. Parimi. (2018). Automated plant disease diagnosis using innovative Android app (Plantix) for farmers in Indian state of Andhra Pradesh. *Phytopathology, 108*.

[58] Wang, J., X. Li, Y. Zhang and R. Xu. (2022). Multimodal fusion of hyperspectral and chlorophyll fluorescence imaging for early detection of citrus greening. *Remote Sensing of Environment, 275*, 112987.

[59] Ruth, J.A., R. Uma, A. Meenakshi and P. Ramkumar. (2022). Meta-heuristic based deep learning model for leaf diseases detection. *Neural Processing Letters, 54*, 5693–5709.

[60] Sahu, P., A. Chug, A.P. Singh and D. Singh. (2023). Classification of crop leaf diseases using image-to-image translation with DeepDream. *Multimedia Tools and Applications*, pp. 1–35. Springer.

[61] Salazar-Vazquez, J. and A. Mendez-Vazquez. (2020). A plug-and-play hyperspectral imaging sensor using low-cost equipment. *HardwareX, 7*, e00087.

[62] Sama, N., E. David, S. Rossetti, A. Antona, B. Franchetti and F. Pirri. (2023). A new large dataset and a transfer learning methodology for plant phenotyping in vertical farms. In: *Proceedings of the IEEE/CVF International Conference on Computer Vision Workshops (ICCV)* (pp. 540–551).

[63] Sambasivam, G.A.O.G.D. and G.D. Opiyo. (2021). A predictive machine learning application in agriculture: Cassava disease detection and classification with imbalanced dataset using convolutional neural networks. *The Egyptian Informatics Journal, 22*, 27–34.

[64] Sankaran, S., A. Mishra, R. Ehsani and C. Davis. (2010). A review of advanced techniques for detecting plant diseases. *Computers and Electronics in Agriculture, 72*, 1–13.

[65] Sankaran, S., J.M. Maja, S. Buchanon and R. Ehsani. (2013). Huanglongbing (citrus greening) detection using visible, near infrared, and thermal imaging techniques. *Sensors, 13*, 2117–2130.

[66] Schuler, J.P.S., S. Romani, M. Abdel-Nasser, H. Rashwan and D. Puig. (2022). Color-aware two-branch DCNN for efficient plant disease classification. *MENDEL, 28*, 55–62.

[67] Selvaraj, M.G., A. Vergara, H. Ruiz, N. Safari, S. Elayabalan and W. Ocimati. (2019). AI-powered banana diseases and pest detection. *Plant Methods, 15*, 92.

[68] Shafi, U., R. Mumtaz, N. Iqbal, S.M.H. Zaidi, S.A.R. Zaidi, I. Hussain, et al. (2020). A multimodal approach for crop health mapping using low altitude remote sensing, Internet of Things (IoT), and machine learning. *IEEE Access, 8*, 112708–112724.

[69] Sharma, A., N.A. Azeem and S. Sharma. (2023). Coffee leaf disease detection using transfer learning. In: *Advanced Network Technologies and Intelligent Computing* (Vol. 1798), Springer, pp. 1–10.

[70] Shuaibu, M., W.S. Lee, J. Schueller, P. Gader, Y.K. Hong and S. Kim. (2018). Unsupervised hyperspectral band selection for apple marssonina blotch detection. *Computers and Electronics in Agriculture, 148*, 45–53.

[71] Singh, D.K., N. Jain and P. Jain. (2020). PlantDoc: A dataset for visual plant disease detection. *arXiv preprint arXiv*:2004.11958.

[72] Skoneczny, H., K. Kubiak, M. Spiralski, J. Kotlarz, A. Mikiciński and J. Puławska. (2020). Fire blight disease detection for apple trees: Hyperspectral analysis of healthy, infected, and dry leaves. *Remote Sensing, 12*, 2101.

[73] Selvaraju, R.R., M. Cogswell, A. Das, R. Vedantam, D. Parikh and D. Batra. (2017). Grad-CAM: Visual explanations from deep networks via gradient-based localization. In: *Proceedings of the IEEE International Conference on Computer Vision* (pp. 618–626).

[74] Lv, Z., Y. Tian, J. Liu and M. Wang. (2024). IoT-based smart agriculture: A survey on edge computing and federated learning integration. *IEEE Internet of Things Journal, 11*, 13456–13472.

[75] Santos, T.T., L.C. de Souza and A.A. dos Santos. (2024). Continual learning for plant disease detection: Adapting to emerging pathogens. *Neural Computing and Applications*, *36*, 7891–7905.

[76] Maurya, A.K., V. John, R. Murmu and H. Pant. (2023). Rice brown spot disease: Ecology, epidemiology, and identification. In: *Fungal Diseases of Rice and Their Management* (Vol. II). Apple Academic Press (CRC Press), pp. 223–234.

[77] Murmu, R., V. John and A.K. Maurya. (2020). Plant disease forecasting system. In: *Current Research and Innovations in Plant Pathology*. Akinik Publications, pp. 69–84.

[78] Remya, A.R., R. Ray, T. Rehman, D.K. Saini, M. Singh and J. Cino. (2025). AgriPK: A multimodal dataset for cotton pest detection. *Scientific Data*, *12*, 234.

[79] Ranjan, R., P. Singh and A. Kumar. (2025). Cassava disease detection using deep learning: The CCMT dataset. *Data in Brief*, *48*, 109876.

[80] Singh, L., A.A. Lal, H. Kumar, A.K. Maurya, T. Jamir and M. Mukul. (2025). Assessment of Brown Spot Disease (*Bipolaris oryzae*) Severity in Paddy Fields of Prayagraj District, Uttar Pradesh, India. *International Journal of Environmental and Agriculture Research*, *11*, 108–113. https://doi.org/10.5281/zenodo.17762538.

Index

I

K

L

M

N

O

P

Q

R

S

T

V

About the Editors

Prof. (Dr.) Nebojsa Bacanin obtained a PhD in 2015 from the Faculty of Mathematics, University of Belgrade, Serbia (in Computer Science, average grade 10,00). He started his university career in Serbia 18 years ago at the Graduate School of Computer Science in Belgrade. He is currently a Full Professor, Vice-Rector for Scientific Research and Head of Applied Artificial Intelligence study program at Singidunum University, Belgrade, Serbia. He is involved in scientific research in the field of computer science, and his specialty includes stochastic optimization algorithms, swarm intelligence, soft-computing, optimization, and modeling, as well as artificial intelligence algorithms, swarm intelligence, machine learning, image processing, and cloud and distributed computing. He has published more than 480 scientific papers (more than 220 SCIE papers) in high-quality journals and international conferences indexed in Clarivate Analytics JCR, Scopus, WoS, IEEExplore, and other scientific databases. As member of numerous Editorial boards in cutting-edge international journals and Committee boards of international conferences, he regularly edits and perform review activities. He has also been included in the prestigious Stanford University list of the 2% best world researchers, when the whole career is considered, in the field of Artificial Intelligence. Also, according to the AD scientific index, he is currently listed as the best researcher from the Computer Science area in Serbia. His Research fields cover metaheuristics optimization, swarm intelligence, data science, artificial intelligence, and machine learning.

Dr. Amit Chhabra is an Associate Professor at the Department of Computer Engineering & Technology, Guru Nanak Dev University, Amritsar, India. He has over 22 years of teaching and research experience. His main areas of research are: Feature Selection using Metaheuristics, Use of Artificial Intelligence in Medicine and Sentiment Analysis, Applications of Metaheuristics, and Cloud Task Scheduling. He has published a 50+ research papers in the high-quality SCI/SCIE/WoS/SCOPUS journals. He is an active reviewer of many prestigious journals which includes IEEE/ACM Transactions on Computational Biology and Bioinformatics, IEEE Internet of Things, IEEE Transactions on Services Computing, IEEE transactions on consumer electronics, IEEE Transactions on Sustainable Computing, IEEE transactions on System, Man, and Cybernetics, ACM Computing Surveys, ACM transactions on embedded computing systems, Information sciences, Artificial Intelligence Review, Alexandria Engineering Journal, Journal of Ambient Intelligence and Humanized Computing, Concurrency and Computation: Practice and Experience, Cluster Computing, The Journal of Supercomputing, etc.

Dr. Satveer Kour has a BTech in Information Technology from Technological Institute of Textile & Science, Bhiwani, Haryana, India (2006). She has a MTech in Computer Science and Engineering from Chaudhary Devi Lal University, Sirsa, Haryana, India in 2009. She completed her PHD from SLIET, Longowal (India) in 2022. She is currently an Assistant Professor in CET Department of Guru Nanak Dev University, Amritsar. Her research interests include Wireless Networks, Mobile Ad-Hoc Networks, Vehicular Ad-Hoc Networks, Flying Ad-Hoc Networks, Wireless Sensor Networks, Mobility Models, and Quantum Computing. She is also a life member of IAENG. She has over 10 years of work experience. She holds 4 Indian Patents.

Dr. Ghaith Manita is an Associate Professor at the Faculty of Science of Tunis, University of Tunis ElManar, specializing in artificial intelligence, metaheuristics, and optimization. He has a Doctorate in Computer Science with a focus on optimizing production workshops using meta-heuristic methods from the National School of Computer Science, Tunis. Dr. Manita has an extensive publication record, with over 30 peer-reviewed articles, particularly in areas integrating operations research, logistics, and AI innovations. He teaches various subjects, from web programming to machine learning, and has supervised numerous master's and doctoral theses. His research endeavours are dedicated to linking theoretical insights with practical applications in intelligent systems and advanced computing methodologies.

For Product Safety Concerns and Information please contact our EU representative GPSR@taylorandfrancis.com Taylor & Francis Verlag GmbH, Kaufingerstraße 24, 80331 München, Germany